POWER ENGINEERING AND INFORMATION TECHNOLOGIES IN TECHNICAL OBJECTS CONTROL

2016 ANNUAL PROCEEDINGS

Power Engineering and Information Technologies in Technical Objects Control

Editors

Genadiy Pivnyak
Rector of the State Higher Educational Institution "National Mining University", Ukraine

Oleksandr Beshta
Vice Rector (Science) of the State Higher Educational Institution "National Mining University", Ukraine

Mykhaylo Alekseyev
Dean of Information Technology Department of the State Higher Educational Institution "National Mining University", Ukraine

CRC Press
Taylor & Francis Group
Boca Raton London New York

CRC Press is an imprint of the
Taylor & Francis Group, an **informa** business

A BALKEMA BOOK

Published by:
CRCPress/Balkema
P.O. Box 447, 2300 AK Leiden, The Netherlands
e-mail: Pub.NL@taylorandfrancis.com
www.crcpress.com – www.taylorandfrancis.com

First issued in paperback 2020

© 2016 by Taylor & Francis Group, LLC
CRC Press/Balkema is an imprint of the Taylor & Francis Group, an informa business

No claim to original U.S. Government works

ISBN 13: 978-0-367-73640-8 (pbk)
ISBN 13: 978-1-138-71479-3 (hbk)

**Visit the Taylor & Francis Web site at
http://www.taylorandfrancis.com**

**and the CRC Press Web site at
http://www.crcpress.com**

Typeset by Alyona Khar', State Higher Educational Institution "National Mining University", Dnipro, Ukraine

Power Engineering and Information Technologies In Technical Objects Control – Pivnyak, Beshta
& Alekseyev (eds)
© 2016 Taylor & Francis Group, London, ISBN 978-1-138-71479-3

Table of contents

Preface

The book provides a comprehensive summary of many aspects of power engineering and information technologies. It is dedicated to vital issues of design and development of technical objects that are extremely important for the professional community. Emphasis is given to recent advances in power engineering, electric drives, transport system, power electronics, cybersecurity and others. It introduces and explains recent developments in the given area of knowledge, includes up-to-date theory and the most recent advances. Readers will find a range of algorithms and techniques for performing comprehensive analysis, as well as modern methods for power system analysis, operation, and control. The authors clearly define concepts and explain the most important details in energy-saving and energy efficiency in industry and in information technologies.

The book provides practical guidance on dealing with modern challenges in technical objects control. Moreover it includes links to a number of industrial applications helping manufacturers to decrease energy consumption and financial expenses.

Genadiy Pivnyak
Oleksandr Beshta
Mykhaylo Alekseyev

Power Engineering and Information Technologies In Technical Objects Control – Pivnyak, Beshta
& Alekseyev (eds)
© 2016 Taylor & Francis Group, London, ISBN 978-1-138-71479-3

Elements of quaternionic matrices calculation and some applications in vector algebra and kinematics

G. Pivnyak, V. Kravets, K. Bas
State Higher Educational Institution "National Mining University", Dnipro, Ukraine

T. Kravets
National University of Railway Transport Named After Academician V. Lazaryan, Dnipro, Ukraine

L. Tokar
State Higher Educational Institution "National Mining University", Dnipro, Ukraine

ABSTRACT: Quaternionic matrices are proposed to develop mathematical models and perform computational experiments. New formulae for complex vector and scalar products matrix notation, formulae of first curvature, second curvature and orientation of true trihedron tracing are demonstrated. Application of quaternionic matrices for a problem of airspace transport system trajectory selection is shown.

1 INTRODUCTION

As (Blekhman, Myshkis, Panovko 1983; Chernyshenko, Rushytskiy 2008) explain, mechanics belongs to engineering sciences in terms of the character of investigated physical phenomena, and to mathematical ones according to analytical approaches applied. Vector calculation is a dominating mathematical tool in mechanics of rigid bodies. Methods and approaches of computational mechanics are used to solve a wide range of engineering and technical problems, in particular those concerned with space flight mechanics including navigation, orientation, stabilization, stability, controllability (Branets, Shmyglevskiy 1973; Onishchenko 1983) as well as dynamics of launcher, aircraft, ship, ground transport etc. (Koshliakov 1985; Kravets, Kravets 2006; Kravets, Kravets, Kharchenko 2009; Larin 2007; Lobas, Verbitsky 1990; Martynyuk 2002). Computer technology application involves the necessity to introduce specific reference system and to reduce vector notation of solution algorithm to coordinate matrix form (Strazhegva, Melkumov 1973; Frezer, Dunkan, Kollar 1950). Matrix calculations in computational experiment provide a number of known advantages. The use of specific mathematical tool in the form of quaternionic matrices calculation is quite sufficient to solve a wide range of problems concerning dynamic design of space, rocket, and aviation equipment, ground transport, robot technology, gyroscopy, vibration

protection etc. in analytical dynamics. R. Bellman (Bellman 1976), and A.I. Maltsev (Maltsev 1970) described some types of quaternionic matrices. Quaternionic matrices were used to control orientation (Ikes 1970; Kravets, Kravets, Kharchenko 2010), in the theory of rigid body finite rotation (Plotnikov, Chelnokov 1981), in the theory of inertial navigation (Onishchenko 1983), and in kinematics and dynamics of a rigid body (Kravets 2001; Chelnokov 2006; Kravets, Kravets 2006; Kravets, Kravets, Kharchenko 2009).

Hence mathematical tool of quaternionic matrices can be applied not only in analytical dynamics in terms of mathematical models development complementing and replacing vector calculation; it also turns to be well-adapted to modern computer technologies to carry out computational experiments concerning dynamics of mechanical systems in spatial motion. Moreover, mathematical models and their adequate algorithms gain group symmetry, invariant form, matrix compatibility, and versatility. This helps accelerate programming, simplify verification of mathematical model and computational process improving the efficiency of intellectual work (Blekhman 1983). The paper aims at systematic substantiation of basic matrices selection as initial and fundamental computation element for quaternionic matrices.

2 PROBLEM DEFINITION

The research objectives are to develop a group of monomial (1, 0, -1) quadric matrices on a set of elements of four-dimensional orthonormal basis and opposite elements, to find out non-Abelian subgroups isomorphic to quaternion group forming the basis for quaternionic matrices, and to place isomorphic matrices to quaternion and conjugate quaternion.

3 A GROUP OF MONOMIAL (1, 0, -1)-MATRICES

A system of four normalized and mutually orthogonal vectors is considered:

$$A_0 = \begin{pmatrix} 1 & 2 & 3 & 4 \\ 1 & 2 & 3 & 4 \end{pmatrix}, A_1 = \begin{pmatrix} 1 & 2 & 3 & 4 \\ 2 & 1 & 4 & 3 \end{pmatrix}, A_2 = \begin{pmatrix} 1 & 2 & 3 & 4 \\ 3 & 4 & 1 & 2 \end{pmatrix}, A_3 = \begin{pmatrix} 1 & 2 & 3 & 4 \\ 4 & 3 & 2 & 1 \end{pmatrix};$$

$$B_0 = \begin{pmatrix} 1 & 2 & 3 & 4 \\ 1 & 2 & 3 & 4^* \end{pmatrix}, B_1 = \begin{pmatrix} 1 & 2 & 3 & 4 \\ 2 & 1 & 4 & 3^* \end{pmatrix}, B_2 = \begin{pmatrix} 1 & 2 & 3 & 4 \\ 3 & 4 & 1 & 2^* \end{pmatrix}, B_3 = \begin{pmatrix} 1 & 2 & 3 & 4 \\ 4 & 3 & 2 & 1^* \end{pmatrix};$$

$$C_0 = \begin{pmatrix} 1 & 2 & 3 & 4 \\ 1 & 2 & 3^* & 4 \end{pmatrix}, C_1 = \begin{pmatrix} 1 & 2 & 3 & 4 \\ 2 & 1 & 4^* & 3 \end{pmatrix}, C_2 = \begin{pmatrix} 1 & 2 & 3 & 4 \\ 3 & 4 & 1^* & 2 \end{pmatrix}, C_3 = \begin{pmatrix} 1 & 2 & 3 & 4 \\ 4 & 3 & 2^* & 1 \end{pmatrix};$$

$$D_0 = \begin{pmatrix} 1 & 2 & 3 & 4 \\ 1 & 2^* & 3 & 4 \end{pmatrix}, D_1 = \begin{pmatrix} 1 & 2 & 3 & 4 \\ 2 & 1^* & 4 & 3 \end{pmatrix}, D_2 = \begin{pmatrix} 1 & 2 & 3 & 4 \\ 3 & 4^* & 1 & 2 \end{pmatrix}, D_3 = \begin{pmatrix} 1 & 2 & 3 & 4 \\ 4 & 3^* & 2 & 1 \end{pmatrix};$$

$$F_0 = \begin{pmatrix} 1 & 2 & 3 & 4 \\ 1^* & 2 & 3 & 4 \end{pmatrix}, F_1 = \begin{pmatrix} 1 & 2 & 3 & 4 \\ 2^* & 1 & 4 & 3 \end{pmatrix}, F_2 = \begin{pmatrix} 1 & 2 & 3 & 4 \\ 3^* & 4 & 1 & 2 \end{pmatrix}, F_3 = \begin{pmatrix} 1 & 2 & 3 & 4 \\ 4^* & 3 & 2 & 1 \end{pmatrix};$$

$$R_0 = \begin{pmatrix} 1 & 2 & 3 & 4 \\ 1 & 2 & 3^* & 4^* \end{pmatrix}, R_1 = \begin{pmatrix} 1 & 2 & 3 & 4 \\ 2 & 1 & 4^* & 3^* \end{pmatrix}, R_2 = \begin{pmatrix} 1 & 2 & 3 & 4 \\ 3 & 4 & 1^* & 2^* \end{pmatrix}, R_3 = \begin{pmatrix} 1 & 2 & 3 & 4 \\ 4 & 3 & 2^* & 1^* \end{pmatrix};$$

$$S_0 = \begin{pmatrix} 1 & 2 & 3 & 4 \\ 1 & 2^* & 3 & 4^* \end{pmatrix}, S_1 = \begin{pmatrix} 1 & 2 & 3 & 4 \\ 2 & 1^* & 4 & 3^* \end{pmatrix}, S_2 = \begin{pmatrix} 1 & 2 & 3 & 4 \\ 3 & 4^* & 1 & 2^* \end{pmatrix}, S_3 = \begin{pmatrix} 1 & 2 & 3 & 4 \\ 4 & 3^* & 2 & 1^* \end{pmatrix};$$

$$T_0 = \begin{pmatrix} 1 & 2 & 3 & 4 \\ 1^* & 2 & 3 & 4^* \end{pmatrix}, T_1 = \begin{pmatrix} 1 & 2 & 3 & 4 \\ 2^* & 1 & 4 & 3^* \end{pmatrix}, T_2 = \begin{pmatrix} 1 & 2 & 3 & 4 \\ 3^* & 4 & 1 & 2^* \end{pmatrix}, T_3 = \begin{pmatrix} 1 & 2 & 3 & 4 \\ 4^* & 3 & 2 & 1^* \end{pmatrix};$$

as well as opposite substitutions:

$$\overline{A_0}, \quad \overline{A_1}, \quad \overline{A_2}, \quad \overline{A_3};$$
$$\overline{B_0}, \quad \overline{B_1}, \quad \overline{B_2}, \quad \overline{B_3};$$
$$\cdots$$
$$\overline{T_0} \quad \overline{T_1} \quad \overline{T_2} \quad \overline{T_3}.$$

Each of them is quadratic monomial (1, 0, -1)-matrix.

Considered biquadratic substitutions and their corresponding (1, 0, -1) –monomial matrices form multiplicative group of 64-power and subgroups of

$\|1 \ 0 \ 0 \ 0\|$, $\|0 \ 1 \ 0 \ 0\|$, $\|0 \ 0 \ 1 \ 0\|$, $\|0 \ 0 \ 0 \ 1\|$, to which finite set of elements e_1 e_2 e_3 e_4 (or 1, 2, 3, 4) and opposite elements e_1^* e_2^* e_3^* e_4^* (or 1^*, 2^*, 3^*, 4^*) are correlated. Opposite vectors of orthonormal four-dimensional basis match opposite elements of the set:

$\|-1 \ 0 \ 0 \ 0\|$, $\|0 \ -1 \ 0 \ 0\|$, $\|0 \ 0 \ -1 \ 0\|$, $\|0 \ 0 \ 0 \ -1\|$.

Note that specific relativity theory, theory of finite rotation, and projective geometry use four-dimensional space.

Set of biquadrate even substitution shown as the total of two transpositions and identity permutations is formed with the help of introduced set of elements (Kargapolov, Merzliakov 1982). Expansion of the required substitution is:

32-, 16-, 8-, 4-, and 2-powers represented by Cayley tables.

4 NON-ABELIAN SUBGROUPS, ISOMORPHIC TO A GROUP OF QUATERNIONS.

Two subgroups of 4-power, seven subgroups of 8-power, twenty-four subgroups of 16-power, and one subgroup of 32-power are separated using the analysis of multiplication table of 64-power group. Initial group order is multiple by the order of any of composed subgroups to match Lagrange theorem (Kargapolov, Merzliakov 1982). Two-power subgroups are not considered due to their triviality.

Four-power subgroups are Abelian. Note that five subgroups of 8-power are Abelian ones and two of them are non-Abelian (marked in Table 1).

Table 1. Subgroups of 8-power monomial matrices

No.	Subgroup element							
1	A_0	A_1	A_2	A_3	$\bar{A}_0$	$\bar{A}_1$	$\bar{A}_2$	$\bar{A}_3$
2*	A_0	R_2	S_3	T_1	$\bar{A}_0$	$\bar{R}_2$	$\bar{S}_3$	$\bar{T}_1$
3*	A_0	S_1	T_2	R_3	$\bar{A}_0$	$\bar{S}_1$	$\bar{T}_2$	$\bar{R}_3$
4	A_0	T_1	S_2	R_3	$\bar{A}_0$	$\bar{T}_1$	$\bar{S}_2$	$\bar{R}_3$
5	A_0	S_1	R_2	T_3	$\bar{A}_0$	$\bar{S}_1$	$\bar{R}_2$	$\bar{T}_3$
6	A_0	R_1	T_2	S_3	$\bar{A}_0$	$\bar{R}_1$	$\bar{T}_2$	$\bar{S}_3$
7	A_0	R_0	S_0	T_0	$\bar{A}_0$	$\bar{R}_0$	$\bar{S}_0$	$\bar{T}_0$

The obtained non-Abelian subgroups are isomorphic to quaternion group. Cayley tables of two non-Abelian subgroups are presented in Table 2.

Table 2. Multiplication tables for non-Abelian subgroups

	A_0	T_1	R_2	S_3	$\bar{A}_0$	$\bar{T}_1$	$\bar{R}_2$	$\bar{S}_3$
A_0	A_0	T_1	R_2	S_3	$\bar{A}_0$	$\bar{T}_1$	$\bar{R}_2$	$\bar{S}_3$
T_1	T_1	$\bar{A}_0$	$\bar{S}_3$	R_2	$\bar{T}_1$	A_0	S_3	$\bar{R}_2$
R_2	R_2	S_3	$\bar{A}_0$	$\bar{T}_1$	$\bar{R}_2$	$\bar{S}_3$	A_0	T_1
S_3	S_3	$\bar{R}_2$	T_1	$\bar{A}_0$	$\bar{S}_3$	R_2	$\bar{T}_1$	A_0
$\bar{A}_0$	$\bar{A}_0$	$\bar{T}_1$	$\bar{R}_2$	$\bar{S}_3$	A_0	T_1	R_2	S_3
$\bar{T}_1$	$\bar{T}_1$	A_0	S_3	$\bar{R}_2$	T_1	$\bar{A}_0$	$\bar{S}_3$	R_2
$\bar{R}_2$	$\bar{R}_2$	$\bar{S}_3$	A_0	T_1	R_2	S_3	$\bar{A}_0$	$\bar{T}_1$
$\bar{S}_3$	$\bar{S}_3$	R_2	$\bar{T}_1$	A_0	S_3	$\bar{R}_2$	T_1	$\bar{A}_0$
*	A_0	S_1	T_2	R_3	$\bar{A}_0$	$\bar{S}_1$	$\bar{T}_2$	$\bar{R}_3$
A_0	A_0	S_1	T_2	R_3	$\bar{A}_0$	$\bar{S}_1$	$\bar{T}_2$	$\bar{R}_3$
S_1	S_1	$\bar{A}_0$	R_3	$\bar{T}_2$	$\bar{S}_1$	A_0	$\bar{R}_3$	T_2
T_2	T_2	$\bar{R}_3$	$\bar{A}_0$	S_1	T_2	$\bar{R}_3$	$\bar{A}_0$	S_1
R_3	R_3	T_2	$\bar{S}_1$	$\bar{A}_0$	$\bar{R}_3$	$\bar{T}_2$	S_1	A_0
$\bar{A}_0$	$\bar{A}_0$	$\bar{S}_1$	$\bar{T}_2$	$\bar{R}_3$	A_0	S_1	T_2	R_3
$\bar{S}_1$	$\bar{S}_1$	A_0	$\bar{R}_3$	T_2	S_1	$\bar{A}_0$	R_3	$\bar{T}_2$
$\bar{T}_2$	$\bar{T}_2$	R_3	A_0	$\bar{S}_1$	T_2	R_3	A_0	$\bar{S}_1$
$\bar{R}_3$	$\bar{R}_3$	$\bar{T}_2$	S_1	A_0	R_3	T_2	$\bar{S}_1$	$\bar{A}_0$

It is known that quaternion is determined as a hypercomplex number:

$1a_0 + ia_1 + ja_2 + ka_3$, where $1a_0$ is scalar, $ia_1 + ja_2 + ka_3$ is vector part of quaternion, a_0, a_1, a_2, a_3 are real numbers, and $1, i, j, k$ are elements of the basis where 1 is a real unit, i, j, k are explained as certain quaternions (hypercomplex units) or as basic vectors of three-dimensional space (Branets, Shmyglevskiy 1973; Maltsev 1970). Specific multiplication rules are adopted for elements of quaternion space basis:

$$i^2 = j^2 = k^2 = -1; ij = -ji = k; jk = -kj = i; ki = -ik = j.$$

The set covering eight elements $1, i, j, k, -1, -i, -j, -k$ (where minus is a distinctive mark) makes a group of quaternions with the known multiplication table (Table 3) (Kargapolov, Merzliakov 1982).

Table 3. Multiplication table for quaternion group

$*$	1	i	j	k	-1	$-i$	$-j$	$-k$
1	1	i	j	k	-1	$-i$	$-j$	$-k$
i	i	-1	k	$-j$	$-i$	1	$-k$	j
j	j	$-k$	-1	i	$-j$	k	1	$-i$
k	k	j	$-i$	-1	$-k$	$-j$	i	1
-1	-1	$-i$	$-j$	$-k$	1	i	j	k
$-i$	$-i$	1	$-k$	j	i	-1	k	$-j$
$-j$	$-j$	k	1	$-i$	j	$-k$	-1	i
$-k$	$-k$	$-j$	i	1	k	j	$-i$	-1

Comparison of multiplication tables for quaternion groups and determined non-Abelian subgroups of 8-power makes it possible to define their isomorphism.

5 QUATERNION MATRICES

Process of juxtaposing elements of quaternion space basis with monomial (1, 0, -1)-matrices of the considered non-Abelian subgroups is not unique. Table 4 represents the list of definite alternatives for juxtaposition of the two non-Abelian subgroups.

Table 4. Alternatives for juxtaposing monomial matrices with elements of quaternion basis

Element of basis	Subgroup elements					
1	A_0	A_0	A_0	$\cdots$	A_0	$\cdots$ A_0
i	S_1	T_2	R_3	$\cdots$	$\overline{S_1}$	$\cdots$ $\overline{R_3}$
j	T_2	R_3	S_1	$\cdots$	$\overline{T_2}$	$\cdots$ $\overline{T_2}$
k	R_3	S_1	T_2	$\cdots$	$\overline{R_3}$	$\cdots$ $\overline{S_1}$
$No.$	1	2	3	$\cdots$	16^*	$\cdots$ 24
1	A_0	A_0	A_0	$\cdots$	A_0	$\cdots$ A_0
i	T_1	S_3	R_2	$\cdots$	$\overline{T_1}$	$\cdots$ $\overline{S_3}$
j	S_2	R_3	T_1	$\cdots$	$\overline{R_2}$	$\cdots$ $\overline{T_1}$
k	R_2	T_1	S_3	$\cdots$	$\overline{S_3}$	$\cdots$ $\overline{R_2}$
$No.$	1	2	3	$\cdots$	10^*	$\cdots$ 24

Alternative No.16 is selected among this variety of alternatives for the first non-Abelian group; alternative No.10 - for the second non-Abelian subgroup, i.e.

$$1 \leftrightarrow A_0, \quad i \leftrightarrow \overline{S}_1, \quad j \leftrightarrow T_2, \quad k \leftrightarrow \overline{R}_3,$$

$$1 \leftrightarrow A_0, \quad i \leftrightarrow \overline{T}_1, \quad j \leftrightarrow R_2, \quad k \leftrightarrow S_3.$$

These juxtaposition alternatives meet the criterion of ordering, or symmetry reflected in the possibility to apply the operation of transposition. In this case, it is expedient to use the definitions of the basis according to Table 5.

Table 5. Basic matrices, isomorphic to quaternion elements

Quaternion elements	Basic matrices		Definitions
1	$A_0 = \begin{Vmatrix} 1 & 0 & 0 & 0 \\ 0 & 1 & 0 & 0 \\ 0 & 0 & 1 & 0 \\ 0 & 0 & 0 & 1 \end{Vmatrix}$		$A_0 = E_0$
i	$\overline{T}_1 = \begin{Vmatrix} 0 & 1 & 0 & 0 \\ -1 & 0 & 0 & 0 \\ 0 & 0 & 0 & -1 \\ 0 & 0 & 1 & 0 \end{Vmatrix}$	$\overline{S}_1 = \begin{Vmatrix} 0 & -1 & 0 & 0 \\ 1 & 0 & 0 & 0 \\ 0 & 0 & 0 & -1 \\ 0 & 0 & 1 & 0 \end{Vmatrix}$	$\overline{T}_1 = E_1, \ \overline{S}_1 = {}^t E_1$
j	$R_2 = \begin{Vmatrix} 0 & 0 & 1 & 0 \\ 0 & 0 & 0 & 1 \\ -1 & 0 & 0 & 0 \\ 0 & -1 & 0 & 0 \end{Vmatrix}$	$T_2 = \begin{Vmatrix} 0 & 0 & -1 & 0 \\ 0 & 0 & 0 & 1 \\ 1 & 0 & 0 & 0 \\ 0 & -1 & 0 & 0 \end{Vmatrix}$	$R_2 = E_2, \ T_2 = {}^t E_2$
k	$S_3 = \begin{Vmatrix} 0 & 0 & 0 & 1 \\ 0 & 0 & -1 & 0 \\ 0 & 1 & 0 & 0 \\ -1 & 0 & 0 & 0 \end{Vmatrix}$	$\overline{R}_3 = \begin{Vmatrix} 0 & 0 & 0 & -1 \\ 0 & 0 & -1 & 0 \\ 0 & 1 & 0 & 0 \\ 1 & 0 & 0 & 0 \end{Vmatrix}$	$S_3 = E_3, \ \overline{R}_3 = {}^t E_3$
-1	$\overline{A}_0 = \begin{Vmatrix} -1 & 0 & 0 & 0 \\ 0 & -1 & 0 & 0 \\ 0 & 0 & -1 & 0 \\ 0 & 0 & 0 & -1 \end{Vmatrix}$		$\overline{A}_0 = I$
-i	$T_1 = \begin{Vmatrix} 0 & -1 & 0 & 0 \\ 1 & 0 & 0 & 0 \\ 0 & 0 & 0 & 1 \\ 0 & 0 & -1 & 0 \end{Vmatrix}$	$S_1 = \begin{Vmatrix} 0 & 1 & 0 & 0 \\ -1 & 0 & 0 & 0 \\ 0 & 0 & 0 & 1 \\ 0 & 0 & -1 & 0 \end{Vmatrix}$	$T_1 = {}^t E_1^t, \ S_1 = E_1^t$
-j	$\overline{R}_2 = \begin{Vmatrix} 0 & 0 & -1 & 0 \\ 0 & 0 & 0 & -1 \\ 1 & 0 & 0 & 0 \\ 0 & 1 & 0 & 0 \end{Vmatrix}$	$\overline{T}_2 = \begin{Vmatrix} 0 & 0 & 1 & 0 \\ 0 & 0 & 0 & -1 \\ -1 & 0 & 0 & 0 \\ 0 & 1 & 0 & 0 \end{Vmatrix}$	$\overline{R}_2 = {}^t E_2^t, \ \overline{T}_2 = E_2^t$
-k	$\overline{S}_3 = \begin{Vmatrix} 0 & 0 & 0 & -1 \\ 0 & 0 & 1 & 0 \\ 0 & -1 & 0 & 0 \\ 1 & 0 & 0 & 0 \end{Vmatrix}$	$R_3 = \begin{Vmatrix} 0 & 0 & 0 & 1 \\ 0 & 0 & 1 & 0 \\ 0 & -1 & 0 & 0 \\ -1 & 0 & 0 & 0 \end{Vmatrix}$	$\overline{S}_3 = {}^t E_3^t, \ R_3 = E_3^t$

These definitions reflect the possibility to transform basic matrices by introducing transposition operations: complete (permutation of each row and column), external (permutation of the first row and column), and internal (permutation of kernel elements, i.e. excluding the first row and column). Each quaternion and conjugate quaternion is correlated to two matrices of the ordered structure:

$$\begin{Vmatrix} A \\ {}^tA^t \\ {}^tA \\ A^t \end{Vmatrix} = \begin{Vmatrix} E_0 & E_1 & E_2 & E_3 \\ E_0 & {}^tE_1^t & {}^tE_2^t & {}^tE_3^t \\ E_0 & {}^tE_1 & {}^tE_2 & {}^tE_3 \\ E_0 & E_1^t & E_2^t & E_3^t \end{Vmatrix} \begin{Vmatrix} a_0 \\ a_1 \\ a_2 \\ a_3 \end{Vmatrix}$$

Respectively, expansion is as follows:

$$A = \begin{Vmatrix} a_0 & a_1 & a_2 & a_3 \\ -a_1 & a_0 & -a_3 & a_2 \\ -a_2 & a_3 & a_0 & -a_1 \\ -a_3 & -a_2 & a_1 & a_0 \end{Vmatrix}; \quad {}^tA = \begin{Vmatrix} a_0 & -a_1 & -a_2 & -a_3 \\ a_1 & a_0 & -a_3 & a_2 \\ a_2 & a_3 & a_0 & -a_1 \\ a_3 & -a_2 & a_1 & a_0 \end{Vmatrix};$$

$$ {}^tA^t = \begin{Vmatrix} a_0 & -a_1 & -a_2 & -a_3 \\ a_1 & a_0 & a_3 & -a_2 \\ a_2 & -a_3 & a_0 & a_1 \\ a_3 & a_2 & -a_1 & a_0 \end{Vmatrix}; \quad A^t = \begin{Vmatrix} a_0 & a_1 & a_2 & a_3 \\ -a_1 & a_0 & a_3 & -a_2 \\ -a_2 & -a_3 & a_0 & a_1 \\ -a_3 & a_2 & -a_1 & a_0 \end{Vmatrix}.$$

6 TRANSPOSITION

The composed matrices are characterised by apparent ordering as they are transformed into each other by the proposed operations of complete, external, and internal transposition. Matrix ${}^tA$ is formed from matrix A as a result of the first row and first column permutation; matrix ${}^tA^t$ is transposed relative to matrix A and matrix A^t is transposed relative to ${}^tA$, i.e. the following transposition rules are applied:

$$ {}^t(A)^t = {}^tA^t, \qquad {}^t(A) = {}^tA, \qquad (A)^t = A^t, $$

$$ {}^t({}^tA^t)^t = A, \qquad {}^t({}^tA^t) = A^t, \qquad ({}^tA^t)^t = {}^tA, $$

$$ {}^t({}^tA)^t = A^t, \qquad {}^t({}^tA) = A, \qquad ({}^tA)^t = {}^tA^t, $$

$$ {}^t(A^t)^t = {}^tA, \qquad {}^t(A^t) = {}^tA^t, \qquad (A^t)^t = A. $$

7 VALIDATION OF QUATERNION MATRICES

In equivalent formulation, the quaternion matrices represent basic operations of vector algebra in a particular case when the scalar part of quaternion is equal to zero. It is quite evident that the following correlations of scalar and vector product of several vectors and multiplicative compositions of quaternion matrices are true (Kravets, Kravets, Kharchenko 2010):

- for two vectors $\overline{a}, \overline{b}$

$$ \begin{Vmatrix} \overline{a} \cdot \overline{b} \\ 0 \end{Vmatrix} \leftrightarrow \frac{1}{2}(A_0 + A_0^t) \cdot b_0, \qquad \begin{Vmatrix} 0 \\ \overline{a} \times \overline{b} \end{Vmatrix} \leftrightarrow \frac{1}{2}(A_0 - A_0^t) \cdot b_0; $$

- for three vectors $\overline{a}, \overline{b}, \overline{c}$

$$ \begin{Vmatrix} \overline{a} \cdot (\overline{b} \times \overline{c}) \\ 0 \end{Vmatrix} \leftrightarrow \frac{1}{4}(A_0 + A_0^t)(B_0 - B_0^t) \cdot c_0, \quad \begin{Vmatrix} 0 \\ \overline{a} \times (\overline{b} \times \overline{c}) \end{Vmatrix} \leftrightarrow \frac{1}{4}(A_0 - A_0^t)(B_0 - B_0^t) \cdot c_0; $$

- for four vectors $\bar{a}, \bar{b}, \bar{c}, \bar{d}$

$$\left\|\begin{array}{c} \bar{a}\cdot\left[\bar{b}\times\left(\bar{c}\times\bar{d}\right)\right] \\ 0 \end{array}\right\| \leftrightarrow \frac{1}{8}\left(A_0+A_0^t\right)\left(B_0-B_0^t\right)\left(C_0-C_0^t\right)\cdot d_0,$$

$$\left\|\begin{array}{c} 0 \\ \bar{a}\times\left[\bar{b}\times\left(\bar{c}\times\bar{d}\right)\right] \end{array}\right\| \leftrightarrow \frac{1}{8}\left(A_0-A_0^t\right)\left(B_0-B_0^t\right)\left(C_0-C_0^t\right)\cdot d_0,$$

$$\left\|\begin{array}{c} 0 \\ \left(\bar{a}\times\bar{b}\right)\times\left(\bar{c}\times\bar{d}\right) \end{array}\right\| \leftrightarrow \frac{1}{8}\left[\left(A_0+A_0^t\right)\left(B_0+B_0^t\right)-\left(B_0+B_0^t\right)\left(A_0+A_0^t\right)\right]\left(C_0-C_0^t\right)\cdot d_0,$$

and other correlations.

8 KINEMATICS

To illustrate the above mentioned correlations application, kinematic problem of selecting motion trajectory for reusable airspace transport system is considered (Panov, Gusynin, Serdiuk, Karpov 1999). Development of completely reusable airspace transport system in Ukraine involves solution of a number of specific problems; territorial limitations being one of them. Traditional flight trajectory for rocket-and-space systems in the form of sloping lines in a shooting plane turns to be unacceptable due to impossibility to ensure safety exclusion area. Using airspace system in which AH-225 launch plane is the first stage and the aircraft-spacecraft with supersonic combustion ramjet is the second stage helps to implement innovative payload deployment trajectories in the form of a spiral line (Fig. 1).

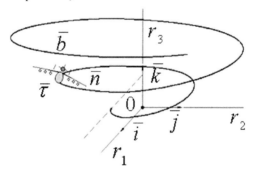

Figure 1. Payload deployment trajectories in the form of a spiral line

Let us assume that the spatial trajectory is given in a fixed coordinate system of the ground complex (air facilities) by a hodograph

$$\bar{r} = \bar{i}\left(\rho_0+\rho_1 t+\rho_2 t^2+\rho_3 t^3\right)\cos\omega t+\bar{j}\left(\rho_0+\rho_1 t+\rho_2 t^2+\rho_3 t^3\right)\sin\omega t+\bar{k}\left(h_0+h_1 t+h_2 t^2+h_3 t^3\right)$$

where $\omega, \rho_i, h_i\,(i=0,1,2,3)$ are variable trajectory parameters determined by the set boundary conditions and apparent dependence: $\omega t_k = 2\pi n$ where n is the number of spiral turns and t_k is staging time.

The following kinematic trajectory parameters are used as initial data: $H(0)$, $L(0)$, $H(t_k)\cdot L(t_k)$ are altitude and distance from air facilities in horizontal plane up to the point of first stage entry into spiral

trajectory $(t = 0)$ and in a staging point $(t = t_k)$; $v(0), w(0), v(t_k), w(t_k)$ are horizontal and vertical components of launch plane velocity in a point of spiral trajectory $(t = 0)$ and a moment of aircraft-spacecraft separation $(t = t_k)$. The specified initial data are used to develop two independent systems of algebraic equations relative to variable parameters of the required trajectory – both linear and non-linear. These systems allow the following analytical solution

$$h_0 = H(0), \qquad h_1 = w(0),$$

$$h_2 = 3\frac{H(t_k)-H(0)}{t_k^2} - \frac{w(t_k)+2w(0)}{t_k}, \quad h_3 = \frac{w(t_k)+w(0)}{t_k^2} - 2\frac{H(t_k)-H(0)}{t_k^3},$$

$$\rho_0 = L(0), \qquad\qquad \rho_1 = \sqrt{v^2(0)-\omega^2 L^2(0)},$$

$$\rho_2 = 3\frac{L(t_k)-L(0)}{t_k^2} - 2\frac{\sqrt{v^2(0)-\omega^2 L^2(0)}}{t_k} - \frac{\sqrt{v^2(t_k)-\omega^2 L^2(t_k)}}{t_k},$$

$$\rho_3 = \frac{\sqrt{v^2(t_k)-\omega^2 L^2(t_k)}}{t_k^2} + \frac{\sqrt{v^2(0)-\omega^2 L^2(0)}}{t_k^2} - 2\frac{L(t_k)-L(0)}{t_k^3}$$

under apparent condition $v^2(t) \geq \omega^2 L^2(t)$.

The basic kinematic parameters below are determined analytically for the obtained spiral trajectory: single vectors of tangential $\overline{\tau}$, principal normal $\overline{n}$, binormal $\overline{b}$ as well as camber K, torsion χ, the matrix of directional cosine angles connecting axes of the moving trihedron and fixed axes of the ground complex, i.e.

$$\overline{\tau} = \frac{\dot{\overline{r}}}{V}, \qquad \overline{n} = -\frac{\dot{\overline{r}}\times(\dot{\overline{r}}\times\ddot{\overline{r}})}{V^4 K}, \qquad \overline{b} = -\frac{\dot{\overline{r}}\times\left[\dot{\overline{r}}\times(\dot{\overline{r}}\times\ddot{\overline{r}})\right]}{V^5 K},$$

$$K = \frac{\left|\dot{\overline{r}}\times\ddot{\overline{r}}\right|}{V^3}, \qquad \chi = -\frac{\dot{\overline{r}}\cdot(\ddot{\overline{r}}\times\dddot{\overline{r}})}{\dfrac{1}{V^2}\begin{vmatrix}\dot{\overline{r}}\cdot\dot{\overline{r}} & \dot{\overline{r}}\cdot\ddot{\overline{r}} \\ \dot{\overline{r}}\cdot\dot{\overline{r}} & \dot{\overline{r}}\cdot\ddot{\overline{r}}\end{vmatrix}\cdot\begin{vmatrix}\dot{\overline{r}}\cdot\dot{\overline{r}} & \dot{\overline{r}}\cdot\ddot{\overline{r}} \\ \dot{\overline{r}}\cdot\dot{\overline{r}} & \dot{\overline{r}}\cdot\ddot{\overline{r}}\end{vmatrix}}, \qquad V^2 = \dot{\overline{r}}\cdot\dot{\overline{r}}.$$

It is convenient to calculate complex vector products used in the given formulae with the help of the proposed matrix algorithms adapted to computer implementation:

$$\dot{\overline{r}}\times\ddot{\overline{r}} \leftrightarrow \frac{1}{2}(\dot{R}-\dot{R}^t)\ddot{r},$$

$$\dot{\overline{r}}\cdot\dot{\overline{r}} \leftrightarrow \frac{1}{2}(\dot{R}+\dot{R}^t)\dot{r}, \qquad \dot{\overline{r}}\times(\dot{\overline{r}}\times\ddot{\overline{r}}) \leftrightarrow \frac{1}{4}(\dot{R}-\dot{R}^t)^2\ddot{r},$$

$$\dot{\overline{r}}\cdot(\ddot{\overline{r}}\times\ddot{\overline{r}}) \leftrightarrow \frac{1}{4}(\dot{R}+\dot{R}^t)(\ddot{R}-\ddot{R}^t)\ddot{r}, \qquad \dot{\overline{r}}\times\left[\dot{\overline{r}}\times(\dot{\overline{r}}\times\ddot{\overline{r}})\right] \leftrightarrow \frac{1}{8}(\dot{R}-\dot{R}^t)^3\ddot{r}.$$

Note also that

$$\begin{vmatrix}\dot{\overline{r}} & \ddot{\overline{r}} \\ \dot{\overline{r}}\cdot\dot{\overline{r}} & \dot{\overline{r}}\cdot\ddot{\overline{r}}\end{vmatrix}\cdot\begin{vmatrix}\dot{\overline{r}} & \ddot{\overline{r}} \\ \dot{\overline{r}}\cdot\dot{\overline{r}} & \dot{\overline{r}}\cdot\ddot{\overline{r}}\end{vmatrix} \leftrightarrow \frac{1}{16}\ddot{r}^t(\dot{R}-\dot{R}^t)^4\ddot{r}.$$

Where $\dot{R}$, $\dot{R}^t$, $\ddot{R}$, $\ddot{R}^t$ are quaternion matrices with zero scalar part.

Thus, we obtain the following formulae in quaternion matrices:

$$\overline{\tau} \leftrightarrow \frac{1}{V}\dot{r}, \quad \overline{n} \leftrightarrow -\frac{1}{4V^4 K}(\dot{R}-\dot{R}^t)^2\ddot{r}, \quad \overline{b} \leftrightarrow -\frac{1}{8V^5 K}(\dot{R}-\dot{R}^t)^3\ddot{r},$$

$$\chi = -\frac{4V^2(\dot{R}+\dot{R}^t)(\ddot{R}-\ddot{R}^t)\ddot{r}}{\ddot{r}^t(\dot{R}-\dot{R}^t)^4\ddot{r}}, \qquad K^2 = \frac{\ddot{r}^t(\dot{R}-\dot{R}^t)^2\ddot{r}}{4V^6}, \qquad V^2 = -\frac{1}{4}\dot{r}^t(\dot{R}+\dot{R}^t)\dot{r}.$$

The dependence between the moving trihedron orts and the basic reference system in quaternion matrices is:

$$\left\| 0 \ \bar{\tau} \ \bar{n} \ \bar{b} \right\| = \frac{1}{4 \, V^4 K} \left\| 0 \ \bar{i} \ \bar{j} \ \bar{k} \right\| \left\| 0 \left| 4 \, V^3 K \, \dot{r} \right| - \left(\dot{R} - \dot{R}^t \right)^2 \ddot{r} \left| - \frac{1}{2V} \left(\dot{R} - \dot{R}^t \right)^3 \ddot{r} \right\| .$$

The matrix of directional cosine angles can be derived directly from it.

9 CONCLUSIONS

It is proposed to apply mathematical technique of quaternion matrices for analytical and computational mechanics, which is sufficient both for mathematical model development and for computational experiments. Algorithms in quaternion matrices are adapted for computer technology. Calculation of quaternion matrices is isomorphic to algebra of quaternions and vector algebra in three-dimensional space (Branets, Shmyglevskiy 1973; Onishchenko 1983).

The basis for the introduced collection of four quaternion matrices on a set of elements of four-dimensional orthonormal space and opposite element in the form of monomial (1, 0, -1)-matrices making two non-Abelian subgroups of 8-power is determined. Isomorphism of elements of quaternion space basis and the developed collections of basic matrices is shown. Symmetry of quaternion matrices is reflected in three transposition operations and appropriate nomenclature. The results are the basic computational elements for quaternion matrices. Complex vector and scalar products used in mechanics are represented in the equivalent formulation by the considered quaternion matrices. The problem of determining basic kinematic parameters of spiral trajectory of reusable airspace transport system is solved analytically. Calculation algorithms are represented in the form of quaternion matrices providing convenient computer implementation.

REFERENCES

Bellman R. 1976. *Introduction into matrix theory* (in Russian). Moscow: Nauka: 352.

Blekhman I.I., Myshkis A.D., Panovko Y.G. 1983. *Mechanics and applied mathematics: Logics and peculiarities of mathematics application* (in Russian). Moscow: Nauka. Glavnaia red. phis.-mat. lit.: 328.

Branets V.N., Shmyglevskiy I.P. 1973. Mechanics of *space flight: Application of quaternions in problems of rigid body orientation* (in Russian). Moscow: Nauka: 320.

Ikes B.P. 1970. *New method to perform numerical calculations connected with operation of orientation system control based on quaternion application* (in Russian). Raketnaia Tekhnika I kosmonavtika, Issue 8, Vol.1:13−19.

Chernyshenko I.S., Rushytskiy Y.Y. 2008. *Timoshenko S.P. Institute of Mechanics of NAN of Ukraine (1918-2008). − 90th anniversary of the Institute (History. Structure. Information aspects) /Under general editorship of A.N. Guz* (in Russian). Kyiv: Litera LTD: 320.

Kargapolov M.I., Merzliakov Y.I. 1982. *Foundations of group theory* (in Russian). Moscow: Nauka: 288.

Koshliakov V.N. 1985. *Problems of rigid body dynamics and applied theory of gyroscopes. Analytical methods* (in Russian). Moscow: Nauka: 288.

Kravets T.V. 2001. *On the use of quaternion matrices to describe rotating motion of rigid body in space* (in Russian). Tekhnicheskaia mekhanika, Issue 1: 148-157.

Kravets V.V., Kravets T.V., Kharchenko A.V. 2010. *Representation of multiplicative compositions of four vectors by quaternion matrices* (in Russian). Vostochno-Evropeiskiy zhurnal peredovykh tekhnologiy, Issue 54 (47): 15-29.

Maltsev A.I. 1970. *Foundations of linear algebra* (in Russian). Moscow: Nauka: 400.

Onishchenko S.M. 1983. *Hypercomplex numbers in a theory of inertial navigation. Autonomous systems* (in Russian). Kyiv: Naukova dumka: 208.

Panov A.P., Gusynin V.P., Serdiuk I.I., Karpov A.S. 1999. *Identifying kinematic parameters of airspace system stage motion* (in Russian). Tekhnicheskaia mekhanika, Issue 1: 76-83.

Plotnikov P.K., Chelnokov Y.N. 1981. *Quaternion matrices in the theory of rigid body finite rotation* (in Russian). Moscow: Vysshaia shkola: Collection of scientific papers in theoretical mechanics, Issue 11: 122 − 128.

Strazhegva I.V., Melkumov V.S. 1973. *Vector and matrix method in flight mechanics* (in Russian). Moscow: Mashinostroienie: 260.

Frezer R., Dunkan V., Kollar A. 1950. *Matrix theory and its application for differential equations in Dynamics* (in Russian). Moscow: Inostrannaia literatura: 445.

Chelnokov Y.N. 2006. *Quaternion and biquaternion models and rigid body mechanics method and their application. Geometry and kinematics of motion* (in Russian). Moscow: Fizmatlit: 512.

Kravets V.V., Kravets T.V. 2006. *On the nonlinear dynamics of elastically interacting asymmetric rigid bodies*. International Applied Mechanics, Issue 42, Vol. 1: 110 – 114.

Kravets V.V., Kravets T.V., Kharchenko A.V. 2009. *Using Quaternion Matrices to Describe the Kinematics and Nonlinear Dynamics of an Asymmetric Rigid Body*. International Applied Mechanics, Issue 45, Vol. 2: 223 – 231.

Larin V.B. 2007. *On the Control Problem for a Compound Wheeled Vehicle.* International Applied Mechanics, Issue 43, Vol. 11: 1269–1275.

Lobas L.G., Verbitsky V.G. 1990. *Quantitative and Analytical Methods in Dynamics of Wheel Machines* (in Russian). Kyiv: Naukova Dumka: 232.

Martynyuk A.A. 2002. *Qualitative Methods in Nonlinear Dynamics: Novel Approaches to Liapunov's Matrix Functions.* New York: Basel: Marsel Dekker: 301.

Power Engineering and Information Technologies In Technical Objects Control – Pivnyak, Beshta & Alekseyev (eds)
© 2016 Taylor & Francis Group, London, ISBN 978-1-138-71479-3

Computer analysis of photobiological utilizer parameters of solid oxide fuel cells emissions

O. Beshta, V. Fedoreyko, A. Palchyk, N. Burega & R. Sipravskyy
State Higher Educational Institution "National Mining University", Dnipro, Ukraine

ABSTRACT: The purpose of the article is justification of automation methods and analysis of microbiology systems in the autonomous power supply complex based on solid oxide fuel cells for intensifying the flow of renewable energy, as well as reducing greenhouse gas emissions. Investigation of guaranteed power supply processes are based on the laws of mass conservation, electrical engineering, electrochemistry, biological processes of photosynthesis, by using the data obtained from simulation and physical modeling. The research analysed biological characteristics of algae during their use in autonomous power supply system based on solid oxide fuel cells. The necessity for automatic analysis of photobioreactor's biological systems for effective utilization of carbon dioxide and biomass generation has been justified. The algorithm control of the photobioreactor and the methods of computer segmentation of Chlorella Vulgaris biological parameters were elaborated. The algorithm of segmentation of microalgae biological parameters was tested against the existing methods. Using biological technology photosynthesis module for automatic analysis of microorganisms in the autonomous power supply system based on fuel cells for selection and accumulation of energy and utilization of renewable sources of carbon dioxide was suggested for the first time. Control algorithms and state analysis of photosynthesis subsystem of the autonomous power supply complex based on solid oxide fuel cells were elaborated.

1 INTRODUCTION

Energy consumption in the world increases, the subsoil resources of the biosphere constantly decrease due to significantly adverse human activity. Environmental problems largely arise from the interference of technogenic emissions by means of large amounts of carbon dioxide in the generational process of electrical, mechanical or thermal energy.

Hence, the modern science has focused on increasing power generation efficiency and developing methods for recycling carbon dioxide.

One of the actively developing technologies is Solid Oxide Fuel Cell (SOFC) as the future replacement of conventional power plants run on methane. Considerable interest in this technology is explained by the fact that its efficiency is about 70%, while the classic power plant's efficiency is - 45%. Although the efficiency is actually twice as high, the system will generate a significant amount of carbon dioxide (Beshta, Fedoreyko, Palchyk, Burega 2015).

In order to fully utilize the carbon dioxide as a result of the electricity generation we should use photobioreactors for the cultivation of unicellular microalgae.

Using photobioreactors to grow microalgae for the purpose of recycling CO_2 and biofuelgeneration is one of the technologies involving renewable sources, which are actively developing. The technology has several advantages associated with the possibility of full utilization of CO_2 and its conversion into lipids as a feedstock for biodiesel and enzymes, proteins for food and pharmaceutical industries. The efficiency of this technology is about 15% compared to solar panels; the control ability of biological processes of cell lipids allows to receive the output up to 60% taking into account their dry weight.

2 PURPOSE

Justification of automation methods and analysis of microbiology systems in the autonomous power supply complex based on solid oxide fuel cells for intensifying flow of renewable energy, as well as reducing greenhouse gas emissions.

3 METHODOLOGY

Investigation of guaranteed power supply processes based on the laws of conservation, electrical engineering, electrochemistry, biological processes of photosynthesis, by using data obtained from simulation and physical modeling.

4 RESULTS

The technology of photobioreactors usage is sufficiently investigated in the laboratory where a culture of microalgae and its condition and purity can be constantly monitored and controlled to ensure the optimal conditions for its growth. It is necessary to constantly monitor the temperature of the environment, pH level, lighting, carbon dioxide concentration and the number of cells (biomass) with relation to volume.

During the transfer of technology of microalgae growth onto the large-scale production, the Control Measuring Module(CMM) performs the main control function; it provides the necessary photobioreactor mode by using the information from sensors (Fig. 1a, b).

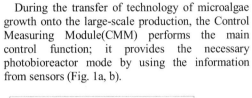
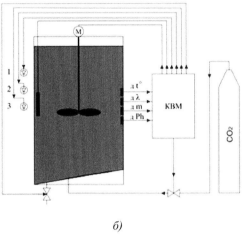

a)   *б)*

Figure 1. a) General view of a flat vertical photobioreactor which has a working volume of 50 liters
b) Scheme of the flat vertical Photobioreactor: 1,2,3 – ribbon panel on the base of white, red and blue LEDs; дt°– sensor of ambient cultivation temperature; дλ – light level sensor of the photobioreactor; дm – microalgae concentration sensor; д pH – pH level sensor

Photobioreactors are built as a flow system in which the cultivation and supplement of the breeding environment takes place continuously (Fig. 2a), or cyclic environment with limited growing period and full restart of microcultures as well as food environment. (Fig 2b) (Hrubinko, Handzyura 2008).

Regular collection of photobioreactors microculture and its review in Goryaeva camera take place in the laboratory to establish the rational concentration of cells for screening and receiving biomass of microalgae. However, this approach is not feasible for mass production due to significant operating time required for cell counting and impossibility to organize the process on the required scale.

The object of the laboratory research was an analogous pure culture of green microalgae Chlorella Vulgaris Beij., which was cultured in the Fitzgerald conditions with modifications Zehnder and Gorham №11, at 22-25 ^{O}C and fluorescent lighting (intensity of 2500 lux) for 16 hours a day (Topachevsky 1975).

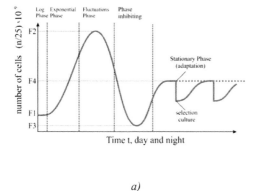
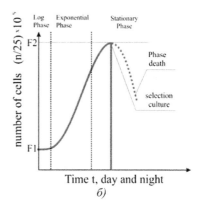

a)   б)

Figure 2. a) Change in the cells number in a mode of continuous cultivation where: F1 - addictive period; F2 -
a sharp increase in the cells concentration (10-100 times) due to energy resources; F3 – decrease in the
functional activity below the initial level; F4 - restructuring plastic and energy supply and accessing to
stationary phase;
b) cells growth in the reactor of the accumulation mode (one production cycle)

Therefore, one of the necessary stages of microalgae cultivation technology development will be designing photobioreactors which can support the optimum parameters, calculate and control the amount and condition of the cells automatically. The main task of the algorithm is to determine the internal temperature of the reactor, pH acidity, light level, the volume of carbon dioxide gotten and utilized by the reactor and maintaining these parameters within normal limits by means of controlling heating, gas supply, lighting and mixing of the substrate. This process takes place according to the time of day as well as time and state of cultivation and quantity of biomass. The algorithm provides time for the culture rest (8 hours) and the selection of the substrate to hold the number of cells in the stationary phase by an automatic segmentation of images during automatic review with the help of a microscope.

One of the algorithm components is tracking the number of cells for which the appropriate mathematical apparatus and software was developed based on processing and segmentation of the received images through the automated system of the microscope.

Real images which are obtained from the microscope, except cells and their fragments, contain a lot of noises, irregularities, impurities and foreign objects whose presence is to be identified. The image gotten from the microscope is photographed to recognize the contours of the cells (Fig. 3a).The algorithm converts the image into the grayscale, converting the color components into monochrome images appropriated for their brightness (Fig. 3b), and then the image contrast is increased by means of stretching the intensity values of the dynamic range (Fig. 3c) (Gonzalez & Woods 2009).

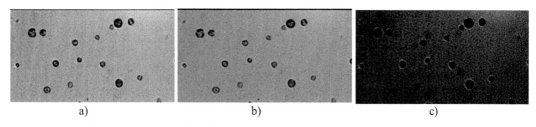

a)   b)   c)

Figure 4. Image processing microalgae cells: a) original image; b) grayscale image;
c) images with increased contrast X1000

Segmentation is achieved by applying the boundary operator to the image for reducing the amount of the processed data. Filtered part of the data is less important, but the most important structural parts are stored. As a result of the

boundary selection the set of connected curves is formed, which indicates the boundaries of objects.

Methods of image segmentation were investigated and represented in the form of masks to calculate the number of cells. The principle of these methods

13

is based on the difference between brightness of elements and background images. This difference is represented as derivatives approximating the gradient. Gradient operators were made like matrix masks of Roberts, Previtt, Sobel, and Laplacian-Gauss operators.

The Roberts cris-crossed operator has a high performance, is easy to use, but is very sensitive to interferences and noises (Fig. 4a). This operator uses four values of the image brightness (Lukyanitsa & Shishkin 2009).

$$G_{i,j} = \sqrt{(E_{(i+1),(j+1)} - E_{i,j})^2 + (E_{(i+1),j} - E_{i,(j+1)})^2}, \quad (1)$$

or

$$G_{i,j} = |E_{(i+1),(j+1)} - E_{i,j}| + |E_{(i+1),j} - E_{i,(j+1)}|, \quad (2)$$

Where E - matrix of the original image; i, j - coordinates of the image element (pixel).

Sobel operator uses twice more brightness values of the processed image (Fig. 4 b).

$$G_{i,j} = \sqrt{G_x^2 + G_y^2}, \quad (3)$$

Where G_x and G_y — two images, where each point contains partial derivatives with respect to x and y respectively.

Matrices of this operator are as follows

$$G_x = \begin{bmatrix} -1 & 0 & +1 \\ -2 & 0 & +2 \\ -1 & 0 & +1 \end{bmatrix} \cdot E \text{ и } G_y = \begin{bmatrix} -1 & -2 & -1 \\ 0 & 0 & 0 \\ \mp 1 & +2 & +1 \end{bmatrix} \cdot E, \quad (4)$$

Where E –the matrix of the original image.
Software of the image is presented as follows:

$$G_{i,j} = \sqrt{G_{i,j(x)}^2 + G_{i,j(y)}^2} \text{ або } G_{i,j} = |G_{i,j(x)} + G_{i,j(y)}|, \quad (5)$$

$$G_{i,j(x)} = [E_{(i-1),(j-1)} + E_{(i-1),j} + E_{(i-1),(j+1)}], \quad (6)$$

$$G_{i,j(y)} = [E_{(i-1),(j-1)} + E_{i,(j-1)} + E_{(i+1),(j-1)}] - [E_{(i-1),(j+1)} + E_{i,(j+1),j} + E_{(i+1),(j+1)}]. \quad (7)$$

Laplacian operator is based on second-order derivatives whose task is to search the border of places where the derivative changes the sign of the brightness function. It is also very sensitive to noise and can also trigger doubling of contours.

In order to reduce the impact of noise, this operator is used in conjunction with smoothing by Gaussian method and this combination is named Gauss-Laplacian (Laplacian Of Gaussian) (Fig. 4c).

Operator mask is represented by the formula:

$$LoG(x,y) = -\frac{1}{\pi \cdot \sigma^4} \cdot \left(1 - \frac{x^2 + y^2}{2\sigma^2}\right) \cdot e^{\frac{x^2 + y^2}{2\sigma^2}}, \quad (8)$$

Where σ– standard deviation of Gaussian distribution.

Filter mask is represented by the formula:

$$LoG(x,y) = \frac{1}{1+a} \begin{bmatrix} -a & a-1 & -a \\ a-1 & a+5 & a-1 \\ -a & a-1 & -a \end{bmatrix}, \quad (9)$$

Where a – parameter at the range $[0, 1]$

Canny operator has such features as an increased signal / noise ratio, good localization and a single response per border (Fig. 4d).

These results of the operator are achieved because it is a very close approximation to the first Gaussian derivative and has the form:

$$f(x) = -x \cdot \exp\left[-\frac{x^2}{2s^2}\right]. \quad (10)$$

To study image segmentation algorithm, its results were compared with standard cell counting method and it was determined that the quality of segmentation depends on the work of selected boundary operators. It was determined that Canny operator satisfies the required accuracy of cells counting constituting 97.3% of the number of cells counted in the Goryaev chamber by the standard method (Fig. 4i).

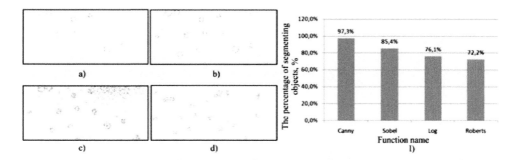

Figure 4.The operator selection borders: a) Roberts; b) Sobel; c) Laplacian-Gaussian; d) Canny;
i) Accuracy of counting cells operators regarding to the selection borders

In order to eliminate some accidental drawbacks of images fragmentation (boundary curves are not connected to each other, all borders absence in some places or the presence of erroneous boundary) a build-up function is used.

The algorithm fills the holes in the binary image for further image processing. With this approach, we get an array of disk-shaped structure of the investigated binary fields (images) , outline of its path and we can calculate the radius of the object (Fig. 5).

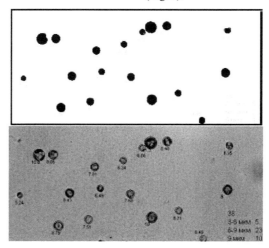

Figure 5. The result of cell counting

According to the algorithm of the biological parameter identification, the sample of microalgae culture taken from photobioreactors of Goryaev camera was studied. The study was conducted in 7 visual fields and after that the results were treated by a computer program. The obtained data were matched with the standard method of counting cells.

5 CONCLUSIONS

1. The structural and technological schemes of power generation based on solid oxide fuel cell and photobioreactor subsystem were created and the laboratory stand for the technology of growing microcultures Chlorella Vulgaris according to the

operating modes of the fuel cell was made.

2. Expediency of automatic analysis of biological systems by computer photobioreactor image segmentation was grounded and developed via appropriate algorithm and software.

3. It is established that the segmentation algorithm satisfies the required accuracy of counting cells constituting 97.3% of the number of cells counted in the Goryaev chamber by the standard method.

REFERENCES

Beshta O., Fedoreyko V., Palchyk A., N. Burega V. 2015. *Independent power supply of manage*

15

object based on biosolid oxide fuel. UK: London: Taylor & Francis Group: Power Engineering, Control and Information Technologies in Geotechnical Systems: 33-39.

Hrubinko V.V., Handzyura V.P. 2008. *The concept of harmfulness in ecology* (in Ukrainian). Kyiv - Ternopil: ITNPU named after B. Hantyuk: 144.

Topachevsky A.V. (ed.) 1975. *Methods of physiological and biochemical studies of algae in hydrobiological practice* (in Russian). Kyiv: Naukova Dumka: 247.

Gonzalez R.C., Woods R.E. 2009. *Digital Image Processing using MATLAB.* Pearson Prentice Hall: 827.

Lukyanitsa A.A., Shishkin A.G. 2009. *Digital image processing* (in Russian). Moscow: "I-S-S Press": 518.

The submission backward trigonometrically functions as periodical functions

O. Aziukovskyi, A. Golovchenko
State Higher Educational Institution "National Mining University", Dnipro, Ukraine

ABSTRACT: In this article is offer for the submission impulse periodical functions with a kink, apply a reverse trigonometrically (circular) functions and their combination with trigonometrically functions. At the same time not necessary in a lot of summands of unknown function for accuracy submission a kink, of original function. In the paper is described, examples of submission and graph more simply periodical circular functions with a kink. The time t is selected as argument of the function.

1 INTRODUCTION

The periodical functions y(ωt), are often, submission as sum of trigonometrically function with different period T=2π/ω, as trigonometrically polynomial. On accuracy of submission function y(ωt) is influence her view and number of its members of polynomial. Especially a lot of members of polynomial is necessary use when submission are impulse periodical function, which include a kink. A lot of members in polynomial is create a difficulties for obtain a solver (Bhattachryya, Datta, Keel 2009; Zhang 2012).

2 FORMULATING THE PROBLEM

Same times, the impulse periodical functions with a kink submission no periodical functions. For example, this is may be algebraic polynomials (Bhattachryya, Datta, Keel 2009; Dudgeon, Mersereau 1988; Zhang 2012). But in this case, is necessary a limit of their by the logical operator.

This method, in case his applying to periodical trigonometrically functions, interconnected with necessary to consider a lot of polynomial and conditions of their existence. No need for a large number of summands of the required analytical functions for accurate description of breaks of the original function. The paper presents the description and graphics of the simplest periodic functions with characteristic circular breaks in which the argument selected time t.

3 MATERIALS UNDER ANALYSIS

Consider bipolar isosceles triangles, which may be submissions periodical circular function (fig.1):

$$F(\omega t) = H \frac{2}{\pi} \arcsin \left[\sin \left(\omega t - \varphi \right) \right], \qquad (1)$$

were: H, ω, φ – amplitude (the height of the triangle); angular frequency; initial phase of periodical function.

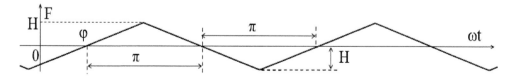

Figure1. Periodical bipolar isosceles triangular

The periodical right triangle is submission next functions (fig.2):

$$F = H \frac{2}{\pi} arctg \left(tg \frac{\omega t - \varphi}{2} \right) \qquad (2)$$

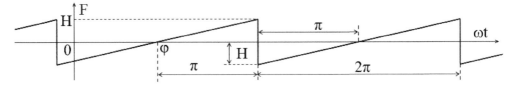

Figure 2. Periodical right triangular.

In the case of periodical bipolar rectangle we apply periodical circular function, which show as (fig.3):

$$F(\omega t) = H\frac{2}{\pi}\left[arctg\left(tg\frac{\omega t-\varphi}{2}\right) - arctg\left(tg\frac{\omega t-\varphi+\pi}{2}\right)\right]. \tag{3}$$

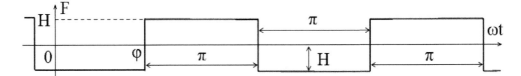

Figure 3. Periodical bipolar rectangular.

Next periodical circular function is submission of periodic bipolar trapezium (fig. 4):

$$F(\omega t) = \frac{H}{\pi-\beta}\left\{arcsin\left[sin\left(\omega t-\varphi-\frac{\beta}{2}\right)\right] - arcsin\left[sin\left(\omega t-\varphi+\frac{\beta}{2}-\pi\right)\right]\right\}, \tag{4}$$

where $0\leq\beta=\pi-2\tau<\pi$ - width horizontal part of impulse in angular measurements; τ – width inclined part of impulses in angular measurement. Value τ is not to be less then deal minimum number, for example $\tau\geq10^{-11}$.

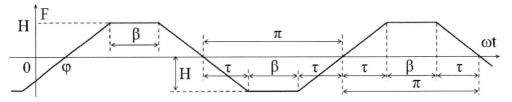

Figure 4. Periodical trapezium with variable inclined part

On the base of formula (4) we may describe periodical impulse such as isosceles triangular (fig.1) if $\beta=0$. Also, formula (4) may be applied when если $\beta\rightarrow\pi$ (periodic rectangular signal rectangular (fig.3).

In addition, in practical work with electrotechnical equipment we needed analyze periodical bipolar rectangular with time delay (fig 5). Like this signals may be submissions next periodical circular function:

$$F(\omega t) = H\frac{1}{\pi}\left[\begin{array}{l}\left(arctg\left(tg\frac{\omega t-\varphi-\beta+\pi}{2}\right) - arctg\left(tg\frac{\omega t-\varphi+\pi}{2}\right)\right)\\ \left(-arctg\left(tg\frac{\omega t-\varphi-\beta}{2}\right) + arctg\left(tg\frac{\omega t-\varphi}{2}\right)\right)\end{array}\right] \tag{5}$$

were $0<\beta\leq\pi$ - width impulse in angular measurement; $p=\pi-\beta$ – width between impulses in angular measurement.

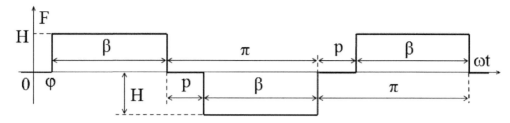

Figure 5. Periodical bipolar rectangular with variable pause

Electrotechnical and information systems also generate and process periodical bipolar trapezium with a time pause. This class signals we offer submission next periodical circular function (fig. 6):

$$F(\omega t)=\frac{H}{2\tau}\left\{\begin{array}{l}\arcsin\left[\sin\left(\omega t-\varphi-\frac{\pi}{2}\right)\right]-\arcsin\left[\sin\left(\omega t-\varphi-\tau-\frac{\pi}{2}\right)\right]+\\ \arcsin\left[\sin\left(\omega t-\varphi-\beta-2\tau-\frac{\pi}{2}\right)\right]-\arcsin\left[\sin\left(\omega t-\varphi-\beta-\tau-\frac{\pi}{2}\right)\right]\end{array}\right\},\qquad(6)$$

were $0\leq\beta=\pi-p-2\tau<\pi$ – width horizontal part of signals in angularly measurement; τ – width of inclined part of signal. This value not be mast less, them some very small number, for example $\tau\geq10^{-11}$; $p=\pi-d-2\tau$ – width time pause between signals in angular measurement.

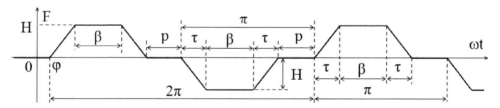

Figure 6. Periodical trapezium with inclined part and time pause

Draw attention, that periodical trapezium with time pause (fig. 6) may be transform to periodical isosceles triangular (fig.1), when condition $\beta\to0$ and $\tau\to\pi/2$ is right. When $\beta\to\pi$ и $\tau\to0$ the periodical trapezium with time pause may be transform to rectangular without time pause (fig.3).

And in conclusions, when $\beta+2\tau\to\pi$ to trapezium without time pause (fig. 4); when $\tau\to0$ to rectangular with time pause.

Periodical one polar rectangular submissions next periodical circular function (fig.7):

$$F(\omega t)=H\frac{1}{\pi}\left[arctg\left(tg\frac{\omega t-\varphi-\beta+\pi}{2}\right)-arctg\left(tg\frac{\omega t-\varphi+\pi}{2}\right)+\frac{\beta}{2}\right]\qquad(7)$$

where $0<\beta=2\pi-p<2\pi$ – width of impulse in angular measurement; $p=2\pi-\beta$ – width of time pause in angle measurement.

19

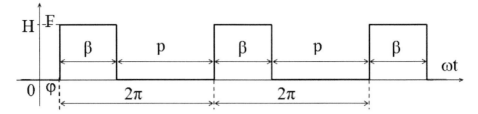

Figure 7. Periodical one polar rectangular signal with time pause

Various combination of periodical circular function is provide a forming of various periodical signals. As a more simple example for that signals consider nonsymmetrical triangular (fig. 8) and symmetrical triangular with time pause (fig.9):

$$F(\omega t)=H\frac{2}{\pi}\left\{arcsin[sin(\omega t-\varphi)]+arctg\left(tg\frac{\omega t-\varphi+\pi/2}{2}\right)\right\} \tag{8}$$

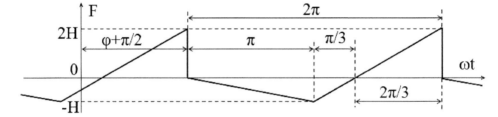

Figure 8. Periodic nonsymmetrical rectangular

$$F(\omega t)=H\frac{1}{\pi}\{arcsin[sin(\omega t-\varphi)]+arctg[tg(\omega t-\varphi)]\} \tag{9}$$

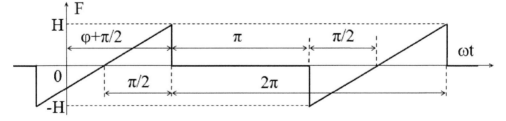

Figure 9. Periodical symmetric rectangular with time pause

The needed form of periodical function we may submission as sequence periodical circular function with the even period and even type. But this is may be only when they different in amplitude and phase.

Example the application, which base on usage periodical one polar rectangle (7) for the imaging the signal, show in formula (10) and fig. 10.

$$F(\omega t)=\frac{1}{\pi}\sum_{k=1}^{3}H_k\left[arctg\left(tg\frac{\omega t-\varphi_k-\beta_k+\pi}{2}\right)-arctg\left(tg\frac{\omega t-\varphi_k+\pi}{2}\right)+\frac{d_k}{2}\right] \tag{10}$$

20

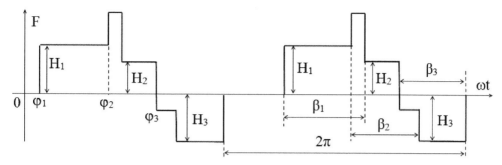

Figure 10. The combinations of the periodic one polar rectangular with use the overstrike

More interesting is a case of periodical one polar rectangular (7) with unity amplitude (H=1). They have a specific property in period and step to change from 1 to 0 and back with the course of time "t". This property we may be called "property of periodical logical commutator". This commutator can be used with another periodical function, as multiplicand. For example, on the base of combination of sine function, show the capability to create elements of sine function, which is relevant to impulse devices (11), (fig. 11):

$$F(\omega t) = H \frac{\sin(\omega t - \psi)}{\pi} \left\{ \begin{array}{c} arctg\left[tg\left(\omega t - \varphi - \beta + \frac{\pi}{2} \right) \right] - \\ - arctg\left[tg\left(\omega t - \varphi + \frac{\pi}{2} \right) \right] + \beta \end{array} \right\},$$ (11)

were - ψ start phase of the sine function; φ - beginning phase of the logical commutator.

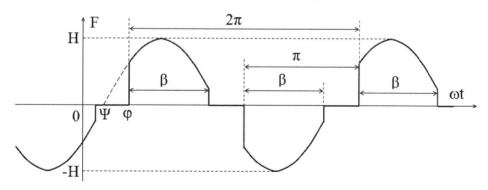

Figure 11. Periodical impulse as part of the sine function

4 CONCLUSIONS

In this paper a new mode of analytic submissions of continuous periodic functions is described. This offer is based on usage of impulse periodical function with typical for their kinks. This kink is created by the backward circular functions. This circular function is not having, a limitation on parameter "t" without use logical operator.

The considered periodical circular function may be, in some case, alternative for classic trigonometrical functions. All of this are formed background for solve a new tasks. In the sequence of impulse periodic functions more interesting for submission periodic signal is a one polar rectangular impulse. This signal may be considered as periodic logical commutator. Such commutators which combine with another periodical functions, is condition a capability for creation periodical signal another forms. Also this mode of solve allow to describe a continuous periodic function, as sequence periodical impulse with some period repeat and differ on amplitude, beginning phase and form.

Also, in a case, when we need to create difficult periodical signals for different processes investigation in electrotechnical devices, we also may to use this method to describe periodical signals.

Particular attention should be paid to the possibility of describing asymmetrical periodic signal that is characteristic for the analysis of processes in communication systems. Also the consideration of electromagnetic processes in electromechanical systems with frequency converters is relevant. Thus, the signals described by periodic bipolar rectangles are frequent in transformative technology. These signals are applied as state control signals of the power transistors. At the same time, periodic trapezoid with variable inclined sides is typical for providing soft switching mode.

Proposed an alternative way to describe periodic signals opens up new possibilities for their further processing. The main advantage in comparison with classical approaches is the possibility to describe arbitrarily complex periodic signal change laws.

With that, the magnitude of the signal does not have to be variable. In communication systems, telemetry periodic signals of complex shape are applied. These signals often have a break. Investigations of the processes that occur in these systems are faced with the complexity of generating these signals in real-world models. The proposed mathematical tool for the reproduction of such signals provides the opportunities for research processes.

REFERENCES

Bhattachryya S.P., Datta A., Keel L.H. 2009. *Linear control theory (Structure, robustness and optimization)*. Boca Raton: CRC Press: 911.

Dudgeon D.P., Mersereau R.M. 1988. *Multidimensional Digital Signal Processing* (in Russian). Moscow: Mir: 487.

Zhang W. 2012. *Quantitative Process Control Theory*. Taylor & Francis Group: CRC Press: 443.

Power Engineering and Information Technologies In Technical Objects Control – Pivnyak, Beshta
& Alekseyev (eds)
© 2016 Taylor & Francis Group, London, ISBN 978-1-138-71479-3

Peculiarities of experimental data reading for DC power supply of the stator in the problem of parameters identification of induction motor equivalent circuit

O. Beshta, A. Siomin
State Higher Educational Institution "National Mining University", Dnipro, Ukraine

ABSTRACT: Substantiation of the value of the signal sampling frequency in experimental data reading in the problem of parameters identification of induction motor equivalent circuit. Methods for describing electromagnetic processes known from the theory of electrical machines and a method for solving differential equations are used. The solutions obtained make it possible to determine the value of sampling frequency in experimental data reading while using the corrective factor which depends on the power of the motor. The results presented in the paper allow specialists to make a reasonable choice of sampling frequency for experiments with DC power supply of the stator . Using of the proposed corrective factor will better the results of parameters identification. Proposed recommendations can be used for the development of algorithms and software for microprocessor systems of AC drives with induction motors for the identification of equivalent circuit parameters.

1 INTRODUCTION

Considered in (Beshta, Siomin 2014) method of parameter identification involves asymmetrical (two-phase) DC power supply of induction motor (IM) stator, in which experimental data of current transient are received. The results of experimental data processing are used to define the parameters of equivalent circuit of IM. Efficiency of data receiving and processing affects the result of identification. This applies fully to the value of sampling frequency of the signal, the justification of which requires further consideration. Let us examine this question in more detail.

The aim of the paper is to get substantiation of the value of sampling frequency of the signal in experimental data receiving that will contribute to effectiveness of identification process.

2 THE USE OF CORRECTIVE FACTOR

The process of the current change with time with two phase DC power supply of the stator is described by the following differential equation written in relative units (Beshta, Siomin 2014):

$$\left(\frac{x_1 x_2 - x_{12}}{r_2}\right)\frac{d^2 i_S^|}{d\tau^2} + \left(\frac{x_1 r_2 + r_1 x_2}{r_2}\right)\frac{d i_S^|}{d\tau} + r_1 i_S = u_S^| + \frac{d u_S^|}{d\tau}\frac{x_2}{r_2}, \tag{1}$$

where $u_S^|$, $i_S^|$ – stator voltage and current in relative units (r.u.);

r_1, r_2 –stator resistance and rotor resulting resistance, r.u.;

x_1, x_2, x_{12} – inductive resistances of stator, rotor and mutual inductance, r.u.
$x_1 = x_{12} + x_{1\sigma}$, $x_2 = x_{12} + x_{2\sigma}$;

$x_{1\sigma}, x_{2\sigma}$ – the stator and rotor leakage inductive resistances, r.u.;

τ – dimensionless time.

The second term in (1) in case of DC voltage supply is equal to zero. Equation (1) is heterogeneous second order differential equation.

It is known (Piskunov 1985) that the general solution of the heterogeneous equation is the sum of the general solution of the corresponding homogeneous differential equation y_0 and a particular solution y_1 of heterogeneous equation:

$$y = y_0 + y_1.$$

Finding a particular solution in the form of $y_1 = A \cdot i_S^| + B$ and substituting it in (1), we obtain that $y_1 = \dfrac{u_S^|}{r_2}$. To find the general solution of the homogeneous differential equation we ought to calculate the roots of the auxiliary characteristic equation:

$$\left(\frac{x_1 x_2 - x_{12}}{r_2}\right)\cdot k^2 + \left(\frac{x_1 r_2 + r_1 x_2}{r_2}\right)\cdot k + r_1 = 0$$

The assumption $x_{1\sigma}\cdot r_2 - x_{2\sigma}\cdot r_1 \approx 0$ allows us to obtain the value of the discriminant as:

$$D = \frac{x_{12}^2 (r_2 + r_1)^2}{r_2^2}.$$

If we assume that

$$x_{1\sigma}\cdot x_{2\sigma} \ll x_{12}\cdot(x_{1\sigma} + x_{2\sigma})$$

and

$$x_{1\sigma}\cdot r_2 + x_{2\sigma}\cdot r_1 \ll 2x_{12}\cdot(r_2 + r_1)$$

which is consistent with the typical parameter values of IM (Ivanov-Smolenskiy 1980), the roots of the characteristic equation are:

$$k_1 = -\frac{x_{1\sigma}r_2 + x_{2\sigma}r_1}{2x_{12}(x_{1\sigma} + x_{2\sigma})} = \frac{-1}{T_1}; \quad k_2 = -\frac{r_2 + r_1}{x_{1\sigma} + x_{2\sigma}} = \frac{-1}{T_2}.$$

Then the general solution of the heterogeneous differential equation will be:

$$i_S^| = \frac{u_S^|}{r_1} + C1^|\cdot e^{-\tau/T1^|} + C2^|\cdot e^{-\tau/T2^|}. \tag{2}$$

The integration constants C1 and C2 considering zero initial conditions are defined as:

$$C_1 = \frac{u_S^|}{r_1}\frac{T_1}{(T_2 - T_1)}, \quad C_2 = -\frac{u_S^|}{r_1}\frac{T_2}{(T_2 - T_1)}.$$

For identification of parameters (Beshta, Siomin 2014), area of the figure, enclosed between the steady-state value of the current and current transient curve is used. This area is equal to:

$$S = \frac{u_S^|}{r_1}(T_1 + T_2). \tag{3}$$

Justification of assumptions made in the determination of (2) was performed by comparing the area of a figure calculated on the basis of (3) with the numerical determination of the area based on the original equation (1) using the trapezoidal method. We used the typical values of the equivalent circuit parameters (Ivanov-Smolenskiy 1980) taken for motors of various power. The ratios of area of the figure calculated by means of (3) to the value of the area, calculated using (1) are shown in Figure 1.

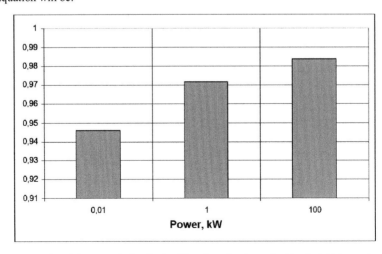

Figure 1. Ratios of the area calculated using (3) to the value determined by the initial equation (1)

According to Figure 1, we can make conclusions on the expediency of using the corrective factor, the value of which can be taken as 0.95 for motors having power approximately 0.01 kW, 0.97 – for 1 kW motors and 0.985 for motors of about 100 kW.

3 THE CHOICE OF THE SAMPLING STEP

When receiving the experimental data it is important to determine the correct value of a sampling step.

As a criterion we have chosen the value of the area ratio, calculated for a given sampling step to the value of this area, calculated using the expression (3). Similarly, the information about the specific values of the equivalent circuit parameters of the motors of various capacities was used.

The calculation results are shown in Figure 2 for 100 kW motor. Calculations for lower power machines were practically identical and so are not presented in this paper. For better presentation the abscissa was taken as the ratio of the time constant T1 to the value of the sampling step.

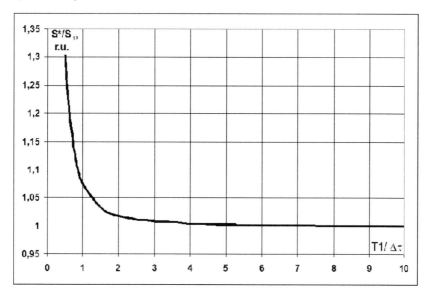

Figure 2. The dependence of the accuracy of determining the area depending on the sampling step for 100kW motor

5 CONCLUSIONS

– assumptions made in the paper for finding the analytical solutions of the differential equation (1) are acceptable. This is illustrated by Figure 1. In addition, we recommend to use the corrective factor which depends on the power of the motor;

– for subsequent use of the area value (after the experimental determination) for identification of parameters by the method presented in [1], it is advisable to use a corrective factor, whose value can be taken as 0.95 for motors having power approximately 0.01 kW, 0.97 - 1 kW motors and 0.985 for motors of about 100 kW;

– dependence of the calculations accuracy on the sampling step is of nonlinear character, approaching uniformity with an increase of $\dfrac{T1}{\Delta\tau}$;

– acceptable accuracy of calculations using the method of trapezoids (<2%) is achieved with the ratio $\dfrac{T1}{\Delta\tau} \geq 5$. In relative units the sampling step has

to be $\Delta\tau \leq 80...4\ r.u.$ (higher values relate to motor power of about 100kW, lower values – to motor power of about 0.01 kW). In absolute units $\Delta t = \dfrac{\Delta\tau}{\omega_{\acute{A}}} = \dfrac{\Delta\tau}{2\pi f_H} \leq 0,255...0,013\ c.$ If we take the value of the sampling step equal to $\Delta t = 0,005\ sec.$, it will suffice for the entire range.

REFERENCES

Beshta O., Siomin A. 2014. *Parameters estimation of induction motor equivalent circuit using asymmetrical stator supply* (in Russian). Kremenchug: KrNU: Elektromekhanichni ta energozberigayuchi systemy. Schokvartalniy naukovo-virobnichiy zhurnal, Issue 2/2014 (26): 10-16.

Piskunov N.S. 1985. *Differential and integral calculus for technical universities, vol.1* (in Russian). Moscow: Nauka: 432.

Ivanov-Smolenskiy A.V. 1980. *Electric machines* (in Russian). Moscow: Energiya: 928.

25

Power Engineering and Information Technologies In Technical Objects Control – Pivnyak, Beshta
& Alekseyev (eds)
© 2016 Taylor & Francis Group, London, ISBN 978-1-138-71479-3

Estimation of electromagnetic compatibility of traction frequency converters for mine transport

M. Rogoza, Yu. Papaika, A. Lysenko & L. Tokar
State Higher Educational Institution "National Mining University", Dnipro, Ukraine

ABSTRACT: The article discusses the electromagnetic compatibility of traction frequency converters. Determined voltage harmonics. The analysis of losses of active energy in the cable lines and transformers. Additional losses in the electrical equipment are defined. The method of calculating the damage from the non-sinusoidal voltage. For a practical evaluation of the quality of electricity damage zone proposed which depend on the level higher harmonics.

1 INTRODUCTION

Higher harmonics (HH) and voltage unbalance (VU) have become the key problem in achieving electromagnetic compatibility of traction frequency converters (TFC) with the mains while launching a transport system with inductive power delivery in the context of gas-hazardous and dust-hazardous coal mines (Zhezhelenko 2004; Zhezhelenko, Shidlovskiy, Pivnyak 2012). As underground mine mains are of limited values of short circuit (SC) power due to specific norms and requirements for power-supply systems of coal mines (spark safety, individual electric supply of underground consumers, and large extension of cable lines), higher harmonics and unbalance within underground mains are issues of concern. It is virtually impossible to lower levels of the specified electromagnetic disturbance using standard mains techniques common in industrial power-supply systems. It should also be noted that for more than a decade, an active discussion of problems connected with interharmonics generated into mains by converters has been going on in specialized literature (Zhezhelenko, Saenko, Baranenko, Gorpinich, Nesterovich 2007; Pivnyak, Shidlovsky, Kigel, Rybalko, Khovanska 2004; Pivnyak, Zhezhelenko, Papaika 2014). Thus, the problem of TFC electromagnetic compatibility in view of new facts and state of current power-supply systems is still topical and has to be solved.

2 MATERIALS FOR RESEARCH

TFC circuit (Fig. 1) uses 6-pulse bridge diagram where 5th, 7th, 11th, and 13th harmonics, called classical ones, prevail; their levels (in relation to 1st harmonic) being inversely proportional to the number; i.e. 1/5, 1/2, 1/11, 1/13 etc.

Calculation practice of HH of different valve converters considers rectifier as a source of classical HH of current whose level is inversely proportional to the number of harmonics. That corresponds to a case when direct-current circuit inductance is $L_d = \infty$, and commutation inductance is $L_k = 0$ (Zhezhelenko, Saenko, Baranenko, Gorpinich, Nesterovich 2007). Curves of linear currents are of rectangular-stepped configuration with 120 electrical degree interval. In actual practice considering L_d - L_k ratio, the curves are deformed due to availability of pulsating dc component (Fig. 2).

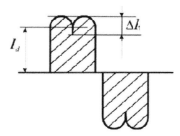

Figure 2. Shape of HH line current for $L_d \neq \infty$

Pulsations affect the value of classical harmonics in proportion to the intensity of pulsations characterized by pulsation factor:

$$\lambda_I = \frac{I_{6m}}{I_d}.$$

while we obtain $I_1 = I_d (1.10 + 0.14 \lambda_1)$ for fundamental harmonic.

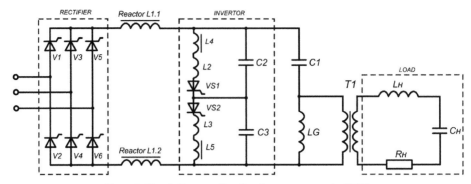

Figure 1. Circuit diagram of traction frequency converter (TFC)

For the amplitude of classical harmonics we get:

$$I_n = 100 \left(\frac{1}{n} + \frac{6,46\,\lambda_I}{n-1} - \frac{7,13\,\lambda_I}{n} \right)(-1)^k \text{ for } n = k_p - 1;$$

$$I_n = 100 \left(\frac{1}{n} + \frac{6,46\,\lambda_I}{n+1} - \frac{7,13\,\lambda_I}{n} \right)(-1)^k \text{ for } n = k_p + 1.$$

In the context of certain λ_I values, HH relative values can be determined according to (Zhezhelenko, Shidlovskiy, Pivnyak 2012; Zhezhelenko, Saenko, Baranenko, Gorpinich, Nesterovich 2007) expressions:

n	5	7	11	13
$I_n^{(\lambda)}$	$0,2 + 0,9\,\lambda_I$	$0,14 + 0,21\,\lambda_I$	$0,091$	$0,076 - 0,0087\,\lambda_I$

It is possible to represent expression for $\lambda1$ as follows:

$$\lambda_1 = \frac{U_{d0}K_{d6}\cos\alpha}{6\omega_1 L_d I_d},$$

where U_{d0} is the amplitude of rectified voltage; and K_{d6} is the coefficient related to the voltage of 6th harmonic.

Fig. 3 represents function curves of four classical harmonics in the mains and two harmonics of direct current.

By analysing the curves represented in Fig. 3, we can conclude that inductance increase L_d (or increase in commutation circuit inductance Lk which is possible due to decrease in power of transformer feeding TFC or decrease in power of short circuit within the connection node of distribution reducing substation DRS-6 kV) results in substantial increase in the mains current 5th harmonics and decrease of 7th harmonic; while levels of 11th and 13th harmonics experience minor changes. In similar conditions, increase in 6th

harmonic level within pulsating current network causes comparable increase of HH level in curves of mains current.

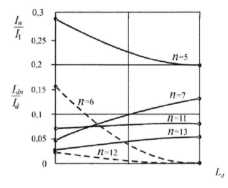

Figure 3. Curves of relationships between mains current harmonics, load current and L_d/L_k ratio

The obtained relationships make it possible to estimate intensity of TFC primary current pulsation in terms of traction converter parametrization as well as mine power supply system state. It will allow to estimate and forecast simultaneous operation modes of TFC and power mains when electromagnetic compatibility experiences critical disturbance; besides, unstable performance of converter is possible.

Higher harmonics in electric mains are undesirable due to a number of consequences for power-supply system of a mine: they have negative effect on technical state of electric facilities impairing economic parameters of their performance. This is the reason for extra loss in energy which worsens thermal conditions of electric equipment complicating reactive power compensation with the help of static capacitor banks. This fact shortens the life of electric machines and apparatus due to accelerated ageing of

28

insulation; in addition, errors of networking and system automatic facilities as well as remote control equipment take place.

Transformers, engines, generators aerial and cable power supply lines demonstrate the greatest value of active power loss resulting from higher harmonics. Increase in active resistance of the components along with frequency increase are proportional to the value $\sqrt{v}$ despite the fact that the approximation is not accurate to some extent (Zhezhelenko, Shidlovskiy, Pivnyak 2012). In some cases extra loss may result in inadmissible overheat and electric equipment failure. Value of active power extra loss is determined by the operation mode of electric equipment and main level of higher harmonics. Calculation of electric power extra loss within electrical distribution system of a mine has been performed using calculation of higher harmonic current level within terminals of traction converter.

Value of active power extra loss within cable overhead lines is determined as follows:

$$\Delta P_{add\ v} = 3\sum_{v=3}^{n} I_v^2 R_v, \qquad (1)$$

where I_v is current of vth harmonic; R_v is active resistance of overhead line within vth harmonic.

Generally, it is re ommended to determine active resistance of overhead lines using the dependence (Zhezhelenko, Saenko, Baranenko, Gorpinich, Nesterovich 2007):

$$R_v = R_2 K_r K_{rv}, \qquad (2)$$

where $R_2 = r_0 l$ is active resistance of reverse sequence of overhead line; $K_r = \sqrt{v}$ is the coefficient related to changes in active resistance in respect to frequency; K_{rv} is correcting coefficient related to parameter distribution within equivalent circuit (as cable lengths within longwall are shorter than 1 km, then parameter distribution may be neglected. That is why $K_{rv} = 1$).

Tables 1 and 2 show calculation results of extra loss within both overhead line cables and transformers. Harmonic current spectrum has been analyzed; 5th, 7th, 11th, and 13th have been taken into consideration.

Table 1. Calculation of extra loss levels in transformers

Transformer type	ΔP_{SC}, kW	Loss in terms of standard operation mode ΔP_i, kW	Extra loss resulting from HH, ΔP_{add}, kW
ТСВП-160	1.9	1.216	0.057
ТСВП-250	2.8	1.792	0.084
ТСВП-400	3.6	2.304	0.108
ТСВП-630	4.7	3.008	0.141

Table 2. Calculation of extra loss levels in cable lines

Cable type used in underground mine workings	The most typical cross-sections of electric conductors	r_0, Ohm/km	Extra loss resulting from HH per 100 cable meters δP_{add}, kW
КШВЭБбШв ВЭБбШв СБН	50	0.363	0.65
	70	0.26	0.5
	95	0.191	0.45
	120	0.151	0.3

If we assume that the same section cable has been applied for each site of underground cable main, it is possible to develop a coefficient taking into consideration electric remoteness of electromagnetic disturbance source in the main. In the context of electromagnetic compatibility level calculation, the coefficient is reduced to the length of cable line detecting zones of reliable and unstable operation of TFC depending upon voltage quality. Table 2 determined estimation levels of active power extra

loss within power cables reduced to 100 m length of cable line.

To perform qualitative estimation of electromagnetic damage within power supply system of coal mine resulting from voltage anharmonicity it is required to determine values of anharmonicity factors at various stages of power energy distribution within a mine (Fig.4). It is proposed to specify the following typical load nodes to estimate quality indices:

- Low voltage side of underground mobile substation– point *K1*;
- High voltage side of underground mobile substation– point *K2*;

- Collecting buses **DUS**-6 kV – point *K3*;
- Collecting buses **CUS**-6 kV – point *K4*;
- Collecting buses 6 kV **MRS** – point *K5*.

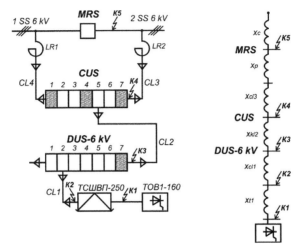

Figure 4. Electric circuit of coal mine underground mains (a) and the replacing circuit (b), where MRS – main reducing substation, CUS - central underground substation, DUS - distribution underground reducing substation, TOB - converter, CL - cable line

Electric power supply system of a mine is a large dynamic system why time constants of basic components are rather high; that is why they can be considered as wide constant components with parameters in frequency spectrum of the power envelope line (0.1 – 15 Hz) being practically constant. Thus, they can be considered as linear ones. While developing equivalent circuits of power components (transformers, reactors, most loads) transfer function module can be represented by transfer coefficient with the general expression

$$K = \frac{x_c}{x_c + x_p}, \qquad (3)$$

where x_c is inductance of PSS prior to component;

x_p is inductance of the power component.

Resistances are calculated according to the replacing circuit of PSS performed to calculate SC currents. Depending on the estimated values, there may be different values of coefficient of voltage curve distortion K_U in anharmonicity modes calculation.

Table 3 shows values of transfer coefficients K for standard transformers of underground substations with $U_{nom} = 6$ kV and various values of PS power. In each case, values of nonlinear load within buses of anharmonicity source (connection of TOB-1-160 converter) are the initial calculation point (node) using the Tables.

Table 3. Values of transfer coefficients for standard underground mobile substations with $U_{HV} = 6$ kV

$S_{nom.t}$, kV·A	S_{scs}, MV·A	Transfer coefficient		
		$K_{TOB-DUS}$	$K_{DUS-CUS}$	$K_{CUS-MRS}$
160	50	0.122	0.488	0.224
	75	0.104	0.386	0.177
	100	0.095	0.323	0.153
250	50	0.153	0.604	0.292
	75	0.125	0.502	0.228
	100	0.112	0.433	0.195
400	50	0.195	0.706	0.234
	75	0.156	0.613	0.187
	100	0.136	0.546	0.143

Adequate value of electromagnetic loss resulting from voltage anharmonicity within power supply system calls for a new idea of frequency converter (FC) as a component of power supply system. Innovative approach involves classical harmonics in a spectrum of converter initial current and analysis of inverter interharmonics calculated using commutation function.

Modern frequency regulated electric drive in coal mines of Ukraine is a powerful tool to control technological processes of coal mining and transportation. Basic technological production stages (rock and coal hoisting, water pumping, and ventilation of mine workings) are mostly equipped with frequency converters operating according to the converter circuit with direct current segment (including high-frequency underground transportation facilities). Application of frequency regulated electric drive is expedient when it is required to control machines and mechanisms in a wide range of working values. In this context, optimum power consumption with minimum specific expenses is achieved (Pivnyak, Zhezhelenko, Papaika 2014). However, it should be noted that powerful frequency converters are sources of higher harmonics of current and voltage which affects voltage quality in electric power supply systems and negatively influences electromagnetic compatibility of electric facilities. Classical higher components and side frequencies (interharmonics) are available within a distortion spectrum depending upon rectifier circuit and control laws; they may reach 40% of the main harmonic current in relative units [80]. A number of national and foreign scientists (I.V. Zhezhelenko, Yu.L. Saenko, Yu.S. Zhelezko, O.G. Grib, A.V. Volkov) have analysed and synthesized the curve of input current of frequency converters in terms of various loading modes (Zhezhelenko 2004; Zhezhelenko, Shidlovskiy, Pivnyak 2012; Zhezhelenko, Saenko, Baranenko, Gorpinich, Nesterovich 2007; Pivnyak, Shidlovsky, Kigel, Rybalko, Khovanska 2004; Pivnyak, Zhezhelenko, Papaika 2014). Representation of FC as a power supply system component, substantiation of mathematical apparatus to analyze operation mode and higher harmonics generation, joint assessment of system parameters of power supply, converter and other complex load in quality indices simulation in the load node of power interconnections are topical tendencies of the research into power supply and electromagnetic compatibility. These issues can only be solved through complex approach to scientific problems and the results validation in practice.

Electromagnetic component of damage due to voltage anharmonicity is characterized by increase in active power loss; growth of active and reactive power consumption; acceleration of electric equipment insulation ageing; restriction of application area of capacitor banks with the view to increasing power coefficient.

Thus, annual damage by electric power loss (electromagnetic component of loss) is:

$$D_e = \Delta P_{a.add} C_e T_b , \qquad (4)$$

Where $\Delta P_{a.add}$ is additional loss of active power resulting from HH; C_e is cost of one kWh of electric power in terms of specified enterprise (assume current tariff by the National Committee for Regulation of Power Energy (NCRPE) for 2nd class enterprises); T_b is operation time for the biggest distortion loads.

The research allowed to obtain boundary values of electromagnetic damage for the basic electric facilities (transformers and asynchronous motors). The capacity of transformers of mobile underground substations was taken within $160...630$ kV·A; the power of asynchronous motors was taken within $10...100$ kW.

Statistic data obtained while carrying out scientific investigations in coal mines of "DFEC Pavlogradugol" served as the basis for estimation of extra loss levels used for the first time to determine graph indices of higher harmonic current behaviour. The coefficients explain HH behaviour during a shift, a day, a year. Thus, it is possible to say that the average shift and daily timing of the biggest distortion loads was validly determined by the method of active power loss determination.

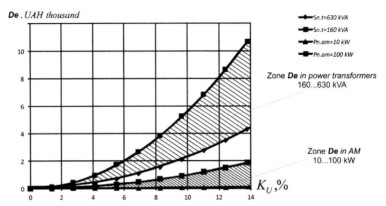

Figure 5. Boundary dependences of annual electromagnetic damage in transformers and asynchronous motors (zone of electromagnetic damage in power transformers of 160-630 kVA; the zone of electromagnetic damage in asynchronous motors of 10-100 kW)

The relations in Fig.5 are based on the information on HH current behaviour obtained experimentally. Scientifically substantiated interconnections between HH current and technological algorithm of a power consumer have not been available before. Thus, accurate determination of electromagnetic damage within power supply system, estimation of the possible connection of transport TFC in any point of electric main as well as possibility to meet the requirements of electromagnetic compatibility of other nonlinear consumers should involve technical features of certain segments with frequency drive.

3 CONCLUSIONS

The conducted research allowed us to establish the following regularities:

1. Specific character of transients while starting TFC of contactless transport has helped highlight the field of non-solved scientific problems dealing with electromagnetic compatibility and spark safety of exterior contours. Taking into account the fact that TFC itself is a damage source energizing from anharmonic voltage, it is required to estimate total level of extra loss within electric power supply system. That is why values of electromagnetic disturbances have been analyzed taking into consideration features of underground mine mains – low level of short-circuit power in energizing circuit of underground sites; the level is responsible for high values of extra power loss resulting from HH in transformers and cable lines.

2. Analysis of TFC mode parameters as a component of electric power supply system should involve total complex node load (in terms of coal

mines). Such settin of the simulation problem helps to detect interrelations of mode parameters of the power system, linear load, and powerful nonlinear loads of TFC as well as estimate levels of electromagnetic compatibility of load nodes. Mathematical modeling has made it possible to identify zones of annual electromagnetic damage resulting from anharmonicity voltage in basic electric facilities (UAH 3.5…8.0 thousand for transformers of mobile substations with 160…630 kV·A power respectively and UAH 0.3…1.8 thousand for asynchronous motors with 10…100 kW power respectively).

REFERENCES

Zhezhelenko I.V. 2004. *Higher harmonics within systems of power supply of industrial enterprises* Moscow: 254.

Zhezhelenko I.V., Shidlovskiy A.K., Pivnyak G.G. 2012. *Electromagnetic compatibility of consumers.* Moscow: 350.

Zhezhelenko I.V., Saenko Yu.L., Baranenko T.K., Gorpinich A.V., Nesterovich V.V. 2007. *Certain problems of anharmonic modes within electric mains of enterprises.* Moscow: 296.

Pivnyak G.G., Shidlovsky A.K., Kigel G.A., Rybalko A.Ya., Khovanska O.I. 2004. *Specific operation modes of electric mains.* Dnipropetrovsk: National Mining University of Ukraine: 375.

Pivnyak G.G., Zhezhelenko I.V., Papaika Yu.A. 2014. *Calculation of electromagnetic compatibility indices.* Dnipropetrovsk: National Mining University of Ukraine: 114.

Power Engineering and Information Technologies In Technical Objects Control – Pivnyak, Beshta & Alekseyev (eds)
© 2016 Taylor & Francis Group, London, ISBN 978-1-138-71479-3

The development of electric alkaline electrochemical fuel cell hybrid system

W. Czarnetzki
Esslingen University of Applied Sciences, Germany

S. Sevruk
Moscow Aviation Institute, Russia

W. Schneider
Esslingen University of Applied Sciences, Germany

Yu. Khatskevych
State Higher Educational Institution "National Mining University", Dnipro, Ukraine

K. Pushkin
Moscow Aviation Institute, Russia

I. Lutsenko & A. Rukhlov
State Higher Educational Institution "National Mining University", Dnipro, Ukraine

ABSTRACT: Dissemination of innovative small vehicles with a hydrogen fuel has been recognized by European Commission as one of the preferable directions of transport system development. In the article existing energy fuel cells for e-bikes are analyzed. A new source of energy that combines alkaline Anion Exchange Membrane Fuel Cell and Hydronic Chemical Current Source is proposed. The main characteristics of this fuel cell and current source are shown. The main stages of further development of the proposed energy system are described. It is expected that the energy source with the alkaline Anion Exchange Membrane Fuel Cell and the Hydronic Chemical Current Source will increase capacity and safety of vehicles, will allow widespread this form of transport because no grid is required for charging. Additional application of the system is universal power device, used for charging laptops, cell phones, music players, industrial or medical devices.

1 INTRODUCTION

Today's world is undergoing major changes. One of the main reasons is the continuous "drying up" of traditional energy sources. In the other hand, there is an ever-growing hunger of mankind for more energy. Mechanical systems are increasingly being replaced with electronic ones and internal combustion engine is being replaced with an electric motor.

Due to the depletion of fossil fuels, electrical energy is becoming more and more important. In addition the production of electricity by renewable energy sources is growing. Although electrical energy has many advantages, there is one big disadvantage: how to store it. This is a crucial point for portable and mobile applications as well as for electrifying passenger cars. Therefore one of the main questions is how to store the energy that moves the electric motor in the end. One can say that the state-of-the-art option for this is using lithium ion batteries. Among other reasons, the limited range of battery electric vehicles and the limited availability of lithium are substantially the eliminatory criteria for this storage technology in future. The major car manufacturers also recognized this. That is why almost all of them are going towards a hydrogen driven fuel cell technology in their products. Six years ago it had also has been recognized by the European Commission. In the "REGULATION (EC) No 79/2009 OF THE EUROPEAN PARLIAMENT AND OF THE COUNCIL of 14 January 2009 on type-approval of hydrogen-powered motor vehicles, and amending

Directive 2007/46/EC (Text with EEA relevance)" the following has been stated: "Innovative small vehicles, designated under EC type of approval legislation as L category vehicles, are considered as early users of hydrogen fuel. Introducing hydrogen for these vehicles requires less effort, as technical challenge and level of investment required is not as high as in the case of M and N category vehicles, as defined in Annex II to Directive 2007/46/EC". This article will contribute to introduce hydrogen and fuel cells for these vehicles but with a novel approach.

2 PROBLEM DEFINITION

The decision on the storage technology to be used depends on several factors. In most cases the result of it is an electrochemical System. Such a system can be realized as shown in Fig. 1. Such an ordinary battery electric system (BES) has two big disadvantages: Low energy density and limited cycles per lifetime.

By trying to compensate these disadvantages, one is working on a system that is a combination of a Proton Exchange Membrane Fuel Cell (PEMFC) and a battery (Fig. 2). The power output of the PEMFC and the battery in this Fuel Cell Hybrid

System (FCHS) varies depending on the application. Sometimes it is completely done without a battery.

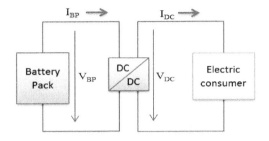

Figure 1. Ordinary BES

By reducing the disadvantages of a pure battery electric system new problems are appearing for a FCHS. PEMFCs are known by being expensive and hydrogen is difficult to store.

In this article a totally new approach will be analyzed to eliminate the problems of the mentioned systems:

BES has	FCHS with am PEMFC
• a short lifetime,	• is difficult to charge (refill),
• a low energy capacity,	• has a short lifetime,
• a low energy density.	• is expensive.

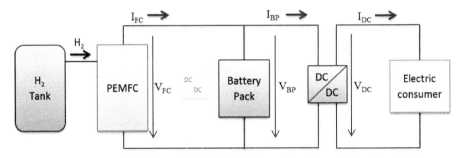

Figure 2: Fuel Cell Hybrid System (FCHS)

This approach comprises both, novel technologies and a novel system that is able to compensate the mentioned disadvantages. Figure 3 shows the initial electrochemical system (IES) which will be analyzed in this work. The core components in it are the alkaline Anion Exchange Membrane Fuel Cell (AEMFC) and the Hydronic Chemical Current Source (HCCS).

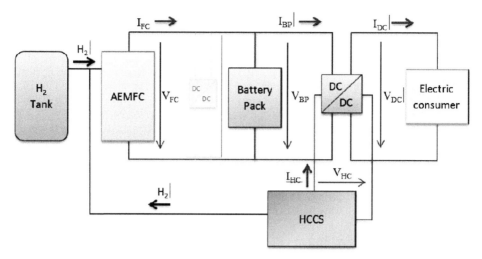

Figure 3: Initial electrochemical system (IES) as start of the examination.

The HCCS is an electrochemical cell that produces electricity, while it sets hydrogen free. Due to this fact the hydrogen tank and the battery pack are unnecessary in such system. It is necessary to analyze to which extent the HCCS can compensate a hydrogen tank and a battery pack, with the final goal of the alkaline electrochemical fuel cell hybrid system (AAE-FCHS) without a hydrogen tank and a battery as shown in Fig. 4. Only then the previous disadvantages of battery system or a FC battery hybrid system will not be present in it.

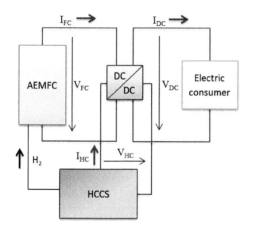

Figure 4. Targeted alkaline all electro-chemical fuel cell hybrid system (AAE-FCHS)

Thus, in this article possibility of development of a new energy supply system for electric transport on the base of the alkaline Anion Exchange Membrane Fuel Cell (AEMFC) and the Hydronic Chemical Current Source (HCCS) is analyzed. For this purpose characteristics AEMFC and HCCS are examined, requirements for power train and energy management system are worked out. As a result the main steps of study and development of electrochemical fuel cell hybrid system (AAE-FCHS) will be defined.

3 THE MAIN CONTENT OF THE WORK

3.1. Hydronic Chemical Current Source

Hydronic Chemical Current Source (HCCS) in essence is an electrochemical cell in which aluminium acts as the anode and an inert material like nickel or molybdenum acts as the cathode; space between these electrodes is filled with an electrolytic solution. Alkaline or neutral (saline) solution may be used in the cell. The HCCS is an innovative environmentally friendly source of electrical current based on the aluminium fuel, and also a controlled hydrogen generator with a wide range of hydrogen discharge speed successfully further developed by the Moscow Aviation Institute (National Research University) (MAI) (s. Fig. 5).

The main feature of the HCCS includes its capability to create a principally new storage system and at the same time a well-controlled hydrogen supply system as well as an additional source of electricity. The HCCS can operate especially well in combination with the Oxygen-Hydrogen Fuel Cells (O_2-H_2 FC), which have a need of hydrogen supply during operation.

The problem of hydrogen storage for O_2-H_2 FC is one of the most crucial problems for today. There are three hydrogen storage options for autonomous usage: metal hydrides, cryogenic and compressed gas tanks. Metal hydride based storage systems are heavy and have a limited flow rate, because the release of hydrogen is an endothermic reaction what actually makes it safer than the other Technologies. Hydrogen is tightly bound in molecules and enters the system in its pure form only as needed. In case of a cryogenic storage, the system faces higher costs and the need for complicated cryogenic equipment, which is not appropriate for the targeted usage. The gas tank storage (hydrogen in a gaseous form is stored at high pressures) substantially reduces the energy-mass ratio of the O_2-H_2 FC-based (FC + hydrogen supply system) power plant (PP). Both compressed gas tank and cryogenic hydrogen storage are highly inflammable and explosive due to pure hydrogen.

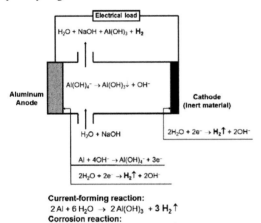

Current-forming reaction:
$2\,Al + 6\,H_2O \rightarrow 2\,Al(OH)_3 + 3\,H_2\uparrow$
Corrosion reaction:
$2\,Al + 6\,H_2O \rightarrow 2\,Al(OH)_3 + 3\,H_2\uparrow$

Figure 5. Basic diagram showing HCCS with an aluminium anode and an alkaline electrolyte

The main consumable substances in the HCCS are the aluminium anode and the water electrolytic solutions, which are well abundant low cost materials and non-toxic agents, whereas hydrogen, aluminium hydroxide $Al(OH)_3$, and electrical power are the by-products of the reaction.

In case of the successful development of the PP based on the O_2-H_2 FC and the HCCS as a hydrogen storage system, this PP will be notable for its environmental friendliness, high specific energy characteristics, and a higher safety of its usage, since pure hydrogen enters the system only as needed and is consumed at once in the FC. Thus the HCCS is an effective and safe solution to the problem of hydrogen storage for its autonomous usage.

Preliminary experimental results used several combinations of working materials for the anode, the cathode and the electrolyte are shown in Fig. 6. It shows that the energy characteristics of the HCCS with an alkaline electrolyte are significantly higher than those of the HCCS with a saline electrolyte. Due to the more positive polarization characteristics of the researched anodic materials the HCCS with a saline electrolyte can work as the source of power only within a short current density range (a short circuit occurs at the cur-rent density of 300-500 A/m^2), which may result in the bigger sizes.

Based on the mathematical model, preliminary experiment (Fig. 7) and test results the estimation of energy characteristics of a combined PP (HCCS + 1 kW PEMFC Ballard Nexa Module) had been done. As it is shown in Fig. 7, the energy characteristics of the O_2-H_2 FC are increased approximately by 40% due to the energy from the HCCS.

According to the obtained mathematical data, this HCCS can operate on the basis of one set of anodes for approximately 10 hours. The operating period of the electrolyte is about 5 hours. Therefore, the electrolyte needs to be replaced every 5 hours and aluminium anode plates need to be replaced every 10 hours.

The size and weight dimensions of the HCCS will be optimized and a higher energy efficiency of the HCCS will be achieved during further research even with the same working materials.

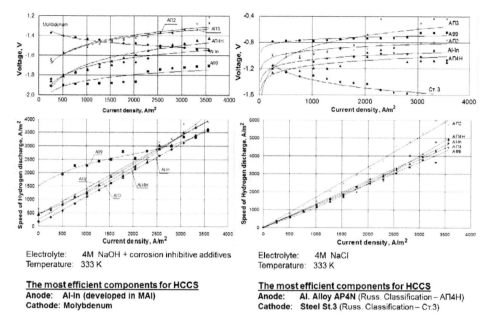

Electrolyte: 4M NaOH + corrosion inhibitive additives
Temperature: 333 K

Electrolyte: 4M NaCl
Temperature: 333 K

The most efficient components for HCCS
Anode: Al-In (developed in MAI)
Cathode: Molybdenum

The most efficient components for HCCS
Anode: Al. Alloy AP4N (Russ. Classification – АП4Н)
Cathode: Steel St.3 (Russ. Classification – Ст.3)

Figure 6. Experimental results of the HCCS

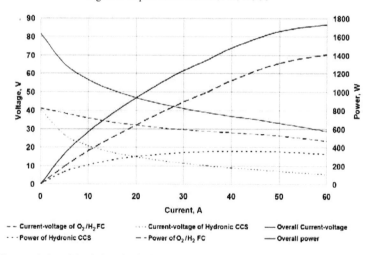

Figure 7. Characteristics of the designed HCCS, Ballard Nexa Module FC and their overall characteristics

3.2. Anion Exchange Membrane Fuel Cell

Since 2010 the Institute for Sustainable Energy Technology (INEM) of the University of Applied Sciences Esslingen (UASE) enhanced research activities in the field of production of the AEMFCs (s. Fig. 8) in addition to the research on the PEMFC. The goal is to drastically reduce the investment costs for fuel cell systems. This is only partly possible when using the PEMFC. Beside other reasons it is because platinum must be used as the catalyst. The research activities in the AEMFC field were funded until 2012 by the Fuel Cell and Battery Alliance Baden-Wuerttemberg (BBA-BW) and 2014 by the Baden-Württemberg Ministry of Science, Research and the Arts (MWK). The AEMFC converts hydrogen directly into electricity and heat according to the following reaction equations (s. Fig. 8):

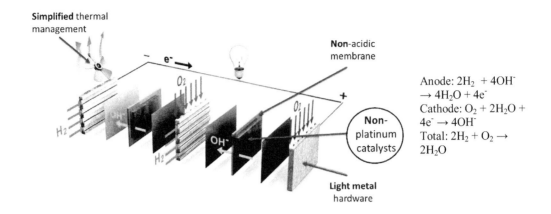

Anode: $2H_2 + 4OH^-$
$\rightarrow 4H_2O + 4e^-$
Cathode: $O_2 + 2H_2O +$
$4e^- \rightarrow 4OH^-$
Total: $2H_2 + O_2 \rightarrow$
$2H_2O$

Figure 8. Scheme of the AEMFC (Gottesfeld S., 2015)

In general, the process is very close to that what is going on in the PEMFC shown in Fig. 9. The main difference is the electrolyte membrane, or to be more precise, its charge carrier. The very aggressive H^+ Proton in the PEMFC is replaced by OH^- ions in the AEMFC.

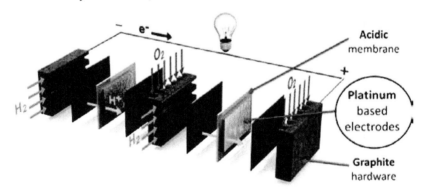

Figure 9. Scheme of the PEMFC (Gottesfeld 2015)

There is a need for expensive materials for the PEMFC due to the aggressive environment caused by the protons. That is why PEMFC Systems are listed with a price around 2,000€/kW. (BOM-based barriers – 90% of stack cost). In addition Platinum as core component underlies a high cost volatility with for example 600€/Oz - 1,500€/Oz in the last 8 years. It can be said that PEMFCs are maturely developed compared to other FC technologies with a solid electrolyte.

In contrast to this AEMFC Systems can be realized with price around 600€/kW, although this technology is in its period of formation. Due to the low corrosion activity of the alkaline electrolyte, less expensive materials can be used (for electrolytes, catalysts, system components), e.g. nickel as catalyst. Worldwide there is only one company that works successfully on AEMFCs. CellEra is a small company produces AEMFCs for telecommunication backup power (Gottesfeld 2015).

INEM has years of experience in the field of battery, hydrogen and fuel cell technology with partners such as Daimler, Ballard, CellEra, NuCellSys, Mahle, HyEt, FumaTech, HyLionTec and others. In the past, electrolyzers and fuel cells for automotive, portable, and stationary applications have been analysed and then integrated. Gradually, the fuel cell and hydrogen laboratory has been established and equipped with the necessary instrumentation and test stands. Currently the fuel cell and hydrogen laboratory can almost universally present the subjects production and analysis of fuel cells systems as well as hydrogen production development and research with test benches as:
1. methanol reformer;
2. methanol synthesis;
3. ethanol reformer;

4. natural gas steam reformer;
5. FC test bench with heat extraction for building heating;
6. FC test bench for analysis of stacks with a power of up to 2kW;
7. FC test bench for the automated operation of cells at different operating conditions;
8. FC test bench for the automated operation of single cells at different operating conditions and their characterization by cyclic voltammetry and electrochemical impedance spectroscopy;
9. hybrid test bench for analysis of EMS of fuel cell battery vehicles;
10. devices for producing membrane electrode assemblies (MEA) for fuel cells and electrolyzers:
   o screen printing machine;
   o spray coating;
   o film applicator.
11. assembly line for producing MEAs for fuel cells and electrolyzers.

In order to build a functional AEMFC highest efforts must be put on the MEA. To create the MEA, catalysts on both sides of the electrolyte must be placed in a way, so that the gases contact with the catalyst, the catalyst itself is in contact with an electrical conductor and the ionic conductor or electrolyte, the so called AEM (Fig. 8 and 10). This represents the catalyst coated membrane (CCM). To get a complete MEA, gas diffusion layers (GDLs) must be added on both sides of the CCM. Producing a stable CCM with the right catalyst ratio and highly active reactive centers is the hardest part. The catalyst layer and the GDL are forming the electrode on both sides.

In the past first usable coating results were obtained by using the spray coating method. This paved the way for homogeneous and porous coatings with a layer thickness accuracy of 1.5µm in the range of 5 to 15µm.

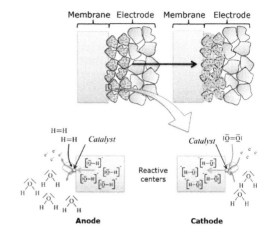

Figure 10. Half MEA and reactive centres

The first coatings were made with inks out of an ionomer solution, demineralized water, 1-propanol, carbon powder and catalyst powder (e.g. anode: Pt based catalysts, 3020 series from Acta and others; cathode: Pt based catalysts, 4020 series from Acta and others). Thereafter reproducible results were also achieved using the screen printing method. Many associated with the coating process parameters had to be evaluated.

Figure 10 only demonstrates how the print speed affects the quality of the coating. There it can be seen that increase of printing speed for instance leads to catalyst layers with a higher density. However, this leads to cracking during drying.

Figure11 clarifies the improvement of the coating during the previous work.

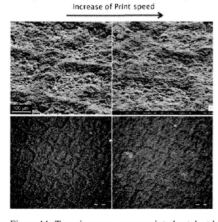

Figure 11. Top view on a screen printed catalyst layer as a SEM picture

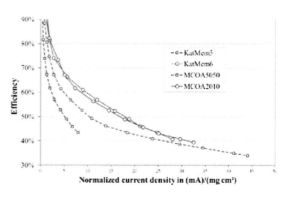

Figure 12. Efficiency curves of produced MEAs

The current density is normalized to the catalyst loading to be able to compare the coating results among one another. KatMem5 and KatMem6 stand for the spray coating. After several optimizing steps, it was only able to achieve minort improvements from KatMem5 to KatMem6 compared to the screen printing results. The MCOA curves stand for the CCMs made by the method of screen printing. With just a few optimizing steps, a clearly visible improvement between MCOA5050 and MCOA2010 was achieved. These are only a small amount of the results earned by working in this field.

Besides focusing on the MEA to achieve affordable AEMFCs, studies on stack design have also been carried out (Fig. 13).

A prototype FCHS, presented in Fig. 2, which is close to the targeted innovative concept is already running at the laboratory of the INEM (Fig. 14). The following picture shows a 100W FCHS in a power unit for a golf trolley that has been invented for a private investor.

The system has already been tested and installed into an electric golf trolley. The battery capacity is much smaller there, but the functionality has been proven and all results will be directly used to continue with a much more favorable AAE-FCHS based on AEM technology compared than those used in PEMFC.

Aluminum AEMFC stack
- Chemically nickel-plated surface
- 5.5x4.5 cm reactive area
- Tests carried out with stacks containing 3-8 cells
- Nickel based catalyst on anode and cathode side

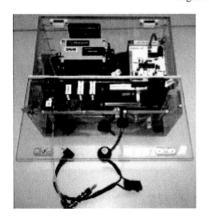

Aluminum steel AEM-Stack
- Galvanically nickel-plated surface
- 50 cm² reactive area
- Tests carried out with stacks containing 3-5 cells
- Nickel based catalyst on anode and cathode side

Investigation of the structural design of electrochemical compressors
- Compressor module with 2 cells and 45 cm² reactive area each
- Examination of various supporting structures for the membrane
- Examination of various supporting principles (metallic nonwovens, foams, ...)

Figure 13. In-house cells and stacks

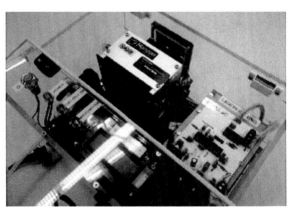

Figure 14. FCHS prototype already running at the laboratory of the INEM

3.3. Energy Management System

The first task of this research package is to develop an EMS to control two electrochemical sources (the HCCS and the AEMFC) in the initial electrochemical system shown in Fig. 3 according to the required parameters of a powertrain and load of the e-bike and the characteristics of electrochemical sources. It is necessary to take into account that the character of the load may change over time or depending on the characteristics of the locality, where e-bike is used. Figure 15 shows characteristics of the 2 most typical driving cycles of e-bike with an example of ARTEMIS European driving cycle (De Luca, Fragiacomo, De Lorenzo, Czarnetzki, Schneider 2015; Michel 2004).

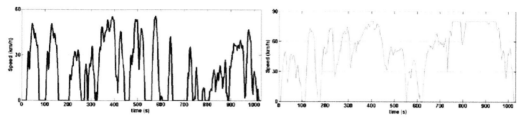

Figure 15. ARTEMIS urban and rural driving cycles

Depending on driving cycle parameters of power consumption and thus required parameters of the electrochemical sources will be obtained.

The parameters of power adjustment over DC/DC converters according to the output power characteristics of the sources and requirements for the parameters of the powertrain have to be justified by the EMS. It will be considered the possibility of using existing converters or working out the converter specifically for such a system taking into account the differences in the output parameters sources.

Different load requirements for various operation modes of on-board generation and storage units call for corresponding electronic power converter configuration and control. The selection of power converter unit for hybrid systems including FC relies on some essential factors and various topologies can be used for the operation of hybrid systems. Based on the analysis of existing topologies for operation of hybrid systems with PEMFC an architecture shown in Fig. 16 (Erdinc, Uzunoglu 2010) can be selected for the target system with two electrochemical sources. An important issue is minimizing the cost of the device. For this purpose multi-input converter based topology also needs to be studied.

Because of different characteristics of multiple power sources, the efficiency and the fuel economy of hybrid systems mainly depend on a proper energy management strategy. Optimization of fuel consumption is the main goal and the global optimum solution can be found within a certain driving cycle. A typical solution for a hybrid system with multiple energy sources that is built according to the common DC bus topology is shown in Fig. 17.

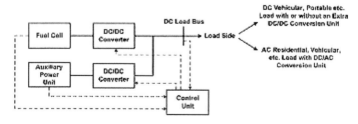

Figure 16. Multiple converter based topology

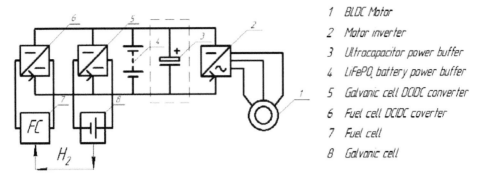

1 BLDC Motor

2 Motor inverter

3 Ultracapacitor power buffer

4 LiFePO battery power buffer

5 Galvanic cell DCIDC converter

6 Fuel cell DCIDC coverter

7 Fuel cell

8 Galvanic cell

Figure 17. DC bus topology for hybrid system

The primary energy source and fuel cell are connected to the bus through DC/DC converters. The best operating efficiency of each component will be provided in the case of 'constant current' control principle.

In the next phase of the research possible ways to control the sources must to be examined in order to determine requirements for the control signals and a hardware for implementation. After determining the basic structural elements of the EMS it is planned to develop a simulation model of the entire electric power system, which will include characteristics of the electrochemical sources (energy characteristics), EMS with H_2-tank and battery pack and powertrain.

The concept of development this model assumes the following: selection of an appropriate software; software emulation system components; study of the system in steady-state and transient conditions; determination of static and dynamic characteristics of the system response to disturbance and control actions.

Initial parameters of the simulation model are:

• current-voltage and load characteristics of both sources

• load curve of the powertrain, i.e. requirements for current and voltage values at the output of the converter and their changes over time;

• dynamic characteristics of the sources;

• requirements for the control signals;

• characteristics of the control system of the hydrogen circuit;

• dynamics of changes in the time-current and load characteristics of sources versus mode and duration of operation (from charge to discharge).

Simulation modeling should be used to analyze system performance, the possibility of reducing the volume of the H_2-tank and power of the battery pack and to create the AAE-FCHS without the hydrogen tank and batteries.

Mobile use of the system for powertrain of two-wheelers suggests that there are three main periods

in the load diagram: acceleration, motion with constant velocity and braking, which succeed each other over time with varying frequency and duration according to the driving cycle. The correlations between vehicle speed, acceleration, required traction force and power during typical intrapezoidal driving cycle are shown in Fig. 18. Dynamics of acceleration of the vehicle imposes stringent requirements on the presence in the energy system of a source that will provide a variable component of the load curve. Therefore problems will arise in such a system that need to be studied.

The results of dimension will depend also on control strategy of the whole energy system. The main issue for Hybrid Electric Vehicle (HEV) design is energy transfer control from sources to the loads with minimum losses of energy which depends on the driving cycles (Bayindir, Gözüküçük, Teke 2011). The main aspect involved in rule-based energy management approach is their effectiveness in real time supervisory control of power flow in a hybrid powertrain. The rules are designed based on heuristics, intuition, human expertise and even mathematical models and generally, without a priori knowledge of a predefined driving cycle. These strategies can be divided into deterministic and fuzzy rule based methods. The main idea of rule based strategies is commonly relies on the concept of load-leveling.

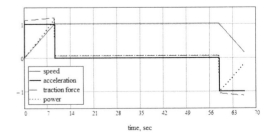

Figure 18. Correlations between vehicle speed, acceleration, required traction force and power

42

The principles described in Fig. 19 should be taken into account in the EMS development. Undoubtedly, each specific load mode should be examined and evaluated by convenient model, e.g. steady-state, dynamic, quasi-static. HEV energy management recommended considering at two control levels: local energy management (real time operating in each subsystem) and global energy management of the power flow for each subsystem coordination. The final goal is management of the entire energy in the whole system.

Thus, the understanding of the function and parameters of each subsystem is crucial. In virtue of above stated requirements and objectives for successful operation of a hybrid vehicle EMS convincing data and dynamic characteristics of each power subsystem are required to achieve and use.

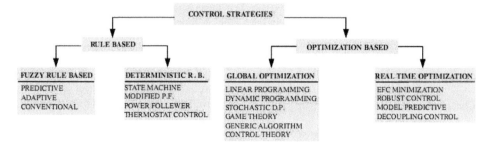

Figure 19. Energy management control strategies for HEV

The obtaining of optimal parameters for each operation mode is highly caused by the load characteristics and type of the vehicle. Thus, to realize the specific load-leveling process, its simulation should be conducted earlier and certain power sources output parameters should be recommended.

According to (Mensler, Joe 2010), the mode control system should select an optimal mode adapted to a driving point of the vehicle. The optimal mode map should be developed for various exploitation parameters for further defining of corresponding mode. This requires previous load simulation. The data obtained as the simulation results should be used for EMS supply aiming the whole power system optimal operation. A simulation model will contain electrochemical energy units, the DC/DC converter, the powertrain and the control unit. The initial data for the development of the model are dependencies and parameters characterizing the electrochemical energy units and powertrain load.

Input characteristics (graphical or analytical) for modeling of the electrochemical energy units are:

1. The output voltage and output power and the possible range of their change;

2. External characteristic curves of the source;

3. Characteristic of the discharge - dependence characterizing the stability of the output voltage in time from charge to discharge at a constant current load. It is necessary to have a set of curves for different load current values;

4. Adjusting characteristics of electrochemical sources.

5. The initial data for modeling of electrical consumer are parameters that indicate the power consumption, instability in time, dynamic characteristics.

Expected simulation results:

1. Based on the characteristics of electrochemical energy units and characteristics of power consumption the parameters and characteristics of DC / DC converter will be determined;

2. It is expected that processing and visualization modeling results will determine the possibility and principles of operation of the process control system;

3. Operational characteristics of the system at different load conditions will be received;

4. Identification of possible fields of application of electrochemical energy units without additional energy storage (Fig. 1.4) and using it will be done;

5. An additional result may be the development of evaluation criteria and requirements for working conditions of technological modules.

Using simulation modeling will be performed:

• study of the limiting possibilities of electrical systems and power sources to define the acceptable parameters of the powertrain load;

• determination of the optimal parameters of the load curve for the most effective use of sources to develop recommendations on operating mode of e-bike;

• development of the EMS with the possibility of adjusting the parameters of the system "sources-converter" and the mode of driving for urban, rural areas and mixed.

- definition of the conditions and parameters of the powertrain load parameters, for which a balance of generation and consumption of energy without the use of additional storage elements (battery and H_2-tank) is possible.

Further improvement of the developed IES in order to exclude the H_2-tank and the battery from the circuit (as shown in Fig. 4) appears to face a number of challenges. As the demand for the traction power varies within a wide range, there will be possibly a need to implement super caps in the system. It allows to reduce the load on the electrochemical power source and the fuel cell, to improve their operation, and to reduce the rating of DC/DC converters.

3.4. Concept and approach

Based on the analysis of the characteristics and requirements in to the operation of the AAE-FCHS system the following steps of the system development are offered.

1. Basic optimization: It is necessary to work out a novel membranes, catalysts and their appropriate coating on the membrane, as well as improvements for the whole MEA and new materials for the HCCS. For the development of EMS and a simulation model for entire concept all necessary processes/system related parameters should be collected. Therefore, this stage should include:
- Improvements on the cell level,
- Material selection,
- Material development,
- Identification of relevant cell parameters for the EMS.

2. Cell/Stack/System design: The research at this stage should be focused on the design of the single cell, stacks and the subsystems, namely, consists of:
- AEMFS single cell design,
- HCCS single cell design,
- Stacking,
- Voltage adaptation.

3. Cell/Stack manufacturing: In this phase the design of the previous phase has to be realized and tested on the device level, as well as EMS and the DC/DC converters.

4. System integration and demonstration: In this phase the integration of all components is focused to form a running system.

4 CONCLUSION

It is proposed the new type of energy source that could be used for e-bykes and other applications –

electro-chemical fuel cell hybrid system. The HCCS is a source for hydrogen and electrical energy. The investigation of the HCCS unit as a stationary and mobile device will reduce necessary size of hydrogen storages and batteries for future systems. Thus, it has a high potential to reduce vehicles weight but only when it is implemented as a combined energy storing unit together with a fuel cell.

Batteries are capable to supply a high power, but if there is a need to increase the range, relying on the batteries is the wrong decision. The energy supply for the electric motor must be delivered from an energy source with a high power and energy density. The batteries have only a high power density. To compensate the lack of high energy density of batteries a low priced AEMFC cell will be developed with a power output of not more than 300 W continuous and 400 W peak. The combination with the HCCS makes the use of the batteries superfluous. In addition the AEMFC will contain no Platinum and bipolar plates will be made of metal or metal plated polymeric material. This is beyond the state of the art as Platinum is commonly used as catalyst and bipolar plates are commonly based on a graphite material.

Analysis of the characteristics of all electro-chemical fuel cell hybrid system shows advantages of it's use in comparison with existing systems:
- Using the HCCS as the only hydrogen source can increase the power of the FCHS approximately by 40%.
- Safe system without gaseous hydrogen.
- For the market introduction, HCCS containers can be sold at gas stations and drug stores.
- No grid is required.

A significant advantage of the AAE-FCHS is a unique selling point and its modularity. It can be used not only as a vehicle component, but also as a standalone charger for all types of e-bikes and pedal electric cycles.

Added values and functions:
- The small fuel cell hybrid module can be used on-board or beside an already existing pure battery electric vehicle;
- Several modules can be combined together for faster charging of already existing pure battery electric vehicle;
- Several modules can be stacked for larger applications;
- Beside the vehicle application, it can be carried as an universal power device for charging laptops, cell phones, music players, industrial or medical devices.

REFERENCES

Bayindir K.Ç., Gözüküçük M.A., Teke A. 2011. *A comprehensive overview of hybrid electric vehicle: Powertrain configurations, powertrain control techniques and electronic control units.* Energy Conversion and Management, Issue 52: 1305–1313.

Erdinc O., Uzunoglu M. 2010. *Recent trends in PEM fuel cell-powered hybrid systems: Investigation of application areas, design architectures and energy management approaches.* Renewable and Sustainable Energy Reviews 14: 2874–2884.

De Luca D., Fragiacomo P., De Lorenzo G., Czarnetzki W.T., Schneider W. 2015. *Development of a Hybrid Electric 2-wheeled Vehicle equipped with Fuel Cell.* 6th International Conference on Fundamentals and Development Fuel Cell, 3 -5 February 2015 Toulouse, France.

Gottesfeld S. 2015. *Anion-Exchange-Membrane Fuel Cells.* Ionomers, Membranes and Catalysts, Seminar, Max-Planck-Institut für Festkörperforschung, Stuttgart 13.03.15.

Mensler M., Joe S. 2010. *Hybrid Vehicle Control System.* United States Patent. Patent US 7,693,637 B2. Jan. 2010

Michel A. 2004. *The ARTEMIS European driving cycles for measuring car pollutant emissions.* Science of the Total Environment, Vol. 334-335: 73-84.

Power Engineering and Information Technologies In Technical Objects Control – Pivnyak, Beshta
& Alekseyev (eds)
© 2016 Taylor & Francis Group, London, ISBN 978-1-138-71479-3

Reactive power and reliability of power supply systems

I. Zhezhelenko, Yu. Papaika
State Higher Educational Institution "National Mining University", Dnipro, Ukraine

ABSTRACT: The article describes the notion of reactive power considering the reliability of a power supply system. The negative consequences of reactive power exchanges in the power network are analyzed. The calculations of the reliability indicators considering the power factor are given. The main indices of electric power systems reliability are considered in the article. The laws of probability for the distribution of these indices are justified. The examples of assessing the annual damage from disruptions in power supply reliability are given.

1 INTRODUCTION

The evaluation of the power coefficient values (tg φ) in power supply systems and the efficient selection of compensating devices is a multifactor task. The influence of various factors was investigated in the known researches. This article describes the influence of the power supply reliability on both the level of reactive power and real power losses in electric power networks.

The concept of reactive power (Neiman, Demirchyan 1966). Reactive power is interchange power as it is a result of the interchange processes between inductive and capacitive elements of power networks. A power supply source is also a part of these processes. The electromagnetic energy interchange takes place in electromagnetic environment i.e. in electromagnetic field surrounding the conductors transporting the electric current (the flow of electrons). It is a way of transporting the electric energy from a source to a consumer. Active power is generated from other types of energy, e.g. mechanical.

2 MATERIALS FOR RESEARCH

Interchange power W is equal to the total of electric $W_Э$ and magnetic W_M fields. The formula for $W_Э$ and W_M is known from physics:

$$W_Э = Cu^2/2 \; ; \; W_M = Li^2/2 \; , \tag{1}$$

where C and L are capacity and induction of circuit sections; u and i are instantaneous values of voltage and current.

The rate of energy accumulation (and output) is equal to:

$$\frac{\partial W_Э}{\partial t} = Cu\frac{du}{dt} \; ; \; dW_M = Li\frac{di}{dt} \; . \tag{2}$$

Consider in details any of these formulas, e.g. $\frac{\partial W_Э}{\partial t}$, to understand the concept of reactive power. This notion is known to be used in the fundamentals of alternating (sine wave) currents.

Instantaneous voltage and current values are equal to: $u = U_{max} \sin \omega t$; $i = I_{max} \sin \omega t$ (phase values are omitted). Make the substitution:

$$\frac{\partial W_Э}{\partial t} = CU_{max} \sin \omega t \; \omega \, U_{max} \cos \omega t = \omega CU^2 \sin 2\omega ti$$

(known proportions $2 \sin \omega t \cos \omega t = \sin 2\omega t$; $U_{max} = \sqrt{2}U$ are obviously considered).

The previous formula allows the following conclusions:

1. $W_{pЭ} = \omega CU^2 = \dfrac{U^2}{1/\omega C}$ is a known formula in electromechanics for the reactive power of a capacitor in the alternating current circuit. The formula for W_{PM} is similar

$$W_{PM} = \omega LI^2 \; , \tag{3}$$

where $1/\omega C$ and ωL are inductive and capacitive reactive resistances.

2. Reactive energy is circulating with double frequency in a power grid, its average value for a period is equal to zero, i.e. reactive energy is not used for its generation.

3. Reactive power formula Q is similar to active power formula:

$P = UI \cos \varphi = S \cos \varphi$;

$Q = UI \sin \varphi = S \sin \varphi$,

where S is gross or apparent power.

Power triangulars S, P, Q and resistance triangulars Z, R, X look similar (fig.1).

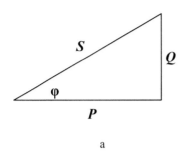

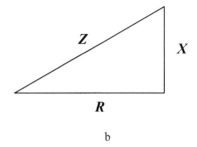

a

b

Figure 1. Power (a) and resistance (b) triangulars

The process of reactive power transmission is characterized by the losses of power and voltage in the power network.

$$\Delta P = \frac{P^2 + Q^2}{U^2} R \; ; \; \Delta U = \frac{PR + QX}{U} .$$

Active and reactive power ratio is defined by the reactive power coefficient $\frac{P}{Q} = \mathrm{tg}\, \varphi$.

The influence of the reactive power level on losses and reliability of electric power networks. Transform the formula for ΔP using different values of tg φ; assume current I as a function tg φ (Zhelezko 2009).

$$I = \frac{\sqrt{P^2 + Q^2}}{\sqrt{3}U} = \frac{P \cdot \sqrt{1 + \mathrm{tg}^2 \varphi}}{\sqrt{3}U} ;$$

$$\Delta P = \frac{P^2 + Q^2}{U^2} R = \frac{P^2 \left(1 + \mathrm{tg}^2 \varphi\right)}{U^2} R .$$

The formulas for I and P evaluate a fraction of a section where current d_{pi} with transmission power $d_{p\Delta p}$ flow, influenced by the reactive power transmission.

$$d_{pi} = 1 - \frac{1}{\sqrt{1 + \mathrm{tg}^2 \varphi}} \; ; \; d_{p\Delta p} = \frac{1}{1 + \mathrm{tg}^2 \varphi} .$$

The values of these indicators considering tg φ are presented below:

Table 1. Indicators considering tg φ

tg φ	0,1	0,2	0,3	0,4	0,5	0,6	0,7
d_{pi}	0,5	1,9	4,2	7,2	10,6	14,3	18,1
$d_{p\Delta p}$	1,0	3,8	8,3	13,8	20,0	26,5	32,9

These results allow the following conclusion to be made: reactive power transmission causes underutilization of some wire and transformer (converter) sections and this results in reducing the possibility of active power transmission (if tg φ = 0,5, it is about 10%). At the same time, power and electricity losses increase (more than 20% with the same tg φ value).

While transporting constant (or almost constant) active power (energy) to consumers, reactive power also changes because of different loads and consumer operating modes. Active losses, heating of conductors, decreasing of insulator life vary depending on the changes of the value proportional to the value $\frac{Q^2}{U_{\text{ном}}^2}$.

So, in extended operations with the increased reactive power consumption additional heating of conducting parts occur causing the reduction of insulator life term. In this case, indicators defining energy consumption reliability suffer changes. If the equipment operates with constant (long term) heating, then the failure rate λ differs from the value λ_0 which is typical for normal conditions:

$$\lambda_\tau = \lambda_0 (1 + \Delta \lambda) = \lambda_0 (1 + d_{p\Delta p}) \qquad \text{(Nepomnyashchy 2013).}$$

The reliability is calculated according to the value of the reliability function R(t), defining the probability of no-failure operation. The probability of failure is equal to $F(t) = 1 - R(t)$.

If the reactive power coefficient tg φ increases, the value of the transmitted active power decreases. In result, the reliability of consumer supply declines and the contravention of the logistics principles takes place.

Multiple calculations made for industrial power supply systems with transformer and reactor equipment allow the following conclusion: with $\pm 12\%$ error and tg φ = $0,1 \div 0,6$, the value $\Delta\lambda \approx d_{\Delta p}$. So, the influence of reactive power can be calculated according to the probabilistic formula $R(t) = f(d_{\Delta p})$.

Probabilistic methods of considering the reliability factor in the calculations of the electric load in industrial networks are practically based on one of three laws: exponential, normal and the Weibull distribution. The easiest (in use) and the most widely spread is the exponential distribution law which is defined by one parameter - failure rate λ. The probability of no-failure operation in this case is equal to $R(t) = e^{-\lambda t} = 1 - \lambda t$ and the failure rate is equal to $F(t) = 1 - R(t) = \lambda$.

Example

Calculate $R(t)$ if tg φ = 0,4 for the transformer 110/10 кВ, $\lambda_0 = 0,03$.

Solution

Find in the table $d_{p\Delta p}(0,4) = 0,133$. According to the formula

$\lambda_\Sigma = 0,03(1 + 0,133) = 0,034$.

Failure rate is equal to $F(t) = \lambda_\Sigma = 0,034$.

Probability of no-failure operation is equal to $F(t) = \lambda_\Sigma = 0,034$.

The known methods of variant calculations are normally based on deterministic approach and assessed values of electric equipment and process parameters. Applying probabilistic approaches in some cases can contribute to obtaining correct results.

In the process of planning electric supply systems for municipal infrastructure or agricultural objects several variants of such systems are normally considered and compared from the point of view of their feasibility. Usually the most appropriate variant is accepted, which ensures the quality and reliability of power supply and efficiency of investments provided by minimizing losses of electricity and creating the conditions for effective operation.

Minimal planned expenses or minimum payback period are the criteria of feasibility. Sometimes these criteria can be equal or close. In this case final decision should be based on comparing the reliability of the suggested variants (Dreshpak, Vypanasenko 2015.).

Two following conditions are necessary to ensure correct comparison of probabilistic calculations:

a) using the same values of failure flow parameters λ;

b) using the same methods of calculation, based on accepting the law of random variables distribution.

Values of failure flow parameters. The author of (Nepomnyashchy 2013) generalized the values of failure flow parameters for electric power lines and $500 - 35$ kV transformers according to Ukrainian and foreign researchers. For $35 - 110$ kV overhead lines which are used in power supply systems, values of failure flow $\lambda_{уст}^{ВЛ}$ lie within 0.75 − 2.2, the maximum likelihood of $\lambda_{уст}^{КЛ} = 0,85$. Average recovery period τ_B is assumed to be 10-3 with planned outage index $K_{пр} \cong 4 - 5 \times 10^{-3}$.

For cable lines laid in canals and gorges average is $\lambda^{каб} = 0,015$ for $\tau_B^{каб} = 1,4 \cdot 10^{-3}$. For transformers and autotransformers $\lambda_{уст}^{т} = 0,04 - 0,1$; prevailing values $\lambda_{уст}^{т} = 0,03$, $\tau_B = 10$-3.

Approximate average values of λ and τ are close to the recommendations given by scientific research institute of electric engineering design ("VNII Proektelectromontazh") (Ovcharenko, Rabinovich 1977).

Distribution laws used for calculating reliability of electric power supply systems.

The laws which require a small number of parameters, such as normal distribution law, Weibull law and the law of exponential distribution are usually applied. The latter is widespread mainly in power electrical engineering and power industry.

Exponential distribution is defined by a single parameter − failure rate λ(t). Reliability parameters (from here on t = 1) are:

failure rate parameter λ(t) = λ;

probability of non-stop operation R(t) = e-λt = 1 − λ;

probability of failure F(t) = 1 − R(t) = 1 − e-λt = λ;

average recovery period τ_B.

As it appears from these equations, if the time of non-stop operation is distributed exponentially, then the ageing of electric equipment doesn't occur, i.e. it is assumed to be in stationary condition. This condition can be violated due to burn-in and ageing/deterioration processes. To describe these processes more complicated distribution laws should be used. During burn-in period the reliability is normally increased because of careful control of manufacturing, assembling and commissioning; during the period of ageing it is provided by proper maintenance. That is why while solving the reliability problems in power engineering failure rate parameter is considered to be constant: $\lambda(t) = \lambda = \text{const}$.

Examples of calculations using reliability indices. Let's consider the algorithm of assessing economic damage of outage in the substation (or part of equipment).

Probable value of power supply limitation $W_{вер}$ in the substation

$$W_{вер} = W_\Sigma q \tau_в \lambda_\Sigma ,$$

where W_Σ is annual power consumption, kWh; q is the share of power consumption of the substation in question; $\tau_в$ – average recovery period; λ_Σ – system failure.

W_Σ can be taken from NERC data for 2014: W_Σ = 171.5 bln kWh. The share of power consumption of the substation is q = 0,05; $\tau_в$ =10-3. The reliability of 12 substations during full power outage was analyzed (Ovcharenko, Rabinovich 1977). This analysis provided the resulting value of failure flow for 35 – 119 kW substations with various electric circuits and types of switches (air, oil, gas-insulated ones), which lie within $\lambda_\Sigma = 0,02 \div 0,12$, prevailing value (more than 70% cases) $\lambda_\Sigma = 0,03$ (Ovcharenko, Rabinovich 1977). A similar result can be obtained from earlier studies.

$W_{вер} = 171,5 \cdot 0,05 \cdot 0,03 \cdot 10^{-3} = 0,257 \cdot 10^6$ kW·h.

The value of the damage У caused by long-term energy outage (Nepomnyashchy 2010):

$$У = W_{вер} y_0 ,$$

where y_0 is specific damage for the given branch of industry, such as metallurgy, mining or other power-consuming manufactures (Nepomnyashchy 2010); for Ukraine

y_0 = 2.0 UAH/kWh

In the given example

$У = 0.257 \cdot 2 = 0.51$m UAH $= 510\,000$ UAH.

The problem of assessing economic damage of the outage in the substations supplying power for separate shops or departments of a plant can be solved in the similar way (Dreshpak, Vypanasenko 2015).

Reliability of various power lines. As it is known, power is supplied through single or double overhead or cable lines. A combination of overhead and cable lines is also used, mainly for supplying power to load-centre substations. Although 35 – 110 kV overhead lines cost significantly less than cable ones, total costs of different variants of usage can be close.

The cost of overhead lines can be significantly increased because of the type of the ground, the width of the line corridors, number and type of turns and surface structures to be crossed and other reasons. The cost of cable lines includes expensive cables and engineering structures (tunnels, canals, ducts) as well as moving already existing facilities (if there is a need for that). The average time of repairing cable lines is significantly longer than the overhead ones. To compare the reliability of using cable and overhead lines in double-circuit variant the failure flow parameter λ_Σ (Nepomnyashchy 2010):

$$\lambda_\Sigma^{в,к} \cong \lambda_в \lambda_к (\tau_в + \tau_к) = (0,85 + 0,015) \cdot (1 + 1,4)10^{-3} =$$
$$= 0,0306 \cdot 10^{-3}$$

A similar calculation for the double-circuit overhead and cable lines resulted in: $\lambda_\Sigma^{zв} = 1,445 \cdot 10^{-3}$; $\lambda_\Sigma^{zк} = 0,0054 \cdot 10^{-3}$.

Thus, the least likelihood of damage F(t) occurs in case of supplying power by two cable lines $F^{2к}(t) = 0,0054 \cdot 10^{-3}$. The maximum likelihood occurs in case of using double circuit overhead line $F^{2в}(t) = 1,445 \cdot 10^{-3}$.

It is obvious that the approximate assessment of reliability in planning and exploitation can be done by the system failure flow λ_Σ = F(t).

3 CONCLUSIONS

The conducted research allowed us to establish the following regularities:

1. Based on the physical concepts, the notion of reactive power for power supply systems is explained.

2. Analytical formula connecting tg ϕ value at the consumer's point with the value of extra losses and the reliability indicators is presented. It is suggested to use the given recommendations while designing the power supply scheme.

3. Approximate assessment of reliability for selecting the optimal variant of power supply is expedient in case when capital costs or total costs are close. Only provided that reliability parameters have been set the described method can be correct.

REFERENCES

Neiman L.R., Demirchyan K.S. 1966. *Theoretical Foundations of Electrical Engineering* (in Russian). Leningrad: 408.

Zhelezko Y.S. 2009. *Electric energy losses. Reactive power. Electric energy quality* (in Russian). Moscow: 447.

Nepomnyashchy V.A. 2013. *The reliability of equipment for power systems* (in Russian). Moscow: 196.

Ovcharenko A.S., Rabinovich M.L. 1977. *The feasibility of electric power systems of industrial enterprises* (in Russian). Kiev: 172.

Nepomnyashchy V.A. 2010. *Economic losses caused by disruptions of power* (in Russian). Moscow: 188.

Dreshpak N., Vypanasenko S. 2015. *Energy Efficiensy improvement of geotechnical systems.* Taylor&Francis Group: CRC Press: 53-59.

Power Engineering and Information Technologies In Technical Objects Control – Pivnyak, Beshta
& Alekseyev (eds)
© 2016 Taylor & Francis Group, London, ISBN 978-1-138-71479-3

Multidimensional simulation of speed controlled induction electric drives with matching reducers and transformers

V. Petrushin, V. Vodichev, A. Boyko & R. Yenoktaiev
Odessa National Polytechnic University, Odessa, Ukraine

ABSTRACT: Operation of different induction motors as a part of controlled - speed electric drives, which carry out the same engineering problem, taking into account turning on of such devices as the transformer and the reducer is examined. A variability of parameters of equivalent circuits of the motors, linked with a modification of magnitudes and frequencies of voltage feeding the motors, and also with saturation of magnetic circuits and displacement of currents in windings of rotors is considered. Comparison of motors characteristics in static and dynamic regimes is carried out. Energy, weight, size and cost parameters are defined. Modeling of a thermal state of motors in static and dynamic regimes is carried out. The analysis of vibroacoustic performance of the motors is made at operation in the certain circuit design of the electric drive on the given control range: Mechanical indexes, characterizing a mechanical state of an induction motor are observed and compared.

1 INTRODUCTION

Using of the controlled-speed electric drive (CED) in all industries and for transport allows to control rationally technological processes at minimization of energy resources consumption. A variety of systems of induction CED is characterized by including such devices as reducers and transformers, and the last can be switched on not only on an entry, but also on an exit of the frequency converter. Using of these devices considerably changes operating characteristics of the CED. The majority of papers is devoted to modeling of the CED without such devices both in static and in dynamic regimes (Shreiner 2001; Baclin, Gympels 2005; Chermalykh, Chermalykh, Maidanskiy 2008; Moshchinskiy, Aung 2007; Khezzar, Kamel Oumaamar, Hadjami, Boucherma, Razik 2009). It is expedient to model operations of the CED including these devices.

2 PROBLEM FORMULATION

To shape models of matching transformers and reducers it is necessary to introduce a row of the initial data defining both functional properties, and mass, weight, dimensional and cost indexes. The last gives the chance to observe economic aspects of the CED. The functional properties are: for the reducer - a gear ratio (i_{red}), for the transformer – a transformation ratio (k_{tr}). For correctness of

energy balance of the electric drive calculation using of efficiency factor is required (η_{red}, η_{tr}).

Taking into account the simulation reducers and transformers in static and dynamic regimes a rotational speed (n_{mech}), the torque on the drive mechanism (M_{mech}), power consumed by the drive (P_{CED}), efficiency of the drive (η_{CED}), power of the mechanism (P_{mech}) are defined. Besides, it is obviously possible to count weight, dimensions and cost parameters of all the CED at use of those or other considered components.

The expressions studying using of the reducer and the transformer in the CED by consideration of a static conditions, look like:

$$n_{mech} = \frac{n}{i_{red}}, \tag{1}$$

$$M_{mech} = M_m \cdot i_{red} \cdot \eta_{red}, \tag{2}$$

$$P_{CED} = P_1 + (1 - \eta_c) \cdot P_1 + (1 - \eta_{red}) \cdot P_1 + \\ + (1 - \eta_{tr}) \cdot P_1, \tag{3}$$

$$P_{mech} = P_m \cdot \eta_{red}, \tag{4}$$

$$\eta_{CED} = \eta_m \cdot \eta_c \cdot \eta_{red} \cdot \eta_{tr}, \tag{5}$$

$$U_2 = \frac{U_1}{k_{tr}}, \tag{6}$$

where M_m – the torque on the motor shaft; P_m – the useful mechanical power on the shaft of the motor; P_1 – a consumed active power of the motor; η_m – efficiency of the motor; η_c – efficiency of the convertor; U_1 – primary voltage of the transformer; U_2 – secondary voltage of the transformer.

The mathematical models (MM) used for examination of the transitive electromagnetic and electromechanical processes in controlled induction motors, are grounded on systems of nonlinear differential equations of equilibrium of voltage and currents in system of converted coordinates (Shestacov 2011; Chermalykh, Chermalykh, Maidanskiy 2010; Chermalykh, Chermalykh, Maidanskiy 2011; Nandi, Toliyat, Li 2005).

$$\frac{d}{dt}\psi_{s\beta} = u_{s\beta}(t) - r_s d(t)[x_r(t)\psi_{s\beta}(t) - \tag{7}$$
$$- x_M(t)\psi_{s\beta}(t)],$$

$$\frac{d}{dt}\psi_{s\alpha} = [-p\omega_r \cdot i_{red}\psi_{s\beta}(t) - \tag{8}$$
$$- r_r(t)d(t)[x_s(t)\psi_{r\alpha}(t) - x_M(t)\psi_{s\alpha}(t)],$$

$$\frac{d}{dt}\psi_{r\beta} = [p\omega_r \cdot i_{red}\psi_{r\beta}(t) - \tag{9}$$
$$- r_r(t)d(t)[x_s(t)\psi_{r\beta}(t) - x_M(t)\psi_{s\beta}(t)],$$

$$\frac{d\omega_r}{dt} = \frac{1}{J}\{\frac{3p}{2}d(t)x_M(t)[\psi_{s\beta}(t)\psi_{r\alpha}(t) - \tag{10}$$
$$- \psi_{r\beta}(t)\psi_{s\alpha}(t)] - M_c(\omega_r)\cdot\eta_{red}\},$$

where $\Psi_{s\alpha}(t)$, $\Psi_{s\beta}(t)$, $\Psi_{r\alpha}(t)$, $\Psi_{r\beta}(t)$ – magnetic flux linkages of the stator and rotor windings of the motor, accordingly on axes α and β; ω_r – an angular speed of the mechanism; p – poles pairs number; J – the total moment of inertia of the drive redused to the shaft of the motor; $M_c(\omega_r)$ – dependence of a moment of resistance of the mechanism on an angular speed; r_s, $r_r(t)$, $x_s(t)$, $x_r(t)$, $x_M(t)$ – active and total reactive resistances of windings of the stator and rotor and resistance of a mutual induction, and all of them, behind exclusion r_s, vary on each integration step; $d(t)$ – an auxiliary variable $d(t) = [x_s(t) \cdot x_r(t) - (x_M(t))^2]^{-1}$; $u_{s\alpha}(t)$, $u_{s\beta}(t)$ – instantaneous values of voltage on axes α and β

which are defined by peak voltage U_m (depending on the law of the frequency control) and an angular position of the generalised voltage vector φ_1

$$u_{s\alpha}(t) = U_m(t)\cdot\cos(\varphi_1), \tag{11}$$

$$u_{s\beta}(t) = U_m(t)\cdot\sin(\varphi_1), \tag{12}$$

thus the system is supplemented with two more differential equations

$$\frac{d}{dt}\varphi_1 = \omega_1 \quad \text{и} \quad \frac{d}{dt}\omega_1 = \varepsilon_1(t), \tag{13}$$

where ω_1 – angular speed, $\varepsilon(t)$ – the dependence in time of angular accelerations of the generalized voltage vector, defined with a drive velocity diagram.

The expressions linking instantaneous values of currents and magnetic linkages, look like the following:

$$i_{s\alpha}(t) = d(t)\cdot[x_r(t)\cdot\psi_{s\alpha}(t) - \tag{14}$$
$$- x_M(t)\psi_{r\alpha}(t)],$$

$$i_{s\beta}(t) = d(t)\cdot[x_r(t)\cdot\psi_{s\beta}(t) - \tag{15}$$
$$- x_M(t)\psi_{r\beta}(t)],$$

where $i_{s\alpha}$, $i_{s\beta}$ – stator currents on axes α and β. The root-mean–square current of the stator

$$i_1(t) = \sqrt{\frac{1}{2}\cdot\left[i_{s\alpha}(t)^2 + i_{s\beta}(t)^2\right]}. \tag{16}$$

A mathematical model constructed on the basis of the above described differential equations, allows calculating of rotation speed, currents in the phases of windings, electromagnetic torque, power losses.

In each of the equations non– linear coefficients take place such as engine parameters varying in each operating point, including due to saturation phenomena of the magnetic system and the displacement current in the rotor winding (Petrushin, Yakimets 2001). One of the approaches to dynamic characteristics of controlled induction motors analysis involves a preliminary determination of these coefficients for the required operating point of the control range. Therefore, to analyse the transient mode steady-state modes calculations are carried out in order to obtain the values of all the equivalent circuit parameters taking into account current displacement in the rotor winding and the magnetic saturation for the required operating point of the control range. MM of steady runs for this purpose are used. In the calculations of dynamic modes changes in each step of the integration of the system are taken into account, i.e. at certain points, the transition from one speed to another, the magnitude and frequency of the supply voltage in accordance with the used law of frequency regulation, equivalent circuit parameters. During the loads of fan and pulling type the load torque also changes, the value of which, corresponding to the angular rotation frequency is determined by the load characteristic. Improving of the adequacy of MM is provided by implementing of this approach.

The conventional active power consumed by the motor in dynamic regimes under condition of a sinusoidal supply voltage can be calculated through virtual values of voltage and currents

$$P_1' = \frac{3}{2}\left[U_{s\alpha} \cdot I_{s\alpha} + U_{s\beta} \cdot I_{s\beta}\right] \qquad (17)$$

Real consumed active power P_1 more than the conventional total on magnitude of not considered losses (in a magnetic circuit, additional, mechanical)

$$P_1 = P_1' + \Delta P_{stbase} + \Delta P_{stadd} + \Delta P_{mech} + \\ + \Delta P_{add}. \qquad (18)$$

Shaft power of the motor can be defined through magnetic linkages and currents with use of meaning of a rotational speed of a rotor

$$P_2 = \omega_r \frac{3p}{2}\left[I_{r\beta}\Psi_{r\alpha} - I_{r\alpha}\Psi_{r\beta}\right] - \\ - \Delta P_{mech} - \Delta P_{add}. \qquad (19)$$

The instantaneous value efficiency is defined by a ratio of instantaneous values of useful power on the shaft of the motor P_2 to consumed active power P_1.

3 MATERIALS FOR RESEARCH

During the modeling of the CED according to system approach principles joint consideration of transformers, motors and loadings, and also reducers and matching transformers (Petrushin 2006) is necessary. Software product DIMASDrive (Petrushin, Riabinin, Yakimets 2001), allowing to realise such model operation is developed at the department of electrical machines in Odessa national polytechnic university.

The CED with the transistor frequency converter with the autonomous invertor of voltage and pulse–width regulating further is observed. The law of the frequency control U/f = const has been observed. Reviewed the draft load with the power of P_{load}= 35 kW with a peak torque 1500 Nm. At the given

stationary magnitude of a loading, a demanded control range (30-250 rpm) in systems of the CED can be ensured by different motors, under condition of using of reducers and transformers.

Three alternatives of the CED are observed at voltage of the main 400 V and a frequency of 60 Hz.

The motor 4A355M12, working with the frequency converter (Mitsubishi FR-A 540 L-G EC, 75 kg, $\eta_c = 0.98$) (Frequency converters. Series FR-A 540) is chosen for direct- drive CED (Fig. 1).

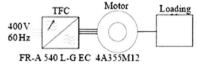

Figure 1. The block diagram of the direct-drive CED

The motor 4A250M4, working with the frequency converter (Mitsubishi FR-A 540 L-G EC, 75 kg, $\eta_c = 0.98$)and the reducer (1ЦУ200, 135 kg, $\eta_{red} = 0.98$, $i_{red} = 6.3$) (Frequency converters. Series FR-A 540; Plant of drive technology. Characteristics of gear) are chosen for the CED with the reducer (Fig. 2)

The motor 4A200L4, the step- up transformer (510 kg, $\eta_{tr} = 0.98$, $k_{tr} = 0.8$), the reducer (1Ц2У-200, 170 kg, $\eta_{red} = 0.98$, $i_{red} = 10$) and the transistor frequency converter (Mitsubishi FR-A 540 EC, 35 kg, $\eta_c = 0.98$) are chosen (Frequency converters. Series FR-A 540; Plant of drive technology. Characteristics of gear; Production of transformers. Characteristics of the trans-shaper) for the CED with the transformer and the reducer (Fig. 3)

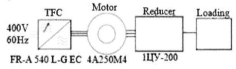

Figure 2. The block diagram of the CED with the reducer

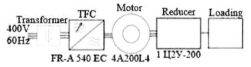

Figure 3. The block diagram of the CED with the reducer and the transformer

The adjusting characteristics representing dependences of a variation of electrical, energy and thermal magnitudes from a rotating speed, can be gained at use of characteristic families, including

55

mechanical, at various parameters of regulating on which characteristics of the load mechanism are superimposed. Fig. 4 depicts the set of speed-torque characteristics and the given loading, matching to the block diagram shown in Fig. 3.

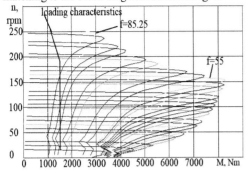

Figure 4. A set of speed-torque characteristics

At such combination of speed–torque characteristics and loadings it is observed three bands which are manifested on character of adjusting characteristics. Within each band the uniform modification of speed-torque characteristics and a load line occurs.

Adjusting characteristics of observed CED are presented in Fig. 5 and Fig. 6.

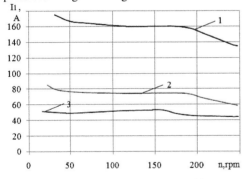

Figure 5. A modification of currents consumed by motors over the speed range: 1 CED without the reducer, 2 CED with the reducer, 3 CED with the reducer and the transformer

The analysis of a thermal state of an induction motor is one of the major problems at examinations of any CED system. Numerous papers (Borisenco, Kostikov, Yakovlev 1983; Goldberg, Gurin, Sviridenco 2001; Bespalov, Dunaikina, Moshinskiy 1987; Petrushin 2001; Petrushin, Yakimets 2002; Petrushin, Yakimets, Cobrin 2003; Kopylov, Clocov, Morozkin, Tokarev 2005) are devoted to the problems linked with the analysis of a thermal state of an induction motor and heat calculations.

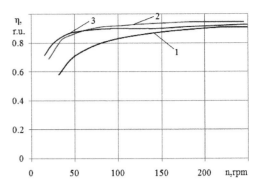

Figure 6. Performances of efficiency of motors over the speed range: 1 CED without the reducer, 2 CED with the reducer, 3 CED with the reducer and the transformer

The thermal state of an induction motor defines level of its operate reliability. The admissible reheat temperature is restricted to a class of thermal classification of applied insulations. The majority of premature fallings out of an induction motor is called by an accelerated ageing of isolation of a winding of the stator owing to excessive heating. This problem is especially actual for the induction motors operated in the CED as additional excessive heating is caused by magnification of losses because of no sinusoidal current, and also a decline of a tap of heat to an environment at lowering of a rotational speed of motors with a self-ventilation.

Heating of a concrete induction motor depends on ambient conditions, magnitude of losses in its constructional devices, intensity of an abstracted heat from these devices. The magnitude of the energy loss and intensity of cooling of controlled induction motors define such operation and technology factors:

• The aspect and parameters of the load mechanism (dependence of a torque of resistance on velocity and a moment of inertia of mechanism) define level of load and accordingly a degree of loses in the motor on a various rotational speed.

• Mode of operation S1 - S8.

• Modification of the electrical machine (enclosed, protected, unclosed), ventilation system (axial, the radial, immixed), a cooling aspect (independent, self-cooling), defining intensity of heat removal from motor to an environment.

Not only a modification of magnitude of a loading, but also parameters of the voltage feeding the motor gained from the semiconductor converter (SC) (frequencies, magnitudes, a spectral distribution) must be considered for an induction motor heat calculation. The last parameters appreciably influence to allocation of losses in active parts and a common thermal state of an induction motor. These parameters depend on structure of the main circuit of the convertor, an aspect of regulating and the law of the control used in it.

The problem of definition of excess of temperatures of various parts of electrical machine over cooling environment temperature is put in a heat calculation. Calculation should be made with the use of the geometrical dimensions and physical parameters of induction motors and information about losses in various parts of the motor, gained as a result of electromagnetic calculation of the motor. An important feature of controlled induction motor is its operation in a various operating points of a control range. Therefore at thermo analysis the calculations of the steady state thermal conditions occurring at long lasting operation in any operating point and also an unsteady thermal conditions at the dynamic transients linked with regulating, starting, a deceleration and reverse of drive are necessary.

Various computational methods of the steady state and unsteady thermal processes: a method of the equivalent heating losses, an analytical method or a temperature pattern method, a method of the equivalent thermal equivalent circuits (ETC) are applied in practice of research and design of induction motors. A simplified heat calculation by the method of heating losses and calculation by means of ETC can be applied for the analysis of a thermal state of controlled induction motors.

Method of ETC is well enough approved in practice of design of series of induction motors. It is grounded on well-designed theory of electrical and thermal circuits and allows to define medial temperatures of parts of a motor. A possibility of various motors designs, a possibility of raise of precision of account for the score of magnification of number of devices of an equivalent circuit and refinement of meanings of thermal conductances are referred to the advantages of this method. Its application for calculation of temperatures of constructive elements of the controlled induction motor should be performed taking into account specificity of operation of the motor in the drive. Following simplifications are accepted at calculation: the cage rotor is considered as one device, cooling of end faces of cores of the stator and rotor is not considered, machine cooling symmetrically and uniformly in a cross-section, thermal conductances are independent of temperature. A variety of constructive solutions of controlled induction motor should be considered at development of mathematical model of a thermal state of the motor. It is necessary to provide a possibility of a heat calculation of motors enclosed (IP44, IP54) and protected (IP22, IP23) modifications both with forced, and with self-cooling with use in system of ventilation of axial and radial ventilating ducts. It is developed ETC for heat calculations of non stationary behaviours of controlled induction motors of the enclosed modification (IP44, IP54) which is presented in Fig. 7 in which some conductances at regulating vary and in the thermal circuit design they are marked out as variables. ETC for heat calculations in a steady conditions will be converted at exclusion of the devices which are representing heat capacities. The mathematical description in appropriate way changes. Calculation of variable conductances is necessary for executing for each operating point. The thermal conductances which have been marked out as variables by a dot line, vary at self-cooling and remain invariable at a dusting the independent ventilating fan. In a heat calculation the problem of definition of excess of temperatures (excessive heating) of various constructive parts of the electrical car over ambient temperature is solved. Constructive parts of the electrical machine possess certain heat capacities which values depend on used materials and their geometrical sizes.

To solve the problem of definition of excess of temperatures of various constructive parts of the electrical machine over environment temperature in an observed equivalent circuit of substitution following constructive parts of an induction motor are introduced:

1. The stator core (fingers and a back) with medial superheat temperature θ_1, heat capacity C_1 and power losses ΔP_1 (magnetic losses in the core taking into account the additional iron loss of the stator).

2. The short-circuited cage of a rotor and fingers of a rotor with medial superheat temperature θ_2, heat capacity C_2 and power losses ΔP_2 (the total of losses of rotor bars, short-circuited rings and the additional losses in fingers and a rotor winding).

3. The grooving part of a winding of the stator with medial superheat temperature θ_3, heat capacity C_3 and power losses ΔP_3.

4. Front parts of a winding of the stator with medial reheat temperature θ_4, heat capacity C_4 and power losses.

$$\Delta P_4 = \Delta P_{el1} - \Delta P_3. \tag{20}$$

5. Interior air (IA) with medial temperature θ_5, heat capacity C_5 and interior ventilating losses ΔP_5.

6. The frame with medial superheat temperature θ_6, heat capacity C_6.

7. End shells with medial temperature θ_7, heat capacity C_7.

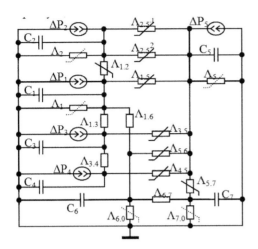

Figure 7. The equivalent thermal equivalent circuit of controlled induction motor of the enclosed modification (IP44, IP54) for the analysis of unsteady thermal processes.

Following thermal conductances are presented in ETC:

Λ_1 – conductance between a package of the stator and a cooling environment at unpackaged modification.

$\Lambda_{1.2}$ – conductance of a stator-to-rotor gap between the core of the stator and a rotor.

$\Lambda_{1.3}$ – shunt conductance of a grooving part of a winding from winding copper to the stator core.

$\Lambda_{1.5} = \Lambda_{rvds} + \Lambda_{avds} + \Lambda_{surf}$ – conductance from a stator package to IA, consists of conductances: Λ_{rvds} the radial, Λ_{avds} axial ventilating ducts of the stator, Λ_{surf} surfaces of the core of the stator to IA.

$\Lambda_{1.6}$ – conductance from a stator package to the frame (for the enclosed induction motors).

Λ_2 – conductance from a rotor to chilling air (a scavenged rotor).

$\Lambda_{2.5}^{1}$ – conductance from front parts of the squirrel cage to IA

$\Lambda_{2.5}^{2} = \Lambda_{rvdr} + \Lambda_{avdr} + \Lambda_{shaft}$ – conductance from rotor core to interior air, consists of conductances Λ_{rvdr} the radial, Λ_{avdr} axial ventilating ducts of a rotor to IA, Λ_{shaft} conductance of a rotor to IA through the shaft.

$\Lambda_{3.4}$ – axial thermal conductance of a winding of the stator.

$\Lambda_{3.5}$ – conductance from a grooving part of a winding of the stator to IA through the radial channels.

$\Lambda_{4.5}$ – conductance from front parts of a winding of the stator to IA

Λ_{5}– the equivalent conductance considering heating of a cooling medium (for the protected motors).

$\Lambda_{5.6}$ – conductance from IA to the blown frame.

$\Lambda_{5.7}$ – conductance from IA to the bearing shields.

$\Lambda_{6.7}$– conductance between the frame and bearing shields.

$\Lambda_{6.0}$ – conductance from a frame surface to chilling air.

$\Lambda_{7.0}$ – conductance from bearing shields to chilling air.

The system of differential equations of a heat account can be made on the basis of offered universal ETC. As for a matrix aspect the system is represented by expression:

$$\frac{d}{dt}\theta = [C]^{-1} \cdot [\Delta P + \Lambda \cdot \theta], \qquad (21)$$

where θ – a matrix-column of medial excessive heating over temperature of a cooling medium,

$$\theta = \begin{bmatrix} \theta_1 \\ \theta_2 \\ \vdots \\ \theta_n \end{bmatrix}; \qquad (22)$$

C – a matrix of heat capacities of matching constructive devices into which the induction motor is conventionally dissected

$$C = \begin{bmatrix} C_1 & 0 & \cdots & 0 \\ 0 & C_2 & \cdots & 0 \\ \vdots & \vdots & \ddots & \vdots \\ 0 & 0 & \cdots & C_n \end{bmatrix}, \qquad (23)$$

Λ – a matrix of thermal conductance

$$\Lambda = \begin{bmatrix} -\Lambda_{1,1} & \Lambda_{1,2} & \cdots & \Lambda_{1,n} \\ \Lambda_{2,1} & -\Lambda_{2,2} & \cdots & \Lambda_{2,n} \\ \vdots & \vdots & \ddots & \vdots \\ \Lambda_{n,1} & \Lambda_{n,2} & \cdots & -\Lambda_{n,n} \end{bmatrix}, \qquad (24)$$

where $\Lambda_{1,1}\Lambda_{1,2}...,\Lambda_{n,n}$ – thermal conductances between motor devices;

ΔP – a matrix-column of Joule heat losses in matching constructive devices of the electrical machine

$$\Delta P = \begin{bmatrix} \Delta P_1 \\ \Delta P_2 \\ \vdots \\ \Delta P_n \end{bmatrix}. \qquad (25)$$

The solution of this first order system, for example by a Runge- Kutt method, allows to observe a modification of temperatures of constructive devices of an induction motor at transients. Adequacy in MM essentially rises at the account of modifications on each integration step of losses.

Fig. 8 shows the calculated temperatures of excessive heating of windings of motors stators θ_s of observed CED for continuous operation within the control range.

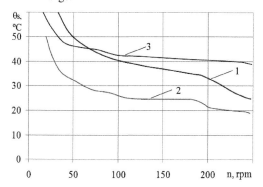

Figure 8. A modification of temperatures of stator windings of motors over the control range: 1 CED without the reducer, 2 CED with the reducer, 3 CED with the reducer and the transformer

The average for the control range design criteria (Petrushin 2001) must reflect energy characteristics of CAM in all control range from n1 to n2 and must be defined as equivalent averaged for this range. The same is applied to the generalized criterion of reduced expenditures of the motor which considers cost of manufacture and an expenditure for maintenance. Because expenditures depend on efficiency and a power factor, the generalised criterion of reduced expenditures has various meanings in different points of a control range and it is advisable to determine the equivalent average value of this criterion for the entire control range.

If the control range energy indicators are calculated as the average for the entire range

$$\eta_{aAM} = \frac{1}{n_2 - n_1} \int\limits_{n_1}^{n_2} \eta_{AM}(n) \cdot dn, \qquad (26)$$

$$\cos \varphi_{1aAM} = \frac{1}{n_2 - n_1} \int\limits_{n_1}^{n_2} \cos \varphi_{1AM}(n) \cdot dn, \qquad (27)$$

then a range criteria of given annual cost of the motor can be determined basing on the following. When you know the full cost of the engine ced criterion value is defined as

$$RC_{AM} = (ced + C_{rpc})[1 + T_s(k_{de} + k_s)] + \\ + CL_{AM}, \qquad (28)$$

where C_{rpc} – cost of expenditures for a reactive power compensation, UAH; CL_{AM} – cost of losses of the electric energy for a year, UAH; T_s – standard pay-back period of the motor, years; k_{de} – a share of expenditures for depreciation expenses; k_s – a share of expenditures for service at motor maintenance.

For controlled induction motors numerical value $T_s = 5$, $k_{de} = 0.065$, $k_s = 0.069$ are accepted the same, as for common industrial induction motors. Then

$$RC_{AM} = 1.67(ced + C_{rpc}) + CL_{AM}, \qquad (29)$$

$$C_{rpc} = C_{cre}P_1(tg\varphi_1 - 0.484), \qquad (30)$$

$$CL_{AM} = C_{cae}P_{1AM}(1.04 - \eta_{AM}), \qquad (31)$$

where C_{cae} – the coefficient considering cost of losses of active energy, representing product of cost 1 kW·h the electric power during life expectancy of the motor ($C = 0.13$ c.u. for kW·h), the number of engine operating hours during the year ($T_{year} = 2100$), number of years of operation before big repair (5) and coefficient of relative motor load ($K_L = 0.8$), C_{cre} – the coefficient considering cost of compensation of a reactive energy and representing product of cost 1 kilovar of a reactive power of compensating devices (10 c.u. for 1 kilovar), coefficient of participation of the motor in a peak load of the system (0.25).

$$RC_{aAM} = \frac{1}{n_2 - n_1} \int\limits_{n_1}^{n_2} RC_{AM}(n)dn. \qquad (32)$$

It is necessary to note that at operation the CAM as a part of up-to-date variable-frequency electric drives because of proximity of a power factor of the drive to 1 a component matching to cost of compensation of a reactive energy may be excluded from expression of measure RC of the electric drive

$$RC_{CED} = cep[1 + T_s(k_{de} + k_s)] + CL_{CED}, \quad (33)$$

where cep – an overall cost of the electric drive,

Numerical values of the coefficients, costs, hours and years are used same, as for definition RC_{AM}

$$RC_{aCED} = \frac{1}{n_2 - n_1} \int_{n_1}^{n_2} RC_{CED}(n)dn. \quad (34)$$

Tab. 1 shows the values of parameters considered CED that include medial band efficiency (η) and reduced expenditures (RC), and also mass, dimensional and cost indexes of motors and drives are resulted.

Account of cost of losses of active energy for year can be executed

$$CL_{ae} = C \cdot T_{year} \cdot K_L \cdot P_{mech} \cdot (1 + 0.04 - \\ - \eta_{CED})/\eta_{CED}, \quad (35)$$

where 0.04 – the relative magnitude of losses in a user supply net.

Comparison of the observed alternatives the CED at cost of losses of active energy for a year (Tab. 2) is executed.

Modeling for each circuit solution of the CED is also executed at operation on set tachogram (starting 1.5 s to 150 rpm, 1 s – 185 rpm) taking into account transients.

The versus time graphs gained at modeling of operation observed CED on set tachogram are presented in Fig. 9, Fig. 10, Fig. 11 and Fig. 12.

Excessive heating characteristics display that at each step of tachogram two sections are observed. The first matches to the transitive electromechanical process, the second is related to heat on an exhibitor with a trend of reaching of final values of superheat temperature.

Forces of a magnetic, mechanical and aerodynamic origin cause vibration and noise of electrical machines.

Table 1. Comparison of various CED indexes

Indexes and parameters	CED Without the reducer and the transformer	CED With the reducer	CED With the reducer and the transformer
η Motor, %	88	93.55	90.34
$\cos\varphi$ Motor, r.u.	0.76	0.89	0.64
η CED, %	86.31	89.9	85.23
RC Motor, c.u.	102926	10908	44690
RC CED, c.u.	205991	108381	118600
Mass Motor, kg	1670	560	325
Volume Motor, dm^3	161.7	75.6	34
Value Motor, c.u.	18039	5437	3294
Mass CED, kg	1745	770	1040
Volume CED, dm^3	290.7	255.6	666
Value CED, c.u.	75039	63437	38762

Table 2. Comparison of costs of losses of active energy for various CED

Indexes and parameters	CED Without the reducer and the transformer	CED With the reducer	CED With the reducer and the transformer
η CED, %	86.31	89.9	85.23
Value CED, c.u	1299	993.36	1398

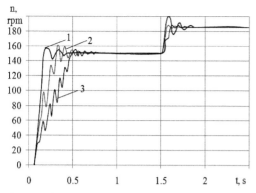

n, rpm

Figure 9. Characteristics of a rotational speed of the mechanism: 1 CED without the reducer, 2 CED with the reducer, 3 CED with the reducer and the transformer

Magnetic sources of vibration and noise are bound to the higher space and time harmonics of a magnetic field which are caused by presence of fingers on the stator and on a rotor, polyharmonic composition of a supply voltage, a stator-to-rotor gap eccentricity, no sinusoidal allocation of MMF of a winding, saturation of the magnetic circuit of the machine and a row of other causes. Mechanical causes are lack of balance a rotor, misalignment and a distortion of mounting faces of the bearing, an aberration in the form of their rings and spread of sizes of a separator, a thermal strain of a rotor, a

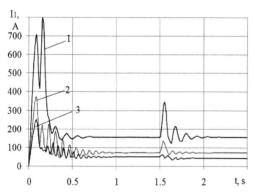

Figure 10. Characteristics of currents consumed by motors: 1 CED without the reducer, 2 CED with the reducer, 3 CED with the reducer and the transformer

shaft bending flexure etc. The aerodynamic noise is formed by the ventilating fan and other details had on a rotor. Vibrations and induction motor noise were observed by many authors (Unified series of asynchronous motors Interelectro, 1990; Shumilov, Gerasimchuk 1997; Shumilov, Chebanuk 1991).

The basic singularities of vibration and acoustic processes at induction motor operation in systems of CED are that in a different operating points of an

observed control range intensity of all three aforementioned sources largely change, and also resonance appearances are observed. In this connection it is necessary the reviewing of noise levels and vibrations in all control range for the purpose of their comparison to allowances. Thus it

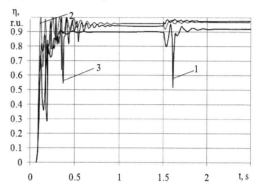

Figure 11. Characteristics of efficiency of motors: 1 CED without the reducer, 2 CED with the reducer, 3 CED with the reducer and the transformer

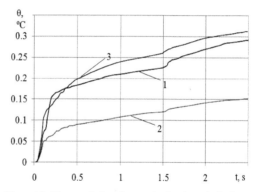

Figure 12. Characteristics of excessive heating of windings of stators of motors: 1 CED without the reducer, 2 CED with the reducer, 3 CED with the reducer and the transformer

is necessary to consider that a source strength of a magnetic origin is in many respects defined by a harmonic composition of voltage feeding motor, i.e. depends on type of the convertor, an aspect of the regulating used in it, a regime of its operation, the applied law of the frequency control and a drive loading. System approach according to which all functional indexes, including vibration and noise levels, are defined by joint consideration of operation of all builders going into the drive, allows generating of complex MM of vibration and acoustic processes of CED, invariant to various convertors and loadings. Account vibration and acoustic indexes of a magnetic origin taking into account the temporary harmonic voltage can be executed on a procedure, designed by Shumilov U.A. and Gerasimchuk V. G. [24], according to it forces of a magnetic origin are divided depending on direction of action into the radial and tangential; vibrations and noise are defined from these components. The basic assumption is the conjecture of linearity of a mechanical system at which frequency of magnetic vibrations and noise is equal to frequency of a magnetic force causes it, and the amplitude of deformations is computed by division of the force operating with given frequency, into rigidity of a construction (taking into account a reinforcement of deformations at a resonance). The end result of calculations is plurality of amplitudes of vibrations on matching frequencies (a vibration spectrum) and a common level of magnetic noise. At the same time at use of this procedure in complex MM for calculation of vibration and acoustic indexes of controlled induction motors it is necessary to consider that in each operating point of a control range parameters of equivalent circuits of controlled induction motors and EMF sections of the magnetic circuit of each observed harmonics vary. An adequacy offered in MM (Petrushin, Riabinin, Yakimets 2001) raises at such account. As a result vibration and acoustic account are defined following vibrations and noise parameters: the relative level of vibrations speed S_v and the level of magnetic noise S_n depending from vibrations speed and the relative radiated power N_{rel}. Their numerical values in a various operating point of a control range of induction motor depend on an aspect of a loading, type of the convertor, an aspect of regulating, the law of the frequency control. When calculating the geometrical sizes and properties of materials of an induction motor, and also magnitude of diameters of the core of the stator, the framework, width of a wall of the stator and the framework, modulus of materials of the stator and the framework and etc. are used.

The given procedure can be used for definition of vibration and acoustic indexes in dynamic regimes. For this purpose it is necessary to use on each step of the solution of simultaneous equations variable meanings of electromagnetic and electromechanical magnitudes in vibration and acoustic calculation.

Rated dependences of vibrations speed $S_v=f(t)$, vibration acceleration $S_a=f(t)$ and magnetic noise $S_n=f(t)$,), gained for different motors in the observed CED are presented in Fig. 13.

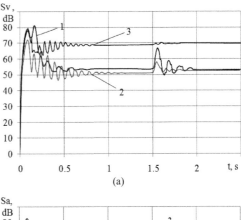

(a)

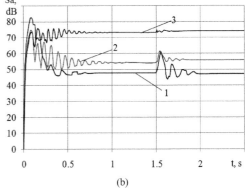

(b)

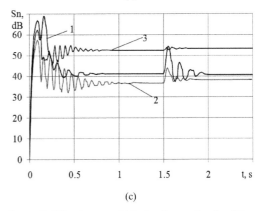

(c)

Figure 13. Vibrations speed (a),vibration acceleration (b) and magnetic noise (c) of motors: 1 CED without the reducer,2 CED with the reducer, 3 CED with the reducer and the transformer

The calculation procedure of ventilating noise of serial induction motors is well completed and confirmed by experimental data (Unified series of asynchronous motors Interelectro, 1990). In induction motors centrifugal fans with various constructions of vanes are applied. Structural features of vanes are considered by introduction in initial data of geometrical sizes of used ventilating fans. Level of ventilating noise is defined for various constructions under different formulas with use of the coefficients which meanings are accepted under tables.

In the induction motors of CED the ventilating fan rotational speed varies over the control range that stipulates an aerodynamic noise modification. The common level of ventilating noise depends also on type of the ventilating fan and its constructive sizes.

Level of this noise in the given control ranges on a known procedure can be defined by means of a program complex for observed alternatives of electric drives (Petrushin, Riabinin, Yakimets 2001). In dynamic regimes (Fig. 14) ventilating noise builds up proportionally to a rotational speed, attaining the meanings matching to a steady run.

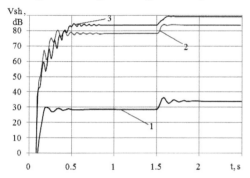

Figure 14. Changes of ventilating noise of an induction motor: 1 CED without the reducer, 2 CED with the reducer, 3 CED with the reducer and the transformer

The known calculation procedure of mechanical vibrations for usual induction short-circuited motors (Unified series of asynchronous motors Interelectro, 1990) is intended for rigid rotors to which rotors of a induction motors of uniform series refer to.

The causes of mechanical vibrations are the residual unbalance at static both dynamic balancing of a rotor and presence of rolling-contact bearings. At calculation of vibrations from rolling-contact bearings it is supposed that on low frequencies the parent of vibrations are an irregularity of manufacture of the bearing on the principal sizes and inaccuracy of mounting, and on frequencies above a 3-fold rotational speed - an irregularity of microgeometry of bearings, and levels of vibrations are maximum on frequencies of an Eigen tone of a rotor. Bearing vibrations have the essential technological spread defined by quality of bearings, and also a construction and manufacturing methods of motors. Indexes of mechanical vibrations of controlled induction motors will vary over the control range. They depend on rotor and machine

masses, and also from a rotational speed. The end result of calculation of mechanical vibrations is: a common level of vibrations speed from imbalance and irregularities of bearings in the radial direction V_r, a common level vibrations speed from imbalance and irregularities of bearings in axial direction V_z. The maximum magnitude of indexes of mechanical vibrations in the given control range in concrete design alternatives can be defined on a known procedure by means of a designed program complex (Petrushin, Riabinin, Yakimets 2001) for comparison of these alternatives. Magnitude of indexes of mechanical vibrations are defined by gyrating masses of the propeller and a rotational speed of a curl and irregularities of manufacture of bearings and inaccuracy of mounting depend on an unbalance.

In dynamic regimes (Fig. 15) mechanical vibrations build up proportionally to a rotational speed, attaining the meanings matching to a steady run. Calculations of indexes of mechanical vibration are executed at the condition of inaccuracy of machining, equalization and an irregularity of manufacture of bearings an equal 1 micron.

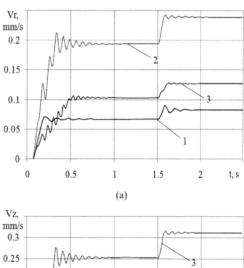

(a)

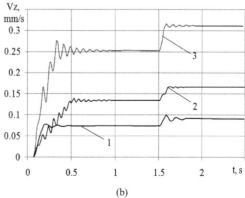

(b)

Figure 15. Vibrations speed in the radial direction V_r (a) vibrations speed in axial direction V_z (b) of an induction motor: 1 CED without the reducer, 2 CED with the reducer, 3 CED with the reducer and the transformer

At mechanical calculations of an induction motor in operation steady runs three factors characterising a mechanical state, - rigidity of the shaft, strength of the shaft and dynamic weight-lifting capacity of bearings (Goldberg, Gurin, Sviridenco 2001; Kopylov, Clocov, Morozkin, Tokarev 2005) are observed. For an estimate of a mechanical state of an induction motor at regulating it is offered to observe the same factors in dynamic modes of operation according to set tachogram.

At shaft calculation on rigidity the mechanical index - resulting bending flexure of a shaft f is being defined. Except the basic bending flexure of the shaft depending on masses of an active steel of a rotor and a short-circuited winding, the additional bending flexure of the shaft which meaning is proportional to a motor torque is observed.

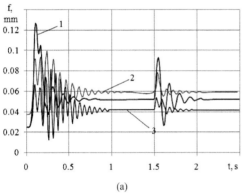

(a)

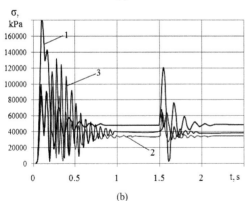

(b)

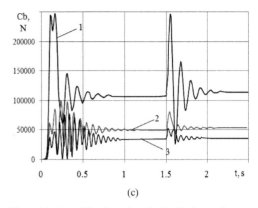

(c)

Figure 16. A modification of mechanical indexes of motors resultant shaft bending flexure (a) reduced mechanical stress (b) dynamic carrying capacity of the bearing (c): 1 CED without the reducer, 2 CED with the reducer, 3 CED with the reducer and the transformer

The shaft bending flexure is called also by forces of a single-sided magnetic attraction which originate at rotor bias. It is possible to count magnitudes of a bending flexure of the shaft while motor speed regulating, using in known algorithm (Goldberg, Gurin, Sviridenco 2001; Kopylov, Clocov, Morozkin, Tokarev 2005) magnitudes of the motor torques, varying throughout a transient process according to set tachogram. At shaft calculation on strength reduced mechanical stress σ, considering combined effect of stress of curving and twisting is defined. Using magnitudes of torque of a motor varying throughout regulating, it is defined reduced mechanical stress at regulating.

Definition of modifications of rated dynamic weight-lifting capacity of bearings C_b at regulating is carried out analogously taking into account type of the bearing and character of a motor load. Designed mathematical models are used in the program (Petrushin, Riabinin, Yakimets 2001) with which help examinations of a mechanical state of the motor at regulating have been executed. In analysing it is accepted joining of motors with actuating mechanisms is performed by means of elastic clutches. Ball-bearings are used in examined electrical machines. In examining character of a loading with moderate jolts is accepted. The overloading coefficient, equal 2.5 for reversible machines is used. The results of calculation of the mechanical indexes set forth above are presented in Fig. 16, a, b, c accordingly to regulation of motors in observed CED.

4. CONCLUSION

1. On the basis of modeling of various CED, inclusive reducer and the matching transformer, which are ensure a demanded control range at a given load, operation of different motors have been investigated, therefore comparison of characteristics of motors in static and dynamic regimes is carried out, energy, mass and dimension parameters of electric drives are defined.

2. Carried out investigations give the chance to justify sampling of the best alternative of the drive depending on the chosen criteria, including measure of active energy losses cost.

3. Magnitudes of efficiency of motors for a static conditions are approximately equal to final values of efficiency in transient regimes accordingly tachogram.

5. Overheating characteristics determined by the increase of heating losses in dynamic mode and the heat capacity of the thermal circuit designs of the motors are observed.

6. Dependences of vibration speed and noise of a magnetic origin for examined motors are placed in one order, unlike dependences of vibration acceleration.

7. Ventilating noise of the motor 4A355M12 is much less than noise of two other motors as it in the core proportionally depends on a rotational speed of the motor shaft.

8. Vibration speed in the radial direction V_r and vibration speed in axial direction V_z is the greatest for the motor 4A250M4 and the least for the motor 4A355M12.

9. The peak value of resultant shaft bending flexure in dynamic regimes for the motor 4A355M12 exceeds, and for two other motors does not exceed a legitimate value (0.1 from magnitude of a stator-to-rotor gap).

10. For all motors the peak value of reduced mechanical stress of shafts in transient regimes do not exceed a legitimate value matching to steels of shafts.

11. The peak value of rated dynamic weight-lifting capacity of the bearing in transient regimes for the motor 4A355M12 exceeds, and for two other motors does not exceed a legitimate value matching to type of the bearing.

REFERENCES

Shreiner R.T. 2001. *Mathematical modeling of AC drives with semiconductor frequency converters* (in Russian). Ekaterinburg: URORAS: 654.

Baclin V.S., Gympels A.S. 2005. *Mathematical modeling of variable frequency induction motor. Electromechanical energy converters* (in Russian). Proceedings of Intern. Scientific and engineering. Conf. -20-22 Tomsk: TPU: 143-146.

Chermalykh V.M., Chermalykh A.V., Maidanskiy I.Ya. 2008. *Investigation of the dynamics and power indicators of asynchronous electric drive with vector control by the method of virtual simulation* (in Russian). Kharkov: Problem of automated electric drives. Journal of the National Technical University "KhPI", Issue 16: 41-45.

Moshchinskiy Yu.A., Aung Vin Tut. 2007. *The generalized mathematical model of a frequency-controlled induction motor with taking into account steel core losses* (in Russian). Moscow: Electricity, Issue 11: 60-66.

Khezzar A., Kamel Oumaamar M. El, Hadjami M., Boucherma M., Razik H. 2009. *Induction Motor Diagnosis Using Line Neutral Voltage Signatures,* IEE Transactions on Industrial Electronic, vol. 56, Issue 11: 4581-4591.

Shestacov A.V. 2011. *A mathematical model of the performance of asynchronous motors with frequency control* (in Russian). Electrical engineering , Issue 2: 23-29.

Chermalykh V.M., Chermalykh A.V., Maidanskiy I.Ya. 2010. *Identification and optimization of electromechanical system parameters by computer simulation* (in Russian). Kharkiv: NTU "KPI": Bulletin of National Technical University "Kharkiv Polytechnic Institute", Issue 28: 45-48.

Chermalykh V.M., Chermalykh A.V., Maidanskiy I.Ya. 2011. *Identification of the parameters of the physical system frequency-controlled induction drive and its simulation model* (in Russian). Electrotechnical and computer systems, Issue 3: 35-37.

Nandi S., Toliyat H.A., Li X. 2005. *Condition monitoring and fault diagnosis of electrical motors-a review*, IEEE Trans. EnergyConversion, vol.20, Issue 4: 719-729.

Petrushin V.S., Yakimets A.M. 2001. *Simulation of dynamic modes of induction motors with frequency control* (in Russian). Kharkov: Problem of automated electric drives. Journal of the National Technical University "KPI", Issue 10: 156-157.

Petrushin V.S. 2006. *Tutorial "Induction motors in the controlled-speed electric drives"* (in Russian). Odessa: Nauka i Technica: 320.

Petrushin V.S., Riabinin S.V., Yakimets A.M. 2001. *The software product "DIMASDrive".The program for analysis, selection and design of asynchronous squirrel cage motors in the variable speed electric drive systems (registration certificate program PA№4065).* Kyiv: State Department of Intellectual Property of liability.

Frequency converters. Series FR-A 540.– http://univolts.ru/trademap/electric/mitsubishi/invertors/fra540

Plant of drive technology. Characteristics of gear. Main technical data of gearboxes.- http://www.reduktor-ptp.ru/.

Production of transformers. Characteristics of the trans-shaper. Main technical data of transformers.– http://transtechno2.ru/produktsiya-3/

Borisenco A.I., Kostikov O.N., Yakovlev A.I. 1983. *Coolingofindustrial electrical machines* (in Russian). Moscow: Energia: 296.

Goldberg O.D., Gurin Y.S., Sviridenco I.S. 2001. *Design of electrical machines* (in Russian). Moscow: Higherschool: 430.

Bespalov V.Y., Dunaikina E.A., Moshinskiy U.A. 1987. *Unsteady heat calculations in electrical machines* (in Russian). Edited by B.K. Clokov. Moscow: Moscow Energy Institute: 72.

Petrushin V.S. 2001. *Design synthesis of high efficiency induction motors up to 400 kW* (in Russian): Dis. Dr. tehn. Sciences: 05.09.01. Odessa: 379.

Petrushin V.S., Yakimets A.M. 2002. *Universal thermal equivalent circuit of induction motors* (in Russian). Odessa: Electrical machinery and electrical equipment, Issue 59: 75-79.

Petrushin V.S., Yakimets A.M. , Cobrin V.L. 2003. *Thermal calculations of unsteady modes of induction motors of controlled drives* (in Russian). Electrical engineering and Electromechanics, Issue 4: 65- 68.

Kopylov I.P., Clocov B.K., Morozkin V.P., Tokarev B.F. 2005. *Design of electrical machines* (in Russian). Textbook for High Schools. Moscow: Higherschool: 767.

Unified series of asynchronous motors Interelectro 1990. Edited by V.I. Radin. Moscow: Energia: 374.

Shumilov Y.A., Gerasimchuk V.G. 1997. *Study of magnetic perturbing forces in induction motor when powered by static frequency converter* (in Russian). Tech. electrodynamics, Issue 4: 44-48.

Shumilov U.A., Chebanuk V.K. 1991. *Induction motors with improved vibration and acoustic performance* (in Russian). Kiev: Technic: 169.

Power Engineering and Information Technologies In Technical Objects Control – Pivnyak, Beshta & Alekseyev (eds)
© 2016 Taylor & Francis Group, London, ISBN 978-1-138-71479-3

A promising approach to the identification and classification of steady-state operation of an iron-ore dressing section

I. Kuvaiev, I. Mladetsky & M. Oriol

State Higher Educational Institution "National Mining University", Dnipro, Ukraine

ABSTRACT: The subject areas which include parameters of an ore-dressing section have been determined and the interconnections between them have been analyzed. The indications which should be established to determine a mode of operation for each stage of iron-ore dressing and each section on the whole have been generalized. The relationship between the perturbation and control actions, which they compensate have been specified. Common approaches to the reduction of a range of process variables identifying and classifying steady-state operation of the production equipment have been grounded.

1 INTRODUCTION

At the mining processing plants of Ukraine, the share of electricity in the production cost is more than 20% (Annual information security issuer 2014: according to the company "Metinvest Holding" LLC; Annual information security issuer 2014: according to the company "Southern Mining and Processing Plant" PJSC; Annual information security issuer 2014: according to the company "Poltava Ore Mining and Processing Plant" OJSC). Of all the technological processes, grinding is the most energy-intensive one, both in mineral dressing and in other industries where there is an operational need for raw grinding before the primary process. Specific energy costs of the grinding process in the first stage is, by an order of magnitude, greater than that of conventional crushing. It is the most energy-consuming one of all the stages of size reduction of the coarse iron ore. A typical iron-ore dressing section consists of several stages connected in sequence, at the head of which a grinding aggregate and a classifying apparatus are necessarily present. These enrichment stages don't differ by the control efficiency function but the design of equipment and methods of classification between the first and subsequent stages are different. Under the spherical grinding, the second and the third enrichment stages are as a rule identical. In the last stage, after vacuum filters, a commercial concentrate is obtained (Fig. 1).

On the block diagram of the iron-ore dressing section (Fig. 1), input parameters in relation to the objects of enrichment technology are marked as "x", output parameters – "y", control actions – "z",

perturbation actions – "w", an executing mechanism – "EM", a measuring device – "MD". By the indices one can determine the enrichment stage to which the parameter or the control action refers.

Executing mechanisms and measuring devices are included into the automated control systems (ACS) and automated control systems of a regulatory type (ACSR), which form the first level of the control system for the dressing section. At each enrichment stage, they control and regulate: optimal filling of a mill with iron ore, ratio of ore-to-water or solid-to-liquid in the mill discharge, classifier drainage density, levels of pulp in technological containers (a sump, a deslimer).

2 TOPICALITY

Analysis of published sources (Solutions for the mining and processing complex: according to the company "Saturn" Data International"; Automated control systems of concentrators processes: according to the company "Promteh"; Iron ore beneficiation solutions: according to the company "Outotec") shows that the control systems of individual aggregates, enrichment sections and ore-dressing plants on the whole have been principally introduced in the domestic and foreign mining industry, the basic ideas of which were worked out be the teams of research and design institutes in the late twentieth century. Their main feature is the task description to the ACS and ACSR by the personnel of the ore-dressing plants. Thus, fundamental scientific research of basic principles for building the top-level of automated control systems of standard production lines, including mining complexes, are relevant..

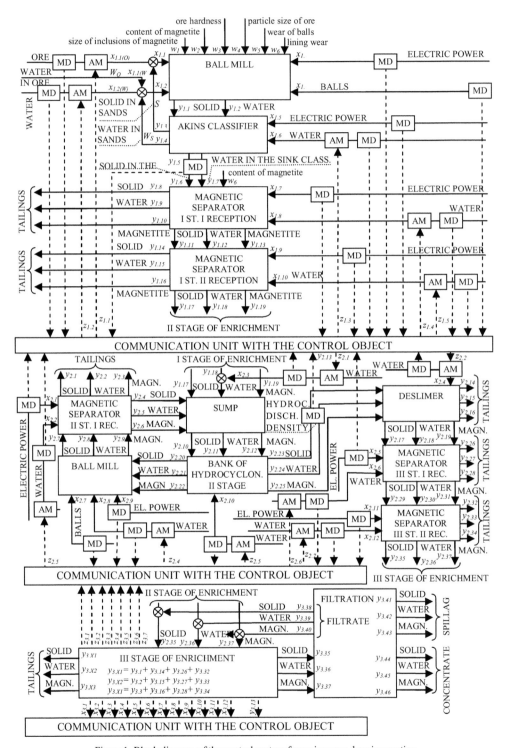

Figure 1. Block diagram of the control system for an iron-ore dressing section

One of the promising directions is the study of general approaches to the application of the method for situational industrial process control (Pospelov 1986) in order to minimize specific costs of electricity at the expense of reduction of technological equipment operation in transient conditions. This direction can be an effective basis for the engineering study of expert systems for situational control that allows the technological equipment operators to be excluded from the control loop, leaving them only the function of monitoring and decision-making in emergency situations. One such approach consists in the reduction of search space of the production situation, describing the steady state operation, which was earlier, and which is similar to the current one emerging at the start search at the controlled object.

3 AIM

The reduction of search space can be ensured at the cost of a minimum number of process variables that describe the technological situation and the selection process of a span value for each of these variables. In this case, it is necessary to identify and ground the possibility of classifying the steady state operation of the enrichment section automatically (without an operator) at the lowest possible number of values of the process variables that are controlled in the real-time technological process. They are included into different subject areas that overlap or are combined with each other. In this regard, in the first enrichment stage (Fig. 2), the decomposition of iron-ore dressing has been performed (Fig. 2).

4 DETERMINATION OF THE SUBJECT AREAS FOR THE FIRST STAGE

This technology makes provisions for getting the "enriched product" with a given value $G.3$ under the actual physical properties of the "original ore", which are determined by the deposit ($A.1 - A.3$) and the results of its size reduction at a given stage of grinding ($A4$). Proceeding from the physical properties of raw materials, the chain and the parameters of the technological devices ($C. *, D. *, F. *$) are determined at the design stage of the ore-dressing plant. Their values remain unchanged except for the parameters of the equipment wear and tear due to the abrasive properties of the raw material ($C.4, C.7, C.8$).

Conversion of the "original ore» (A) into the "enriched product" (G) occurs due to its successive passage through the whole process of "ore concentration" (D) and "magnetic separation» (E). Variations of physical properties of the original ore ($A.1 - A.4$) are compensated by the predeterrmined parameter values corresponding to ACSR and ACS of ore concentration ($B.1 - B.8$). Aim: To ensure maximum recovery of a useful mineral ($E.1 = G.3$) with a minimum specific energy consumption on the enrichment process ($I.7 / B.1$) and its minimal losses in the "depleted product" ($H.1$):

$$G.3 \rightarrow max, \; H.1 \rightarrow min, \; I.7 / B.1 \rightarrow min. \qquad (1)$$

Compensation of changes ($A.1 - A.4$), aimed at the optimum recovery of a useful mineral, should, on the one hand, lead to its maximum recovery into the "enriched product" ($G.3$), and on the other hand – to minimize its losses in the "depleted product" ($H.1$) due to overgrinding of the ready class. It is known that its size is determined by the "drain classifier density" ($B.8$). This suggests that for the first enrichment stage the following static causal relations can be traced:

1) With the unchanged "hardness" and constant "size distribution" of the "original ore", maximum performance of closed circuit grinding for the ready class is equal to the "original ore consumption" into the mill and it directly depends on the "classifier discharge density":

$$(B.8 \uparrow \downarrow) \rightarrow (B.1 \uparrow \downarrow); \; A.1, A.4 - const. \qquad (2)$$

2) The higher the "content of a useful mineral" in the "original ore", the greater amount of the useful mineral can be recovered into the "enriched product", and vice versa:

$$((A.3 \uparrow \downarrow) \rightarrow (G.3 \uparrow \downarrow). \qquad (3)$$

With the constant "content of a useful mineral" in the "original ore", the following dynamic causal relationships can be traced at the first enrichment stage:

1) Deviation of the size of the useful mineral inclusions ($A.2$) in the "original ore" is compensated by the change of the "classifier discharge density" in the direction of deviation $A.2$ to obtain maximum of "useful mineral recovery" in the "enriched product" and minimize its content in the "depleted product":

$$(A.2 \uparrow \downarrow) \rightarrow (B.8 \uparrow \downarrow) \rightarrow (G.3 \uparrow) \rightarrow \\ (H.1 \downarrow); \; A.3 - const. \qquad (4)$$

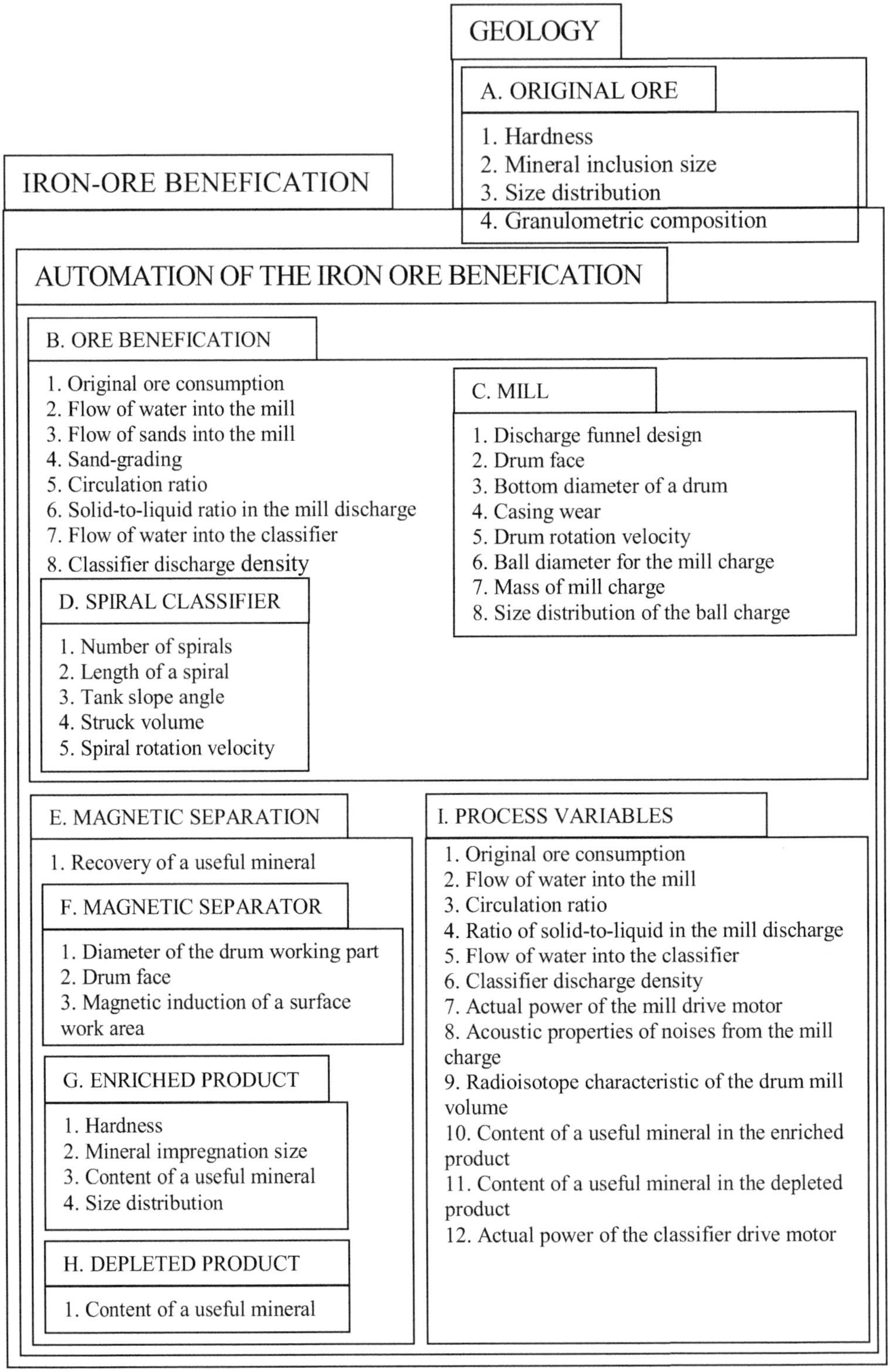

GEOLOGY

A. ORIGINAL ORE

1. Hardness
2. Mineral inclusion size
3. Size distribution
4. Granulometric composition

IRON-ORE BENEFICATION

AUTOMATION OF THE IRON ORE BENEFICATION

B. ORE BENEFICATION

1. Original ore consumption
2. Flow of water into the mill
3. Flow of sands into the mill
4. Sand-grading
5. Circulation ratio
6. Solid-to-liquid ratio in the mill discharge
7. Flow of water into the classifier
8. Classifier discharge density

C. MILL

1. Discharge funnel design
2. Drum face
3. Bottom diameter of a drum
4. Casing wear
5. Drum rotation velocity
6. Ball diameter for the mill charge
7. Mass of mill charge
8. Size distribution of the ball charge

D. SPIRAL CLASSIFIER

1. Number of spirals
2. Length of a spiral
3. Tank slope angle
4. Struck volume
5. Spiral rotation velocity

E. MAGNETIC SEPARATION

1. Recovery of a useful mineral

F. MAGNETIC SEPARATOR

1. Diameter of the drum working part
2. Drum face
3. Magnetic induction of a surface work area

G. ENRICHED PRODUCT

1. Hardness
2. Mineral impregnation size
3. Content of a useful mineral
4. Size distribution

H. DEPLETED PRODUCT

1. Content of a useful mineral

I. PROCESS VARIABLES

1. Original ore consumption
2. Flow of water into the mill
3. Circulation ratio
4. Ratio of solid-to-liquid in the mill discharge
5. Flow of water into the classifier
6. Classifier discharge density
7. Actual power of the mill drive motor
8. Acoustic properties of noises from the mill charge
9. Radioisotope characteristic of the drum mill volume
10. Content of a useful mineral in the enriched product
11. Content of a useful mineral in the depleted product
12. Actual power of the classifier drive motor

Figure 2. Interconnections between subject areas of the first enrichment stage

2) Deviation of the "content of a useful mineral" in the "original ore" simultaneously changes in the same direction "the content of a useful mineral" in the "enriched" and "depleted product" at a constant "size of mineral inclusions" in the "original ore":

$$(A.3 \uparrow \downarrow) \rightarrow (G.3 \uparrow \downarrow) \rightarrow (H.1 \uparrow \downarrow); \quad (5)$$
$$A.2 - const..$$

The parameters $A.2$ and $A.3$ determine the maximum possible value $G.3$. If they and the rest of the parameters, that determine maximum "recovery of a useful mineral" into the "enriched product" under its minimum remains in the "depleted product," are unchanged, then the process is conducted in the steady-state regime. At constant values of the rest parameters that determine maximum recovery of a useful mineral into the magnetic product, an exit from the steady-state regime is possible in three ways: 1) the parameter $A.2$ changes, 2) the parameter $A.3$ changes, 3) both parameters change. Analysis of causal relationships (2) – (5) shows that in any case this is fixed by the change of $G.3$ and $H.1$. In its turn, $G.3 = I.10$ and $H.1 = I.11$, therefore, $I.10$ and $I.11$ are the expectants in a set of process variables, by the values of which an operating regime of the first enrichment stage is determined: steady or transient. This is only partly consistent with the control objective of the first enrichment stage in the automatic mode (1).

To provide the second part of the control objective (1), let us consider "automation of the iron ore уткшсрьуте process" (Fig. 2) from the position of the subject area called "grinding conditions" (Fig. 3). We are going to clearify which of its parameters should enter the totality, enabling us to automatically distinguish the steady and transient modes of operation of the first enrichment stage.

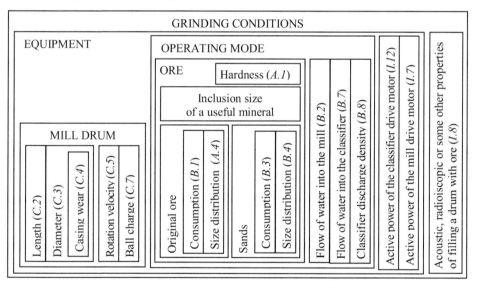

Figure 3. The relationship between subject areas of the ball mill grinding process

The parameters of such subject areas as "Automation of the iron ore benefication process" (Fig. 2) and "Grinding conditions" (Fig. 3) can be divided into three groups. The first group comprises the equipment design parameters, which are the same throughout the whole period of its operation: $C.1, C.2, C.5, D.1 - D.5, F.1 - F.3$.

The second group includes the parameters that are indicators of a change in "grinding conditions": $B.3 - B.5, G.3, H.1$. The deviation values $B.3 - B.5$ inform about the presence of perturbations at the input of the first enrichment stage: $w3$ and $w4$ (Fig. 1). Simultaneous deviation of the values $G.3$ and $H.1$ points at the presence of perturbations $w1$ or $w2$, which correspond to the improvement or deterioration of the useful mineral recovery during the grinding operation. They do not directly affect the grinding conditions, but merely serve as a guide to operations for the personnel of the ore-dressing plant. Therefore, they do not exist in the subject area "Grinding conditions" (Fig. 3).

The third group comprises the parameter values, which, when changed, have a direct impact on the conditions of grinding inside the mill drum: $A.1, A.4, B.1, B.2, B.6 - B.8, C.3, C.4, C.6 - C.8$. Let us classify them from the standpoint of the technological-process automation: perturbation action (*A.1, A.4, C.3, C.4, C.6 – C.8.*) and control action (*B.1, B.2, B .6 – B.8*), compensating the perturbances at the input of the first enrichment stage. A variety of values of these parameters can be determined by direct (*B.1 = I.1, B.2 = I.2, B.6 = I.4, B.7 = I.5, B.8 = I.6*) and indirect (*{C.6, C.7, C.8} ~ {I.3, I.6, I.8, I.9}*) methods in the real-time flow process with an acceptable error. The values of the rest parameters are determined in the laboratory conditions, and thus they are not suitable for the automatic classification of industrial situations,

corresponding to the established mode of operation of the first stage of grinding.

Decisions made by the operator or the upper level of the control system are based on the information about the controlled object. A comparative analysis of the conceptual and content models that are available to the operator and the upper level of the control system, gives an idea of the potential possibilities of both sources for making decisions on the control of the first iron ore enrichment stage (Fig. 4).

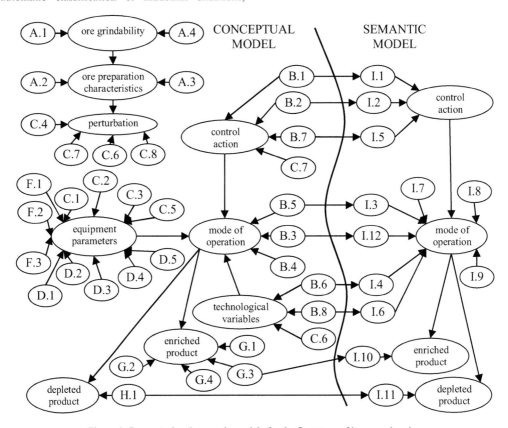

Figure 4. Conceptual and semantic models for the first stage of iron-ore dressing

5 CONCLUSIONS

The following conclusions afford ground for the passage from particulars to generals – from the first enrichment stage to the second and third one as typical structural units which form the dressing section as a whole.

1. The first and subsequent enrichment stages are connected to each other by material flows and they have a number of common features (Fig. 1), which make it possible to apply a single scheme of subject areas, conceptual and semantic models, considered

in depth for the first stage (Fig. 2 – 4). Any material flow, incoming, outgoing and internal one, in the dressing section provides reasonable details within a framework of a single homonymous subject area as well as parameters that describe its properties (Mladetskyy 2006): "hardness", "useful mineral inclusion size", "content of a useful mineral" and "size distribution". For the differentiation of these parameters let us agree on the condition that at the input of any enrichment stage and section as a whole, an "enriched product" is delivered, and at its output two material flows "enriched product" and "depleted product" are obtained. Thus:

– at the section input and at the first enrichment stage, the "original ore" is the "enriched product" (Fig. 2);

– in such subject areas as "Ore preparation" and "Technological variables" the parameter "consumption of enriched product" will be used instead of the parameter "consumption of original product" (Fig. 2).

2. In view of diverse designs of classifying apparatuses used in the first and subsequent grinding stages, alongside with the subject area "Spiral classifier", it is necessary to add the subject area "Hydrocyclone" with the relevant design parameters (Fig. 2): "amount", "diameter of a cylindrical part", "cone angle", "equivalent diameter of a feed opening", "diameter of a drainage hole", "sand hole diameter", "water pressure", "drain size", "productivity". To be able to determine the sand flow in the second and third enrichment stages one can use the parameter $I.12$ "active power of the classifier drive motor." The parameter "active power of a sand pump corresponding to the hydrocyclone battery" will be prescribed there.

A complex analysis of the relationships between subject areas, the process of grinding in a ball mill, conceptual and semantic models of any stage of iron-ore dressing gives reason to make the following conclusions:

1. From the standpoint of the efficiency function (1), the identification and classification of the steady-state regime of iron-ore dressing should be implemented by the following features: recovery of a useful mineral into the enriched and depleted product ($I.10$, $I.11$), grinding conditions inside a mill drum ($I.1 - I.9$, $I.12$), efficiency of the grinding process ($I.7 - I.9$).

2. Static and dynamic causal relationships (2) – (5) show that the use of a full range of process variables controlled in the real-time technological process for the identification and classification of the steady-state regime of iron-ore dressing, unreasonably increases the search space of the production situation similar to the current one which took place earlier at the controlled object.

3. Elimination of information redundancy that enable us to automatically identify and classify the steady state operation of the process equipment, should be carried out after a critical analysis of static and dynamic relationships between the values of process variables by the method of elimination according to the following criteria:

– all the process variables that can't be controlled in the real time process with acceptable errors for the automatic control should be excluded from the totality;

– the least informative process variables should be excluded from the totality of values;

– it is necessary to use specific characteristics of the steady-state operation of the production line.

REFERENCES

Annual information security issuer 2014: according to the company "Metinvest Holding" LLC (in Ukrainian) [See http://ingok.metinvestholding. com/upload/ingok/shareholders/godovoy_otchet_ 00190905_2014.pdf]

Annual information security issuer 2014: according to the company "Metinvest Holding" LLC [See http://sevgok.metinvestholding.com/upload/sevgo k/shareholders/pjsc_-northern-gok_-godovoj_ otchet_2014.pdf]

Annual information security issuer 2014: according to the company "Metinvest Holding" LLC (in Ukrainian) [See http://cgok.metinvestholding. com/upload/cgok/shareholders/00190977- 2014.pdf]

Annual information security issuer 2014: according to the company "Southern Mining and Processing Plant" PJSC (in Ukrainian) [See http://www.ugok.info/images/doc/ report_2013.pdf]

Annual information security issuer 2014: according to the company "Poltava Ore Mining and Processing Plant" OJSC (in Ukrainian) [See http://ferrexpo.ua/system/files/id_reports/fpm_an nual_report_2014.pdf]

Solutions for the mining and processing complex: according to the company "Saturn" Data International" (in Russian). [See http://www.saturn-data.com/img/forall/file/ Mining_rus_mod.pdf]

Automated control systems of concentrators processes: according to the company "Promteh" (in Russian) [See http://www.promtex.ru/ index.php?option=com_content&view=article&id = 38&Itemid=111]

Iron ore beneficiation solutions: according to the company "Outotec" [See http://www.outotec. com/en/Products--services/ Ferrous-metals-and-ferroalloys-processing/Iron/Iron-ore-beneficiation/]

Pospelov D.A. 1986. *Contingency management theory and practice* (in Russian). Moscow: Nauka. Ch. ed. sci. litas.: 288.

Mladetskyy I.K. 2006. *Synthesis of mineral processing technology: Monograph* (in Ukrainain). Dnipropetrovsk: National Mining University: 153.

Power Engineering and Information Technologies In Technical Objects Control – Pivnyak, Beshta
& Alekseyev (eds)
© 2016 Taylor & Francis Group, London, ISBN 978-1-138-71479-3

The mutliphysical processes at strong electric field disturbance by different water micro-inclusions in XLPE insulation

M. Shcherba
National Technical University of Ukraine "Kyiv Polytechnic Institute", Kyiv, Ukraine

ABSTRACT: The multiphysical processes at a strong electric field perturbation in local areas of XLPE insulation near water micro-inclusions of different shapes, sizes and mutual disposition were studied. The water micro-inclusions of shapes: spheres, spheroids, spheres and spheroids with water treeing on the surface were considered. The influence regularities of the inclusions characteristic parameters on the increase degree of maximum values of field strength, electromechanical pressure and total current density in the insulation micro-volumes were determined. Not only sizes and shapes of separate inclusions were taken into account, but also their mutual disposition and orientation with respect to the field lines. The comparison of multiphysical processes of ageing and degradation of XLPE insulation for extra-high voltage cables was made both taking into account its nonlinear properties (in particular, the specific conductivity dependence on the field strength) and without one.

1 INTRODUCTION

Currently, there is an increasing interest in problems of strong electric fields calculation in heterogeneous dielectric mediums. With the development of modern computational methods and their software implementations it became possible to solve the problems not previously available, which are of interest for the development of the electromagnetic field theory (Teixeira 2008). New aspects of the theory are used in important practical problems. One of such problems is the electric fields (EF) calculation in a cross-linked polyethylene (XLPE) insulation of high voltage (HV) and extra high voltage (EHV) cables in the presence of various water micro-defects in this insulation.

Summarizing the results of numerous studies (O'Dwyer 1973; Sletbak & Botne 1977; Dissado & Fothergill 1992; Hvidsten et. al. 1998; Kurihara et. al. 2014) the algorithm for the cable insulation breakdown in a strong electric field in the presence of moisture can be described. In the micron and submicron pores and cracks, which are always present in the cross-linked polyethylene, water is gradually accumulated. Its intrusion can be described by thermodynamics laws due to the difference of chemical potentials (Matsuba & Kawai 1976), by chemistry laws as a result of polyethylene cross-linking agents' disintegration (Sletbak & Botne 1977) and as a result of water molecules dielectrophoresis along the conductor and to the insulation volumes with the highest EF strength (Pohl & Pohl 1978). To a greater or lesser extent, all of these mechanisms appear and contribute to filling of voids in dielectric with moisture.

Filled voids become water micro- and submicro-inclusions, at the poles of which a liquid pressure on the surface of polyethylene is created. In some cases, the nascent pressure magnitude could be comparable or even exceed the mechanical strength of a material and leads to the appearance of new cracks in it (Dissado & Fothergill 1992; Hvidsten et. al. 1998). Moreover, there are regions of increased field strength near the poles of inclusions, in which the new water molecules are drawn in and the inclusions become centers of water trees - thin branched structure forms. In its turn, near the tips of treeing branches the high pressure and field strength also appear, that increase the size of tree structure and promotes further degradation of cross-linked polyethylene.

At close mutual arrangement of water micro-inclusions the EF additionally intensifies in the dielectric gap between them (Shcherba 2013), the electric forces appear, which aimed to converging and coalescence of the inclusions. The appearance of cracks in the polyethylene and germination of water treeing can cause the appearance of conducting channel between the inclusions, which conductivity increases while it is filled with water (Hvidsten et. al. 1998). Ultimately close-located inclusions are combined into one conductive structure, which disturbs EF much more.

Studies of the maximum levels of the electric field strength, which can exceed the breakdown voltage of XLPE insulation, allow to describe fast deterministic processes of dielectric degradation. To describe the slow stochastic processes of its degradation in (Shcherba 2013) it was proposed to evaluate the dimension of tensed volumes areas of XLPE, i.e. those areas where EF tension is though lower the breakdown value E_{break}, but higher then the determined permissible value E_{perm}. The larger tensed volume in insulation, the greater its breakdown probability in any region of this volume.

In addition to solving the electrical and mechanical tasks on the calculation of the EF distribution and pressure in dielectric it is expedient also to solve the thermal task on calculation of heating local areas in insulation during the displacement currents flow as well as heating of water defects areas during the conduction currents flow (Bodega et. al. 2004; Shcherba 2013). These studies are particularly relevant in situations of a sharp change in the cross section of conductive defects: on the inclusion surface turning into the water treeing, at thinning of a treeing channel and near it tips. On the defect - dielectric boundary the conduction currents close through the displacement currents in the XLPE insulation and their high densities can lead to the material heating and thus a reduction of its electrical and mechanical strength (Dissado & Fothergill 1992).

Let's note that the electrical, mechanical and thermal processes in a dielectric are interrelated and for their analysis it is necessary to solve a complex multiphysical problem.

The size, shape and mutual position of the micro-defects determines the electric field disturbance in XLPE insulation. Because of the micro-inclusions and micro-tree configurations can be very different and they vary with time, then for a comprehensive assessment of the specific configuration and for forecasting further insulation resource it is necessary to make a comparative analysis of the occurring electrical processes. It is expedient to determine the levels of the appearing EF strengthening, the values of tensed volumes, pressures and currents in the insulator, as well as to identify the factors with the greatest impact. It is this tasks that given work was dedicated.

Moreover in the works (Tokoro 1992; Boggs 1995) it was demonstrated that in strong electric fields (higher 20 kV/mm) XLPE conductivity becomes a function of the EF intensity and the problem of its calculation becomes non-linear. This fact was also taken into account in this paper and the results of the calculation for both linear and nonlinear problem definitions were compared.

2 PROBLEM FORMULATION

Using a numerical finite element method, it is necessary to calculate and made a comparative analysis of changes in the values of electric field strength, tensed volumes, current densities and pressures occurring in XLPE insulation of high-voltage cables near the water micro-inclusions and micro-treeings depending on their configuration.

3 MATHEMATICAL MODEL FOR CALCULATION

According to the calculation of the field distribution over the cross section of 330 kV cable the average EF intensity is $E_{av} \geq 10$ kV/mm at a distance up to 5 mm from the surface of semi-conductive layer of XLPE insulation. The insulation layer of 1 mm thickness, to which a sinusoidal 10 kV voltage at 50 Hz is applied (see Fig. 1) was simulated.

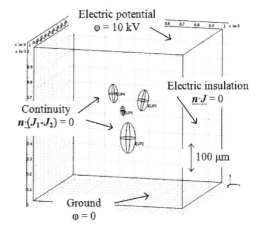

Figure 1. Computational domain of XLPE insulation with water micro-inclusions and boundary conditions

In the layer it was supposed a presence of water micro-inclusions and micro-treeings with different sizes, shapes and arrangement. We considered the most typical micro-inclusions: spheres, spheroids, spheres and spheroids with water treeing on the surface. Also we study the cases of close disposition of such inclusions as well as the situations of appearance of conducting channel between them, treeing branching, fission of large inclusions into many smaller ones and getting of smaller inclusions in electric field of large ones.

Dielectric medium considered was piecewise-homogeneous, isotropic, linear at EF strength lower than 20 kV/mm, and non-linear at strength higher than 20 kV/mm. Under such assumption in (Tokoro

1992) the conductance of XLPE insulation at temperatures 80, 90 and 100 °C and field strength from 37.5 to 60 kV/mm was experimentally measured, and in the work (Boggs, 1995) according to these results the theoretical model was developed and the following approximate equation for specific conductivity σ as a function of EF strength was proposed:

$$\Box(E,T) = \frac{\exp(a+b/T)}{E(T)} \text{sh}\left(\frac{kE(T)}{T}\right). \tag{1}$$

Here a and b stand for estimated constants, $k = 1.38 \cdot 10^{-23}$ J/K Boltzmann's constant, T for absolute temperature.

The final approximate expression for σ(E), which, with over 98 % accuracy converges with the experimental data, is given by:

$$\Box(E,T) = \frac{\exp(2,79 - 6698/T + 1,24 \cdot 10^{-7} \cdot E(T))}{E(T)}. \tag{2}$$

The interrelation of the vectors of the electromagnetic field was described by Maxwell equations (Landau & Lifshitz 2006), the problem was formulated in the quasi-static approximation. The estimated equation for the scalar electric potential φ and for the linear region of XLPE insulation (σ is const) is conveniently written using the method of complex amplitudes (Shcherba 2013):

$$\text{div} \left[-(\gamma + i\omega\varepsilon\varepsilon_0) \cdot \text{grad}\varphi(t)\right] = 0. \tag{3}$$

For non-linear insulation region, taking into account σ(E) dependence, the estimated equation for φ is represented in the following form:

$$\text{div}\left[\Box(E) \cdot \text{grad } \varphi(t) - \varepsilon\varepsilon_0 \frac{\partial \text{grad } \varphi(t)}{\partial t}\right] = 0. \tag{4}$$

In order to obtain the unique solution of the equations (3) and (4) the Dirichlet conditions (values of electric potentials) were set on the upper and lower boundaries of the computational region (see Figure 1), and the Neumann conditions (vanishing of the surface normal derivative of potentials) were set on the side faces of the computational region. At water-XLPE insulation interface (i.e. at inclusion-medium interface) the criteria for potentials and their interface surface normal derivatives were set.

The calculation of the scalar electric potential φ distribution in the computational volume of XLPE insulation was performed using the numerical finite element method implemented in software package Comsol Multiphysics.

As in the work (Shcherba 2013) the tensed volume V_t value for the three-dimensional computational domain was determined according to the equation:

$$V_t = \int_V f(E) dV. \tag{5}$$

Here V stands for the calculated volume of dielectric, $f(E)$ for a function, which for EF strength exceeding the permissible value ($E \geq E_{perm}$) takes the value of $f(E) = 1$, and for $E < E_{perm} - f(E) = 0$.

To calculate the force interactions of water micro-defects with the external EF and the dielectric material the electric Maxwell stress tensor T was used (Landau & Lifshitz 2006). Electrical force F induced by EF has an influence on the conducting inclusions. It can be represented only as a result of forces applied to the inclusion surface. Therefore, to determine the force F the surface integral of the stress tensor T can be used instead the volume integral, according to the expression:

$$F = \iiint f dv = \iint T ds. \tag{6}$$

The equation for the components of the tensor T has the following form:

$$T_{ij} = E_i \cdot D_j. \tag{7}$$

Vector of the total current density J_{tot} was calculated as the sum of the vectors conduction current in the water micro-defects J_{cond} and the displacement current J_{dis} in the XLPE insulation:

$$J_{tot}(t) = J_{cond}(t) + J_{dis}(t) = \Box(E)E(t) + \varepsilon\varepsilon_0 \frac{\partial E(t)}{\partial t}. \tag{8}$$

4 ANALYSIS OF SIMULATED RESULT

The electric field perturbations in the cross-linked polyethylene (XLPE) insulation of high-voltage cables near the water micro-inclusions in the form of spheres, spheroids, spheres and spheroids with water treeing on the surface were simulated. Also the situations of closely located inclusions, the appearance of a conductive channel between them, the treeing branching, the fission of large inclusions into many smaller ones and getting of smaller inclusions in electric field of large ones were simulated.

The maximum values and distribution of the electric field $E(t)$ and the total current density $J_{tot}(t)$ in the XLPE isolation volume, the maximum value and the pressure $p(t)$ distribution on the water inclusions surface, as well as the value of the dielectric tensed volume $V_t(t)$ were also calculated. Since all mentioned values change with time the amplitude values of E, J_{tot}, p and V_t were determined in this work.

77

The values of E and V_t were represented in relative units as the electric field gain coefficient $k_E = E/E_{av}$ and tensed volume coefficient $k_{Vt} = V_t/V$. Coefficient k_E was defined as the ratio of intensity E in the calculated points in dielectric to its average value E_{av} in the computational domain. Coefficient k_{Vt} was defined as the ratio of V_t across the computational domain to the total volume V of all water micro-defects this configuration.

For the analysis of the occurring electrical processes we are most interested in maximum values $k_{E\,max}$, $J_{tot\,max}$, p_{max} which are observed at the poles of micro-inclusions and near the tips of micro-treeing branches. To simplify the notation index "max" will be omitted, implying that the calculated values k_E, p and J_{tot} are the highest values in the computational domain of the XLPE insulation.

4.1 Spherical inclusion

It is known that for a single conductive spherical inclusion in the dielectric the maximum value of the field gains coefficient $k_E = 3$ at the inclusion poles in the EF direction (Landau & Lifshitz 2006). This fact was confirmed by the simulation results for the water micro-inclusions with radius $r = 25$ mm in the XLPE insulation. According to our calculations for such inclusion the tensed volume coefficient $k_{Vt} = 0.6$, the maximum water pressure on polyethylene is $p = 5$ kPa, and the maximum total current density at the inclusion pole is $J_{tot} = 0.5$ A/m^2. These results are in good agreement with the results of other studies (Dissado & Fothergill 1992, Shcherba 2013) that is a check of adequacy of the mathematical model applied.

4.2 Spheroid micro-inclusions

The electric field disturbance in the XLPE insulation near the water micro-inclusions of spheroid forms extending along the field lines was simulated. We considered spheroids with different ratios of semi-axes a − along the EF and $b = c$ − perpendicular to the EF.

As mentioned in paper (Sletbak 1979), for a water droplet in the air the maximum extension along the EF is limited by the $a/b = 1.5$ ratio, higher which the Coulomb forces will rise faster in comparison with the surface tension forces, and by the action of an unbalanced resultant force the droplet is divided into multiple droplets. A solid medium in particular cross-linked polyethylene characterized in that a difference of the Coulomb forces and the interfacial tension forces compensates the elastic reaction forces of medium. In this case, according to the calculation in paper (Sletbak 1979), the most characteristic shape of water micro-inclusions in XLPE

insulation under the influence of external field is a spheroid with semi-axes ratio a/b in the range of 1.5 to 3.

However, since there are micro-cracks of arbitrary configuration in the material the shape of single micro-inclusions may differ significantly from the above-mentioned one. The ratio of the spheroid semi-axes can reach values of 1/10 or higher.

For as much as possible wide research the spheroids with semi-axes ratios from 1.5 to 100 were simulated in this work. The a and b values were chosen in such a way that the volume of liquid in the spheroid did not change and amounted to the volume of liquid in the spherical micro-inclusion of radius $r = 25$ mm. The calculation results of the EF and electric pressure distribution near several spheroids of typical configurations are shown in Fig. 2.

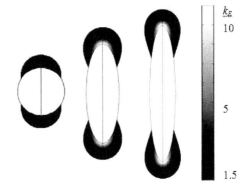

Figure 2. The EF distribution in XLPE insulation near spheroid micro-inclusions.

In Fig. 2 the areas of tensed volume V_t are marked by the dark shading, in which the EF strength E increased by 50 % and more in comparison with the average value E_{av} in the computational domain. The E value is according to the scale on the right.

Distribution of the water pressure p for spheroid micro-inclusion with 1/5 semi-axes ratio is presented on Figure 3 on the left and distribution of the total current density J_{tot} − on the right.

The arrows indicate the direction and the magnitude of the electric pressure that arises at the water-cross-linked polyethylene interface.

The calculated values of the maximum EF gain coefficient k_E, the tensed volume coefficient k_{Vt}, the maximum water pressure p and the maximum total current density J_{tot} at the spheroid poles are shown in the Table. 1.

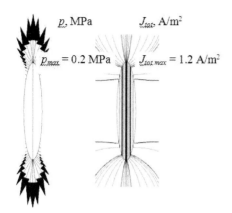

Figure 3. Distribution of the water pressure and the total current density for spheroid micro-inclusion.

Table 1. The values, calculated for the spheroid micro-inclusions in XLPE

$a{:}b$, r.u.	k_E, r.u.	k_{Vt}, r.u.	p, MPa	J, A/m^2
1:1.5	4.3	0.8	0.01	0.3
1:2	7.0	1.5	0.02	0.5
1:3	9.2	2.3	0.05	0.6
1:5	17.9	2.7	0.2	1.2
1:10	47.5	6.4	1.3	3.3
1:20	143	16.4	11.8	9.8
1:100	1275	194.3	936.6	650

Let's note the non-linear increasing of all variables due to the spheroid extension along the EF lines.

Increasing of the tensed volume coefficient k_{Vt} for the same volume of micro-defects with different configuration shows an increasing of dielectric breakdown probability at any point of the tensed volume. If ρ is a probability density of insulation breakdown, than the integral $P = \int \rho \, dV_t$ is equal to the probability of this volume breakdown. If we take $P = 1$ at the critical tensed volume value ($V_{t\,cr}$ 10^{-12} - 10^{-11} m^3 for the XLPE insulation), then for the calculated k_{Vt} and V_t the breakdown probability is in the range from 4 to 50%.

The calculated electrical pressure p, depending on the shape of micro-inclusion, is comparable with the mechanical strength of XLPE insulation (polyethylene breaking stress is 9.8-16.7 MPa). Such force impacts in alternating EF occur with double frequency and depending on changes in the inclusions configuration can increase tens times.

4.3 Micro-inclusion with micro-treeing on surface

Near the micro-inclusion pole in the dielectric the regions, which have the greatest EF strength and the

maximum pressure, arise. It can lead to micron and submicron cracking of the material. Such cracks are gradually filled with water and they turn into water treeing, which initially has an extended unbranched shape. The water treeing appearance in XLPE insulation, in particular on the surface of the conductive micro-inclusions was confirmed by many works, for example (Sletbak 1979; Hvidsten et. al. 1998; Thomas & Saha 2008; Kurihara et. al. 2014).

In the paper the calculation of the EF distribution in the XLPE insulation near the spherical and spheroid micro-inclusions with extended unbranched treeing on the surface has been made. The treeing is simulated as cylindrical tube with radius R_{tr} from 0.5 to 2.5 μm and with rounding at the tip by semisphere or semi-spheroid with different semi-axes ratios. The configuration of such micro-defect is varied so that the total water volume remained unchanged and it was equal to the volume of spherical micro-inclusion with 25 μm radius – 65450 μm^3. It should be noted that since the characteristic dimensions of the micro-inclusion are more than 10 times greater than the treeing diameter, then its volume is about 100 times smaller than the volume of inclusion. Thus, the volume of micro-inclusion with treeing practically equals to the volume of micro-inclusion without treeing, while the EF distribution pattern differs significantly.

The calculation results of the maximum value of the field gain coefficient k_E, the coefficient of tensed volume k_{Vt}, the maximum water pressure p and the maximum total current density J_{tot} near the tip of water treeing on the surface of spheroid micro-inclusion (semi-axes ratio $a/b = 1/2$) are shown in the Table 2.

Table 2. The calculated values for micro-inclusions with treeing on the surface

r_{tr}, μm	r_{tip}, μm	l_{tr}, μm	k_E, r.u.	k_{Vt}, r.u.	p, MPa	J, A/m^2
1	1	25	18.3	1.0	0.2	45
1	1	50	32,6	1.4	0.4	81
1	1	100	79.2	2.0	3.4	195
2	1	100	52.7	1.6	1.5	110
5	1	100	30.0	1.0	0.5	25
1	2	100	71.5	1.9	2.8	195
1	5	100	61.1	1.9	2.0	195

The changes in the calculated values depending on the five dimensional parameters, which characterize the shape of water micro-defect: r_{tr}, the radius of the cylindrical treeing channel, r_{tip}, the rounding radius of the treeing tip, l_{tr}, the treeing length, a and b, the semi-axis of the spheroid micro-inclusion.

The highest values of k_E and p are ob served near the water treeing tip. The increasing of the inclusion and/or the treeing length along the field as well as the decreasing of radius of the tip rounding, intensify the EF disturbance. Whereas increasing of the inclusion size at right angle to the field and increasing of treeing thickness lead to EF disturbance reduction.

The dark shading in the Fig. 4 shows the EF disturbance in the dielectric near the spheroid micro-inclusion (1/2 semi-axes ratio) with water treeing ($r_{tr} = 2$ μm, $r_{tip} = 2$ μm, $l_{tr} = 100$ μm) on the surface.

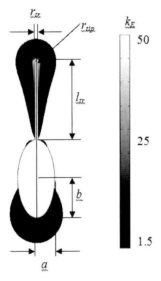

Figure 4. The EF disturbance in the dielectric near the spheroid micro-inclusion with water treeing.

The graph in Fig. 5 represents the dependence of the field gain coefficient k_E of three dimensional parameters r_{tr}, r_{spr} and l_{tr} for the complex inclusion. If the micro-inclusion dimensions along the EF are larger than the micro-treeing size, they are decisive for the disturbance character. If the treeing during its lengthening process becomes larger than the inclusion, then it dimensional parameters become decisive now.

It can be concluded that for micro-defects of considered configuration the appearance of micro-treeing may lead to increasing of the field strength to 80 times, and to the tensed volume V_t becomes twice higher then the inclusion volume. The micro-treeing appearance also has a significant impact on the distribution of the conduction current density J_{cond} along the channel and, as a consequence, on displacement current value J_{dis} in the dielectric at

the treeing tip. If for the spherical inclusion the maximum current density is $J_{tot} = 0.05$ A/m^2, then for the spheroid it is J_{tot} up to the 10 A/m^2, and when the water treeing appears J_{tot} can be increased to 350 A/m^2.

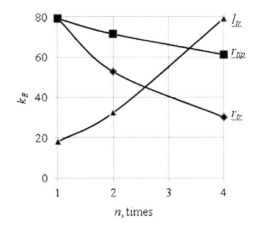

Figure 5. Dependence of the field gain coefficient k_E of three dimensional parameters r_{tr}, r_{spr} and l_{tr}.

The micro-treeing configuration has a influence both on the maximum value of the current density and on the location of the area with the highest density along the treeing channel.

Fig. 6 shows the variation in the density of the total current J_{tot} along the water treeing channel with 50 μm length and in the volume of nearby dielectric depending on the treeing configuration.

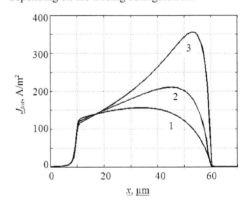

Figure 6. Variation in the total current density J_{tot} along the water treeing channel.

Curve 1 corresponds to the treeing in the form of a cylinder with a spherical rounded tip and curves 2 and 3 correspond to a complex treeing consisting of cone with a different inclination angles α and cylinder with their bases connected. The current density at the treeing base is independent of its configuration, and is about 120 A/m². However current density at the treeing tip with the increasing of angle α of its conical part rises from 70 to 350 A/m².

A non-uniform J_{tot} distribution in the dielectric volume leads to non-uniform thermal energy dissipation in it and as a result to non-uniform insulation heating. With rise of temperature the mechanical strength of XLPE insulation decreases. For example, when the temperature rises from 20 to 60°C the polyethylene breaking stress decreases from 9.8–16.7 MPa to 3.3–8.8 MPa and the material ages more intensively under influence of the pulsating forces owing to micro-inclusions.

4.4 Closely located inclusion

As it was shown in the paper (Shcherba 2014) the EF nearby water micro-inclusions additionally intensifies if the distance between these micro-inclusions less than their characteristic sizes (closely spaced inclusions). On such mutual distances the electrical charges induced on the opposite poles of inclusions are attract and charge density at the poles increase. This leads to an increase in the field strength E, the pressure p and the density of the total current J_{tot} in the dielectric gap between micro-inclusions. Furthermore, due to non-uniform distribution of the surface charge both pressure distribution at the interface inclusion-insulation and current density distribution in the inclusions also are non-uniform. The resultant force F acting on each of the inclusions becomes non-zero and they become mutually attracted.

The inclusions of spherical shape with radius $r = 25$ μm and spheroid shape with different ratios semi-axes, provided equal volumes of water in each inclusions were simulated.

Figure 7 indicates the EF disturbance in the dielectric near two closely located spheroid micro-inclusions with 1/2 semi-axes ratio on different distances.

Dark-shaded areas correspond to the tensed volume areas V_t, the E distribution is according to the scale in Fig. 7 is to the right, the arrows display the vectors of the surface pressure p density.

According to the numerical experiment results the EF in the dielectric gap could increase tens time with decreasing of the distance l between the inclusions. For example when l changes from 50 to 1 μm coefficient k_E increases from 6 to 130, and J_{tot} from 0.2 to 5 A/m². Thus, if the average strength near the core of a high-voltage cable with XLPE insulation equals 10 kV/mm then it amplifies in dielectric gap of 1 mm between two spheroid inclusions 130 times, and E_{max}= 1300 kV/mm i.e. it is more than 30

times higher than the dielectric strength of polyethylene.

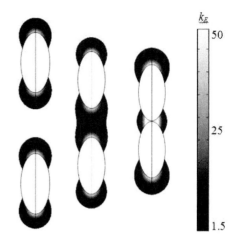

Figure 7. EF disturbance in the dielectric near two closely located spheroid micro-inclusions

Fig. 8 shows the resultant force F of mutual attraction of two spherical micro-inclusions as a function of distance l between them.

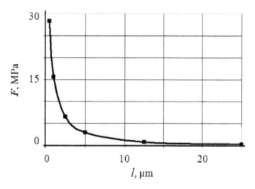

Figure 8. Resultant force F of mutual attraction of two spherical micro-inclusions.

With the l decreasing the F value increases according to the power function law and the approximate equation is F = 36.46 $l^{-1.21}$. Due to the positive feedback, during exploitation of XLPE insulation over a long period of time, its degradation intensifies the process of micro-inclusion approaching and EF intensification.

4.5 The conductive channel

As it was shown in papers (Hvidsten et. al. 1998; Shcherba 2013) because of appearance of the pulsating pressures at the poles of closely located micro-inclusions, which could exceed the polyethylene mechanical strength, the micro-cracks could arise in the dielectric gap between the inclusions. Due to further cracks development a continuous channel connecting the inclusions into a single structure can occur. Over the time this channel is filled with water and its conductivity increases from the polyethylene value $\sigma = 1 \cdot 10^{-14}$ S/m to the water value $\sigma = 1 \cdot 10^{-2}$ S/m.

In this paper both two closely located spheroid micro-inclusions and a group of five such micro-inclusions with a thin connecting channel were simulated. The changes in values k_E, k_{Vt}, p and J_{tot} with the increasing of the channel conductivity were studied.

Thus in Fig. 9 the EF disturbance in the dielectric near five spheroids with 1/2 semi-axes ratio and with a channels length equal to 1.5 inclusions size are shown.

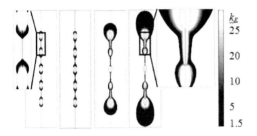

Figure 9. The EF disturbance in the dielectric near five spheroids with connecting channels.

The distribution of the EF gains coefficient k_E in XLPE insulation when the channel conductivity $\sigma_{channel}$ between the micro-inclusions changes from 10^{-14} to 10^{-2} S/m is represented in Fig. 9 (according to the scale on the right). Dark-shaded areas are the tensed volume V_t areas (in which E has raised by 50 % and more in comparison with the average value $E_{av} = 10$ kV/mm in insulation).

The rise of the channel conductivity leads to the increasing in the values of coefficient k_E and the insulation tensed volume V_t. When the channel conductivity reaches the characteristic value for water $\sigma_{water} = 0.01$ S/m the maximum EF strength increases in 4.4 times (up to 250 kV/mm), and the tensed volume $V_t - 55.5$ times (up to 322 μm^3) in comparison with the single water micro-droplets.

It should be note that channel conductivity begins to affect on the EF distribution nature only if its value is equals 10^{-6} S/m and more.

Appearance of water channel between 5 inclusions results in increase the field 25 times, and the tensed volume from 5.8 to 322.2 times.

It can be concluded that connection of closely located micro-inclusions by a conductive channel in a one single water structure is a significant danger to the XLPE insulation because of great EF disturbances.

4.6 Treeing branching

Another important factor, which significantly changes the electric field distribution, is water treeing branching. If the volume of moisture appeared in the XLPE insulation is not enough for the new micro-inclusion formation or configuration of the dielectric micro-pores does not allow such formation, the water micro-droplets could contribute to the origination of water treeing new branches.

Fig. 10 from paper (Shcherba 2014) shows the results of numerical experiment on the EF disturbance calculation near the micro-inclusion with treeing of varying branching degree.

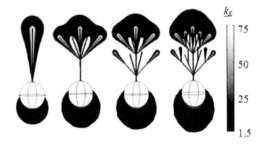

Figure 10. The EF disturbance near the micro-inclusion with treeing of varying branching degree.

The micro-inclusion has spherical shape with a diameter $d = 50$ μm. The micro-treeing has length $l_{tr} = 100$ μm, radius $r_{tr} = 1$ μm of each cylindrical branch and radius $r_{tip} = 1$ μm of tips rounding. Dark-shaded areas correspond to the tensed volume areas V_t, and the EF gain coefficient k_E distribution is determined according to the scale on the right.

If for the unbranched treeing EF could be increased in 72 times then according to the calculations the treeing branching reduces the E on its tip. For branched treeing due to the central branch shielding by other branches the EF strength decreases and its increase could be in a range from 48 to 59 times. It is well agree with the results of other studies (Dissado & Fothergill 1992; Hvidsten et. al. 1998), which describe that intensive treeing branching leads to a slowing of its growth and even to its full stop.

However, in spite of the fact that k_E decreases the value V_t increases. The tensed volume coefficient increases from 0.6 to 2.2 (i.e. 4 times) in comparison with the sphere inclusion. The more intensive is treeing branching, the greater is the volume with increased EF strength in XLPE insulation and, as a result, the greater is the breakdown probability. If the breakdown probability of the dielectric local volume near the branch treeing tip is about 5 %, then in the presence of 17 such branches (case 4 in Figure 10) the breakdown probability of any such local volume becomes $1 - 0.95^{17} = 0.58$, i.e. 58 %.

The number of regions with the increased pressures p as well as with the increased total current density J_{tot} rises at treeing branching that in addition to the electric ageing results in mechanical and thermal degradations of the dielectric.

4.7 Dispersion of the inclusions

Under the influence of electrical forces the volume of conducting liquid in a solid dielectric instead forming of some big inclusions could be distributed over a larger volume in the form of small droplets (of micron sizes). Although the size of each inclusions type is much less than the size acceptable by the quality criteria for XLPE insulation, but the accumulation of group (cloud) of closely located inclusions is no less dangerous (Shcherba 2014).

In the paper it was also studied the dependence of EF amplification from the inclusions dispersion degree – number of inclusions at invariable total volume of conductive liquid. Dark shading in Fig. 11 presents distribution of the field gains k_E and tensed volume V_t area near one, three, five spherical and five spheroid micro-inclusions.

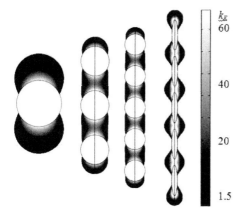

Figure 11. The EF disturbance near one, three, five spherical and five spheroid micro-inclusions.

The sphere diameter and the spheroid semi-axes were chosen so, that the total liquid volume was invariable.

This micro-inclusions configuration in comparison with the two closely located inclusions is characterized by a larger number of areas with increased EF strength E_{max}, pressure p_{max} and current density J_{max}. If for two inclusions there is only one dielectric gap with potential degradation processes

in insulation than for five micro-inclusions there are four such gaps.

4.8 A drop in the inclusion field

We have investigated the EF distribution near small micro-inclusions when they get in the field of a nearby large inclusion. So, if spherical inclusion with 10 µm radius appears 10 µm away from the spherical inclusion with 50 µm radius, then near its poles coefficient k_E increases from 3 (for a single sphere) to 7.7. For this situation the values: $k_{Vt} = 0.6$, $p = 0.25$ MPa, $J_{tot} = 0.5$ A/m^2 increase too.

4.9 Non-linearity

In accordance with experiment results in the paper (Tokoro 1992) it is shown an increasing of a specific conductivity σ of the XLPE insulation with the increasing of the average EF intensity up to 60 kV/mm at the temperatures of 80, 90 and 100 °C. The conductivity value increase from $1 \cdot 10^{-14}$ S/m (characteristic for the XLPE insulation without field) to $7.12 \cdot 10^{-12}$ S/m, which is almost three orders of magnitude greater. In paper (Boggs 1995) it is proposed a theoretical model, which confirms the experimental results and the approximate equation in order to describe the conductivity σ of XLPE insulation as a function of temperature T and the electric field strength E up to 200 kV/mm.

The changes of the EF strength, the electrical pressure and current density in the XLPE insulation near various water micro-inclusions were calculated in this paper using abovementioned dependence. It has been demonstrated that in comparison with the linear problem formulation (when conductivity is taken as a constant), the distinction in EF strength becomes apparent at $σ \geq 10^{-7}$ S/m and this distinction depends on the configuration of micro-inclusions. This value of XLPE insulation conductivity in the operating temperature range corresponds to EF strength 150 kV/mm.

Therefore if the average EF strength in our calculation is 10 kV/mm then change of the conductivity, which has an influence on the EF distribution, occurs only near the inclusions amplifying the EF 15 times or more. That is near only a specific group of inclusions.

Taking into account the dependence of dielectric specific conductivity on the electric field strength makes it possible to consider a significant increase in the conduction current density (three orders of magnitude or more) in the areas of most field amplification near the micro-inclusions. This non-uniform current density distribution promotes the local insulation overheating, deterioration of its electrical and mechanical strength, which leads to the dielectric accelerated degradation.

4 CONCLUSIONS

1. The mathematical simulation and the comparative analysis of the electric field perturbation in the cross-linked polyethylene insulation near water micro-inclusions of the most typical forms: spheres, spheroids, spheres and spheroids with water treeing on the surface have been made.

2. The regularities of the electric field amplification in the XLPE insulation near water micro-inclusions with complex forms and between closely located micro-inclusions of different sizes, shapes and mutual disposition are determined. It is shown how the taking into consideration of the insulation conductivity dependence on the field intensity can influence on the multiphysical processes of insulation degradation.

3. The changes in the calculated values taking into account the XLPE nonlinear conductivity dependence on the electric field strength were determined.

4. It is found that the greatest amplification of the electric field (dozens times) occurs near the poles of water treeing of prolate-shape, which is oriented parallel to the field lines and near micro-inclusions of similar forms. However the greatest field amplification arises in the insulation between closely located inclusions, which are pulled out along the field.

5. For abovementioned configurations of the inclusions there are the most local pressures from water to cross-linked polyethylene (about MPa), which are comparable to the limit of the insulation mechanical strength. And with the field strength increasing, such pressures increase according to the quadratic law.

6. The largest areas of tensed volume in dielectric arise when some micro-inclusions are combined by the conducting channel into a single structure as well as when water treeing branches or near strongly prolate spheroid along the field. The increase of the insulation tensed volume characterizes the stochastic mechanisms its degradation and shows increase of breakdown probability at any point of the tensed volume in the dielectric.

7. Taking into account the dependence of dielectric specific conductivity on the electric field strength allows to consider a significant increase of the conduction current density (three orders of magnitude or more) in the areas of the most field amplification near the micro-inclusions. This non-uniform current density distribution contributes to the local insulation overheating, decreasing of its electrical and mechanical strength. Therefore, if electric field strength in the local micro-volumes of XLPE insulation increases more than 150 kV/mm then in the first place it is necessary to consider the electro-thermal degradation of insulation.

REFERENCES

Bodega R., Montanari G.C., & Morshuis P.H.F. 2004. *Conduction Current Measurements on XLPE and EPR Insulation.* IEEE: Electrical Insulation and Dielectric Phenomena: 101-105.

Boggs S.A. 1995. *Semi-Empirical High-field Conduction Model for Polyethylene and Implications Thereof.* IEEE Transactions: Dielectrics and Electrical Insulation, Issue 2.1: 97−106.

Dissado L.A., & Fothergill J.C. 1992. *Electrical degradation and breakdown in polymers.* UK: London: Peter Peregrinus. Ltd.: IEE Materials and Devices Series 9: 601.

Hvidsten S., Ildstad E., Sletbak J., & Faremo H.A.F.H. 1998. *Understanding Water Treeing Mechanisms in the Development of Diagnostic Test Methods.* IEEE Transactions: Dielectrics and Electrical Insulation, Issue 5(5): 754-760.

Kurihara T., Okamoto T., Kim M.H., Hozumi N., Tsuji T. & Uchida K. 2014. *Measurement of Residual Charge Using Pulse Voltages for Water Tree Degraded XLPE Cables Diagnosis.* IEEE Transactions: Dielectrics and Electrical Insulation, Issue 21(1): 321-330.

Landau L.D., Lifshitz E.M. 2006. *The Classical Theory of Fields. Vol. II* (in Russian). Moscow: Fizmatlit: 534.

Matsuba H. & Kawai E. 1976. *Water Tree Mechanism in Electrical Insulation.* IEEE Transactions: Power Apparatus and Systems, Issue 95(2): 660-670.

O'Dwyer J.J. 1973. *The Theory of Electrical Conduction and Breakdown in Solid Dielectrics.* Oxford: Clarendon Press: 317.

Pohl H.A. & Pohl H.A. 1978. *Dielectrophoresis: the Behavior of Neutral Matter in Nonuniform Electric Fields* (Vol. 80). Cambridge: Cambridge University Press: 579.

Rezinkina M., Bydianskaya E., Shcherba A. 2007. *Alteration of Brain Electrical Activity by Electromagnetic Field.* Environmentalist, Vol. 27, Issue 4: 417-422.

Teixeira E.L. 2008. *Time-domain Finite-difference and Finite-element Methods for Maxwell Equations in Complex Media.* IEEE Transactions: Antennas and Propagation, Issue 56(8): 2150-2166.

Shcherba A.A., Podoltsev O.D., Kucheriava I.M. 2013. *Electromagnetic Processes in 330 kV Cable Line with Polyethylene Insulation.* Technical Electrodynamics, Issue 1: 9-15.

Shcherba M.A., Roziskulov S.S., Vasilyeva O.V. 2014. *Dependence of Electric Field Disturbances in Dielectrics on the Dispersion of Closely Spaced Water Micro-Inclusions.* Technical Electrodynamics, Issue 4: 17-19.

Sletbak J. & Botne A. 1977. *A Study of Inception and Growth of Water Trees and Electrochemical Trees in Polyethylene and Cross Linked Polyethylene Insulations.* IEEE Transactions: Electrical Insulation, Issue 6: 383-389.

Thomas A.J. & Saha T.K. 2008. *A New Dielectric Response Model for Water Tree Degraded XLPE Insulation-Part a: Model Development with Small Sample Verification.* IEEE Transactions: Dielectrics and Electrical Insulation, Issue 15(4): 1131-1143.

Tokoro T., Nagao M. & Kosaki M. 1992. *High Field Dielectric Properties and ac Dissipation Current Waveforms of Polyethylene Film.* IEEE Transactions: Dielectrics and Electrical Insulation, Vol. 27, Issue 3: 482-487.

Energy efficiency optimization of underground electric power networks

V. Trifonov, D. Trifonov, O. Kovalev, I. Koltsov & V. Berdnyk
State Higher Educational Institution "National Mining University", Dnipro, Ukraine

ABSTRACT: On the basis of reactive power compensation is justified method of determining the economic feasibility of reducing electricity losses and a program simulation model of load distribution points underground electrical networks.

1 INTRODUCTION

It is known that for various operation modes of the electrical system the balance of reactive power (RP) is to be observed, i.e. the total generated RP is always equal to the total power consumption. Balance conditions maintain for each node of the electrical network and the entire electrical system. For RP balance, different procedures can be undertaken and they depend on the type of power supply sources. Hence, the problem arises whether the choice of the most economical type of RP sources, their allocation and operation modes both for each network node and for the entire electrical system is technically correct or not.

During the transfer of reactive power from the power plant to the consumers, additional voltage and power losses are observed in the network. The most effective way to reduce these losses is to install the compensating devices (CD)at the consumer'. Since voltage losses in the external network are reduced during RP compensation, the regulated RP source can be used as a means of voltage regulation under certain conditions.

Solving of the RP compensation problem not only contributes to the chance of obtaining optimal operation modes for electric networks by providing a balance of RP in the system as a whole and in its separate units, but also significantly reduces expenditures on the network construction and operation. In addition, it requires no additional capital investments, but is achieved through systematic development and introduction of rational organizational mechanism aimed at implementing the theoretical solutions, and that is why turns out to be particularly relevant on the current stage of electrical power engineering development in the country facing an acute shortage of generating capacity.

2 FORMULATION THE PROBLEM

As far as the coal mines are concerned, it should be noted that reduction of energy losses in the underground cable networks, if one assumes their right choice, is possible only through the total current lowering. It is obtained by increasing the natural power factor, which is not more than 0.6 ... 0.75. However, up to this day there are no science-based guidelines of economically reasonable RP compensation amount in underground networks. It is achieved by economical and mathematical modeling of CD distribution and by the development of an algorithm as for load simulation model with reference to underground DS(Distribution Substation) and CUS (Central Underground Substation), taking into account the parameters of electrical networks. At the same time, optimization of energy losses is achieved at minimal costs spent on measures that ensure their reduction (Rosen & Trifonov 1995).

3 MATERIALS FOR RESEARCH

Thus, the effectiveness increase as for 6 kV underground electrical networks is possible due to the reduction of energy losses through the use of RPC (Reactive Power Compensation) with in certain economically expedient limits. In this regard, development of a computational procedure to determine the economically expedient reduction of energy losses in underground networks based on RPC is essential. This in turn requires both the creation of electricity losses database as for separate links of underground cable networks and the development of the model to optimize these losses.

To ensure the efficiency of the power supply system (PSS), one aims at diminishing of the reactive power transmission in electric networks by reducing its consumption by RP electrical receivers and besides, by the application of special technical procedures of RPC.

The maximum decrease of RP is conditional on the rational allocation of CD on various levels of PSS. At the same time, the most economical effect can be achieved due to the reduction of active power losses in the transmission of reactive power by distribution networks, and besides, the maximal discount payment for electricity is provided.

Significant losses of electricity in underground networks arise due to the low power factor of underground power consumers, which is tgφ = 0,45 - 0,75. Alternate allocation of capacitor batteries in underground mines at the distribution stations provides for maximum effect of compensation. Arrangement of CD in CUS of mines is considered to be economically advantageous if it reduces the required number of shaft cables or increases the underground load in case of the same number of cables (Gitelson 1971).

Active power losses in distribution networks caused by the transfer of active and reactive power are determined by the expression:

$$\Delta P = \frac{P^2 + Q^2}{U^2} \cdot R \cdot 10^{-3}, kW \qquad (1)$$

or

$$\Delta P = \frac{P^2 \cdot R}{U^2} \cdot 10^{-3} + \frac{Q^2 \cdot R}{U^2} \cdot 10^{-3} = \Delta P_a + \Delta P_r, kW; \qquad (2)$$

where P,Q -active and reactive power, kilowatt and kiloVAr; U,R – voltage and resistance of electric mains, kilovolt and ohm; ΔP_a, ΔP_r- losses of active and reactive power, kilowatt and kiloVAr.

Losses ΔPa remain practically unchanged in case of the compensation, so they are not taken into account whenever alternates are considered, and to simplify the notation, ΔP is taken instead of Δ Pr. For radial and main power distribution circuits, the smallest losses are determined if the condition is satisfied:

$$Q = \frac{P^2 + Q^2}{U^2} \cdot R \cdot 10^{-3}, kW$$

$$(Q_i - Q_{ki}) \cdot R_i = (Q - Q_k) \cdot R_e \qquad (3)$$

where $Q_i = Q_1 + Q_2 +$ - total calculated reactive load of DUS-6 (Distribution Underground Substation),kiloVAr; Q_1, Q_2, - calculated reactive loads DUS-6, kiloVAr; $Q_k = Q_{k1} + Q_{k2} +$ - total capacity of underground CD being subject to

distribution, kiloVAr; Q_{ki}, Q_{k2} - optimal capacities of CD being subject to connecting to particular DUS-6, kiloVAr; R_1, R_2 -active resistances of particular lines feeding relevant DUS-6,ohm; R_e - equivalent resistance of the line, ohm, calculated as

$$R_e = \frac{1}{\dfrac{1}{R_1} + \dfrac{1}{R_2} + \cdots} \qquad (4)$$

Nowadays circuits of PSS in mines are complex because they consist of many different elements, i.e. radial, main and mixed. In such cases, to calculate and distribute capacitor banks, it is necessary to implement the equivalenting of mixed circuits (main and radial), and then capacitor banks are distributed.

In this case, the minimum losses in electrical transmission networks of reactive power are determined by the expression:

$$\Delta P_{min} = \frac{(Q - Q_k)^2}{U^2} \cdot R_e \qquad (5)$$

Owing to installed capacitor banks, reduction of active power losses is equal to:

$$\Delta P_r = \frac{\left[Q^2 - (Q - Q_k)^2\right]}{U^2} \cdot R_e = \frac{-Q_k^2 + 2 \cdot QQ_k}{U^2} \cdot R_e = \frac{Q_k \cdot (2 \cdot Q - Q_k)}{U^2} \cdot R_e \qquad (6)$$

For a single connection (DUS-6), by substituting the values in (6) we get:

$$\Delta P_{mi} = \frac{2 \cdot Q P_a (tg\varphi_i - 0.25) - \left[P_a (tg\varphi_i - 0.25)\right]^2 \cdot R_i}{U^2} \qquad (7)$$

where $P_a = \dfrac{T_{max}}{t_n} \cdot P_p$ is the average load per the most loaded shift; P_r -calculated load of underground power consumers (PC); T_{max} =4000-4800 hr - the number of hours of maximum active power; $t_n = N \cdot n \cdot t = 7200 hr$ - the number of labour hours of underground consumers; N=300; n=4; t=6 –the number of working days per year, the number of shifts per day, the duration of a shift, respectively; $tg\varphi_i$ - power factor of the lines; $tg\varphi_i$ =0.25 – the value of amarginal coefficient of reactive power for which there is no extra charge to pay the power system for consumption of reactive power.

To simplify the calculation of underground PC, assume:

T_{max}=4500 hr, U_l=6kV, $tg\varphi_i$=0.25, $R_i = \dfrac{L_i}{\gamma \cdot S_i}$, γ=50

m/Ohm · mm²

Substituting the following values into the formula (6) we get:

$$\Delta P_n = \sum_{i=1}^{n} \frac{P_{ri}^2 \cdot (0.86 \cdot tg\varphi_i^2 + 0.12 \cdot tg\varphi_i - 0.024) \cdot L_i}{1800 \cdot S_i} \qquad (8)$$

Hence, there is reduction in the unit cost of electricity consumption (per 1m line with the cross section of 1mm²) with regard to CD. Results of calculation of specific power losses are shown in Table1.

$$\Delta P_{rn} = \frac{P_{ri}^2 \cdot (0.86 \cdot tg\varphi_i^2 + 0.12 \cdot tg\varphi_i - 0.024)}{1800} \qquad (9)$$

Table 1. Reduction of specific power losses in the networks of underground mines

Parameters	P_r=1000	P_r=1500	P_r=2000	P_r=2500	P_r=3000
$tg\varphi_i - \cos\varphi_i$	$\Delta P_{r.n} \cdot 10^{-3}$	$\Delta P_{r.n} \cdot 10^{-3}$	$\Delta P_{r.n} \cdot 10^{-3}$	$\Delta P_{r.n} \cdot 10^{-3}$	$\Delta P_{r.n} \cdot 10^{-3}$
	кВт	кВт	кВт	кВт	кВт
2,00-0,45	2000	4500	8000	12500	18000
1,4-0,58	1000	2250	4000	6250	9000
1,00-0,7	530	1200	2120	3300	4770
0,7-0,83	270	607	1080	1690	2430
0,44-0,92	100	225	400	625	900
0,25-0,96	0	0	0	0	0

Fig. 1 shows a diagram of the specific reduction of power losses as a function of

$$\Delta P_n = f(P_{ri}, tg\varphi_i)$$

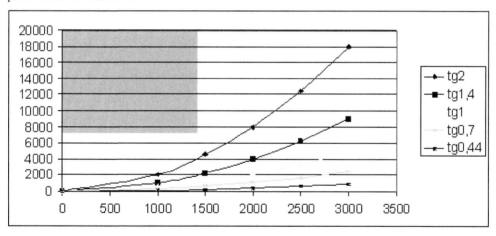

Figure 1. Diagram of reduction of specific power losses in underground networks

Reduction of the overall power losses for the electrical network is given by:

$$\Delta P = \sum_{i=1}^{n} \Delta P_{ni} \cdot \frac{L_i}{S_i} \qquad (10)$$

As a result, the above calculations made possible the development of a program of load simulation model for underground networks, with the help of which the allocation of compensating devices inDUS-6 kWis determined, and the electricity losses in the networks of 6kWare computed. Moreover, the program intended to minimize the losses of electricity was substantiated, and it regards two criteria: the cost of electricity losses, and expenditures on purchase and installation of tools aimed at reduction of energy losses (2).

REFERENCES

Rosen V.P., Trifonov D.V. 1995. *Defining of the feature space that forms energy consumption of mining companies* (in Russian). Coal of Ukraine, Issue 6: 29-30.

Gitelson S.M. 1971. *Economic decisions on designing the power supply of industrial enterprises* (in Russian). Moscow: Energy.

Konstantinov B.A., Zaytsev G.Z. 1976. *Reactive power compensation* (in Russian). L.: Energy: Biblioteka elektromontera, Vol. 445: 104.

Electrical drives automated control systems of open-mine roller-cone drilling rigs

V. Khilov

State Higher Educational Institution "National Mining University", Dnipro, Ukraine

ABSTRACT: The article deals with scientific and applied problems of automated electric drives theory and design aimed at the creation of drive systems for modernized boring rigs as well as the development of the active consecutive correction method with the use of indistinct control in the systems of electric drives regulation in complex electromechanical devices.

1 INTRODUCTION

In open cast mining at the quarries of Krivoy Rog basin (Ukraine) general technology the cost of drilling and blasting operations takes about 16-36% of the total costs in the destruction of hard rocks. The rise of energy, electricity and cost-based materials prices makes the problem of energy-saving technologies in the extraction of rocks application increasingly important. Nowadays the rigs used in Ukrainian quarries have completely exhausted their rated resource. This state of things presupposes not only the modernization of the existing rigs, but also the introduction of the new generation of drilling rigs (DR) with longer barbells and multi-mass system of flight tripping and displacement onto the borehole bottom.

The efficiency of blast holes drilling by the means of the roller-cone method is directly determined by the perfected level of electric drive systems. The introduction of fast-action AC transistor drive into rotation mechanism of the existing rigs, as well as into the rotating, tripping and displacement mechanisms of the new generation rigs instead DC thyristor drive with moderate action, is characterized by the increased vibratory loads on all mechanical parts of the rigs. These loads lead to the energy release flow not only in the area of the bottom zone where the destruction of the rock is performed, but also in the structural elements of the machine directly, increasing mechanical stress and material fatigue of the machine, as well as conduces to accidental breakdowns of the rigs units (derrick, as a rule). Frequency characteristics of the drive systems must be compatible with the dynamic characteristics of multi-mass mechanisms with distributed and lumped masses with changing attached masses. It will provide the release of

mechanical power flow in the area of the bottom zone, as well as energy and resource efficiency of blasting boreholes drilling. Therefore, the solution of scientific and applied problems of establishing patterns of influence of electromechanical and hydro mechanical parameters of the rigs to the rotation drive, displacement, STF as well as the further related development of the theory of operation of fast-action automated electric open-mine drilling rigs now acquires special importance.

2 PROBLEM SATEMENT

Blastholes drilling is one of the most power and material intensive modern processes of the existing iron ore open-pit mining technology in Ukraine. The improvement of rock breaking efficiency by roller bits is associated with the perfection of the level of rigs drive systems. The works by Quang 2010, Novotny 1996, Boldea 1992, Boldea 2006 and many others are devoted to the development of the theory of electromechanical systems operation.

The aim of our study is obtaining energy and resource efficiency in the process of blastholes drilling via focusing the flow of mechanical power in the area of the bottom zone and supporting minimum dynamic loads in a drilling machine in the presence of elastic connections and changing the natural oscillation frequencies in multi-mass transmission with lumped and distributed parameters.

To achieving this aim we had to cope with the following tasks:

• to develop the way of the drilling process control with the formation of mechanical characteristics with the variable stiffness on cone bits, depending on the drilled rock hardness through the influence on the motor rotation, displacement and STF rod for bottom zone;

• to evaluate the effect of forced and natural oscillations of rod on the work dynamics of the rigs with thyristor DC drive (DR RCDR-250MN-32) and to introduced, with the participation of the author, the thyristor and transistor AC drives (rigs БС СБШ-250МН-32 and СБШ-250Н);

• to provide further development of the scientific basis of the electric rotation creation system taking into account its technological features and the development of drilling control method;

• to investigate electro-hydro-mechanical (EHM) transmission as the object of automatic electric control of tripping operations and displacement and to install the regularities in the changes of characteristic frequencies oscillations;

• to provide further development of the scientific basis for constructing the drive of lowering / lifting and displacement operations according to the object peculiarities and the developed method of drilling process control;

• to provide further development of the method of the active sequence correction using fuzzy control systems to regulate complex electric electromechanical (EM) and EHM processes in the devices.

The object of our study is the frequency compatibility of electromechanical processes and electro-hydro-mechanical processes in the rotary drilling rod with distributed parameters and the round-trip drive and displacement rod with lumped parameters associated with the changes in the distributed and lumped masses in the transmission.

The subject of research is the elastic properties of transmissions, the frequency response of control systems and their mutual influence on the quality of work in the electrical drive of rotation, round-trip and displacement rod of the rock drilling machine.

3 RESEARCH MATERIALS

3.1 Current status and development prospects of roller-bit drilling rigs for blasthole drilling drive systems.

Medium type rigs БС СБШ-250МН-32 with thyristor DC drives in rotation rod mechanism, with asynchronous drive mechanisms of round-trip operations and rod displacement, rig movement and rig compressor pneumatic transport system are mainly in operation in the quarries of Ukraine.

Ukrainian machine-building plants have created: УСБШ-250А type drilling rig (developer - KZGM, Krivoy Rog, has been in operation at Poltava processing enterprise since 2004; it is equipped with

drive systems, which are similar to the ones of СБШ-250МН-32 type drilling rig), the new generation rigs СБШ-250/ 270-32 (developer - NKMZ Kramatorsk; in operation at Ingulets ore-dressing plant since 1999; equipped with DC thyristor drive in the mechanism of rotation, hydraulic drive of the rod movement / tripping and displacement system) and СБШС-250Н (developer - NKMZ; in operation at the Central processing enterprise since 2003; they are equipped with AC transistor drives in rotation mechanism, STF of rod and movement, which were introduced as recommended by the author).

Along with the creation of the drilling rigs of new generation, there has been performed the modernization of the existing ones, according to the recommendations of the author, in the process of СБШ-250МН-32 type rigs upgrading the AC drive system has been introduced to the mechanisms of rod rotation. These drive systems are in operation at the Central processing enterprise: AC thyristor drive with current source (in operatin since 1998); AC transistor drives with a voltage source (in operation since 2005).

The use of AC transistor drive systems instead of DC thyristor drives resulted in the significant change of the dynamics of the drilling rig in the process blastholes drilling. To establish the quantitative characteristics of the EM rigs system there were experimental studies performed directly in mining and geological conditions of the Central Mine. The dynamics of the head cross-arm of the drill rotation mechanism of the rig rod with four drive systems was under examination: machines for СБШ-250МН-32 – DC thyristor drives, AC currents and AC drive transistor, and machines for machine СБШС-250Н - AC transistor drive.

The harmonic analysis (Fig. 1, 2) of pressure fluctuations of the head of the drill traverses confirmed the well-known fact that the growth rate of bit rotation leads to the increase in the frequency and amplitude of the forced vibration. Together with the known regularities there appeared some new ones: in the spectra of the amplitude-frequency characteristics of the vibration of the machine of the new generation СБШС-250Н harmonic appears at the frequency of $f = 1,02$ Hz, which is independent of the rod speed. Other harmonic components are subjected to the law of the existing СБШ-250МН-32 type rigs. The appearance of harmonic oscillation, which is independent of the rotational speed and which is located in the bandwidth of the control system, indicates the need to consider the action on the dynamics of the drilling process not only as forced oscillations, which are typical of the type of СБШ-250МН-32, but also as the composition of the natural oscillation frequencies of СБШС-250Н machine. The use of transistor frequency-regulated drive in the rigs СБШ-250МН-32 led to the increase of high-frequency components in the frequency of rotation of the shaft (rod) and motor current, Table 1 (where SIT and VIN are standalone inverter current and voltage)

P, kW; Ia, A; n, rev/min; A, micrometer

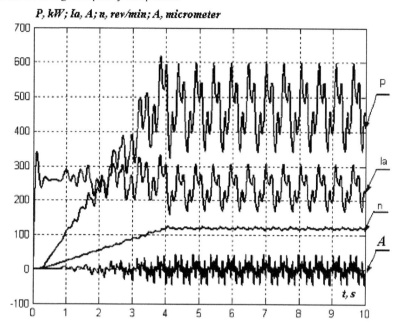

Figure 1. Oscillograms of transients in the rotation drive of the drilling rig СБШС-250МН with the system of frequency converter-induction motor with the inverter voltage: A-vibration displacement of the pressure traverses of the head top (axial pressure on the rod of 250 kN, drilling with three drill rods)

Table 1. Quantitative characteristics of the ripple depending on the type of rod rotation electric drive systems and drilling rigs

The rotation drive types and drilling rigs types	Peak-to peak ripple during drilling, %	
	engine speed (rod)	motor current
Thyristor dc drive, rig СБШ-250МН-32	0,51	28,4
Thyristor dc drive SIT - induction motor, rig СБШ-250МН-32	2,56	25,3
Transistor AC drive SIV - induction motor, rig СБШ-250МН-32	3,51	25,6
Transistor AC drive SIV - induction motor, rig СБШС-250Н	3,02	27,1

Together with general industrial electric drive, which is now implemented at the rigs, there exist object-oriented drives with automatic control systems of holes drilling for roller cone rigs, which are based on the following methods of control: torque stabilization, minimum drilling cost of a borehole meter, maximum drilling speed. However, such drive systems are not widespread in Krivoy Rog basin, as the basic ideas of their foundation were not confirmed during experimental verification. Therefore, to design the electric drive control method for the rod displacement and rotation system during roller cone drilling we took into account another idea. It was the idea of Ivano-

Frankivsk National Technical University of Oil and Gas, partially experimentally tested at Poltava Mining Plant (its geological conditions are similar to those of Krivoy Rog basin quarries): control method with the check of the engine rotation rod power, which assists to increase resilience and to reduce drilling and blasting costs.

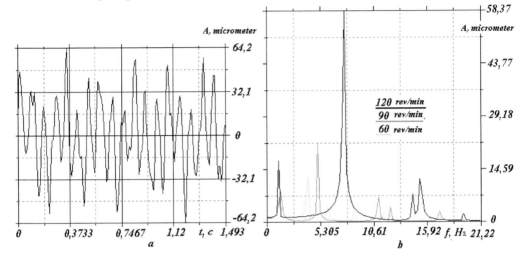

Figure 2. Oscillogram of vibration displacement of the head projectile at 60 rev / min (a) and the amplitude spectrum of vibration frequency characteristics (b) of the machine СБШС-250Н with transistor AC drive system SIV - induction motor (axial pressure on rod 250 kN, drilling with three rods)

To increase the resistance of rock bit by maintaining the flow of power in the region work zone the author has developed the way to control the drilling process by controlling the rotation of electric machinery and STF of the rod, in which there were rigid mechanical characteristics formed, those on the cone bits when drilling in rocks with the rigidity of f≤10-13 by prof. M. M. Protodyakonov scale and soft mechanical characteristics - in more rigid rocks. The implementation of the proposed control method in drilling the rocks with intermittent physical and mechanical properties provides the automatic selection of the mechanical characteristics depending on the strength of the rock. The control method is patented in Ukraine.

The implementation of the developed control method of the drilling process leads to the necessity of control and limitation on the permissible levels in electric drives: rotation rod - mechanical power flow on the cone bits in the work zone; rod tripping and displacement - the pressure in the hydraulic system and the linear velocity of movement of the drilling rod, requiring the further development of the scientific bases for object-orientation drive systems for the new generation of rigs.

Now the problem of mutual influence of the dynamic characteristics of high-speed AC transistor motor and frequency characteristics of the transmission of the drilling rigs is poorly understood and relevant. This is connected with the introduction of new design solutions into new generation drilling rigs and actuators. These solutions have significantly changed the natural frequencies of the EGM systems of the rod tripping / displacement and EM rotation of the rig.

3.2 The analysis of the drill string dynamic properties in the process of drilling as an object of automatic control.

The system of electric drive control rotation, as shown by experimental studies (Khilov 2004; Pivnyak 2005), should consider the frequency characteristics of the controlled object. With this purpose the frequency characteristics of the drill string, used in the quarries of Ukraine in the roller cone machines, were determined.

The drill strings of drilling rigs are from 8 m (machine of УСБШ-250А, СБШ-250МН-32types) to 11 m (СБШС-250Н machine) long. From the design point of view, they are hollow rods of circular cross-section, one end of which is pivotally supported on the bit and the other is moving in sliding fixing of the pressure traverse. The rod is under pressure from the displacement force. Determining natural frequencies, we studied the rod as a linear system with a uniformly distributed mass, in which the longitudinal, torsional and transverse vibrations are independent of one another. The mass of the cone bit in the analytical study was neglected.

As a result of theoretical studies we obtained a mathematical model for the determination of natural frequencies of longitudinal, torsional and transverse vibrations for the drill strings with a single drill rod as well as for the rod with an addition, see Table 2 (where u - longitudinal displacements of current cross section rod; t - time; φ - angle rotation of the section of the bar; x - current distance; y - transverse displacement of the cross section of the rod). Based on the model, the method of natural oscillations calculation was created, the method made it possible to calculate the oscillation frequency of the drill string at a variable number of drill rods - Khilov & Plakhotnik 2004; Pivnyak 2003).

The comparison of natural frequencies of the longitudinal, torsional and transverse vibrations obtained by the calculation with natural frequencies, which are highlighted in experimental studies (Section 3.1), shows that during the drilling with one and two drill rods regardless of the type of the drilling rigs, the natural frequencies of longitudinal and torsional vibrations of the drill string do not fall within the passband of both DC and AC drives. While drilling with three drill rods in the drilling rig СБШ-250МН-32 the low frequency transverse of their own oscillation with the value of 2.36 Hz was experimentally fixed, its calculated value, by the developed technique, is 2.48 Hz. While the drilling with three rods in the drilling rig СБШС-250Н the own low-frequency transverse vibrations of the rod with a cyclic frequency of 1.02 Hz were experimentally fixed. The calculated value of this frequency is 1.05 Hz.

Low frequency components of the drill rod transverse oscillations do not directly affect the dynamic processes in the drive system. However, they lead to inadmissible dynamic loads not only on the drill rod, but also on the drilling rig as a whole, which significantly reduces the mean time of structural components failure significantly limiting the performance of drilling and blasting operations, so the drive system must limit the transverse vibrations on an acceptable level.

Drill rod torsion vibrations are directly affected by the dynamics of the AC drive system provided that its own frequency falls within the pass band of current and speed loops. These conditions are fulfilled in СБШС-250 drill rig AC drive system with transistor independent inverter of voltage pulse-width modulation (SIV-PWM) with a band pass of loop speed 200 rad / s, while drilling with four drill rods.

The operation of the DC voltage drive system with a voltage thyristor converter with the loop bandwidth speed not higher than 50 rad/s in drilling rigs СБШС-250Н and СБШ-250МН-32 does not lead to the ingress of the drill rod torsion vibrations natural frequencies into the bandwidth.

3.3 Drill rod rotating drive system.

In order to obtain the desired hyperbolic relationship between the speed and the rock strength according to the developed drilling method, it is necessary to maintain not only the constant engine speed (bit) at soft rocks drilling, but also the flow of power to the roller cone bit in the process of hard rocks destruction in the area of the work face. In this case, the control system must automatically, without the operator's intervention, select the speed of the roller cone bit at a variable and not a priori known value of the drilled rock strength.

During the drilling with one and two drill rods (Section 3.2) the drill rod lowest frequency torsion oscillations of СБШС-250Н, УСБШ-250А and СБШ-250МН-32 drilling rigs are more than the cutoff frequency of the stator current and speed control loops. Therefore, under such conditions, at the development of the electric drive system of drill rod rotation with controlling electromagnetic power we can disregard the impact of natural torsion vibration on the control dynamics, as it is in the band attenuation outside the bandwidth of the control system.

It is found that the direct application of the external loop of the electromagnetic power leads to the parallel connection of the control loop speed and the current due to the presence of the cross-links in the facility management. For such a control system we created the procedure of loop power dynamic correction control, taking into account the perturbation via the channel of the stator current engine onto the power control loop. The introduction of the online external loop power control effectively maintains set power of the drive system at hard rock drilling, but does not implement hard mechanical characteristics for drilling soft and destroyed rock.

To overcome this shortcoming we proposed and investigated the power control system that uses a nonlinear correction in the back displacement. With this aim in purpose, the reference signal at the input of the speed controller is divided into a signal which is proportional to the moment of resistance on roller cone bits (Fig. 3).

Table 2. Drill rod natural oscillations design equations

Longitudinal oscillations	Torsional vibrations	Transverse vibrations	Notes	
$\dfrac{\partial^2 u}{\partial t^2} = C_1^2 \dfrac{\partial^2 u}{\partial x^2}$	$\dfrac{\partial^2 \varphi}{\partial t^2} = C_2^2 \dfrac{\partial^2 \varphi}{\partial x^2}$	$E \cdot J \cdot \dfrac{\partial^4 y}{\partial x^4} + R \cdot \dfrac{\partial^2 y}{\partial x^2} + m \cdot \dfrac{\partial^2 y}{\partial t^2} = 0,$ $y = X_y(x) \cdot T(t).$	Oscillation equation	
$C_1 = \sqrt{E/\rho}$; E – elastic modulus; ρ – density	$C_2^2 = G/\rho$; G - shear modulus	m – drill rod distributed mass intensity, $E \cdot J$ – drill rod flexural rigidity; J – inertia axial moment; R - displacement force	Denoted	
$X_u = C_u \sin \dfrac{p_u}{C_1} x +$ $+ D_u \cos \dfrac{p_u}{C_1} x$	$X_\varphi = C_\varphi \sin \dfrac{p_\varphi}{C_2} x +$ $+ D_\varphi \cos \dfrac{p_\varphi}{C_2} x$	$\dfrac{\ddot{T}}{T} = -p^2; \dfrac{E \cdot J}{m} \cdot \left(X_y^{IV} + a^2 X_y^{II} \right) = p^2;$ $X_y = C_1 \cdot sh(S_1 \cdot x) + C_2 \cdot ch(S_1 \cdot x) +$ $+ C_3 \cdot \sin(S_2 \cdot x) + C_4 \cdot \cos(S_2 \cdot x)$	The solution of the equation	
C_u and D_u – arbitrary constants	C_φ and D_φ – arbitrary constants	$S_1 = \sqrt{\sqrt{0{,}25 \cdot a^4 + k^4} - 0{,}5 \cdot a^2}$; $S_2 = \sqrt{0{,}5 \cdot a^2 + \sqrt{0{,}25 \cdot a^4 + k^4}}$; $- C_1', C_2', C_3', C_4'$ - integration constants; a, κ – constant coefficients	Denoted	
p_u – longitudinal oscillations natural frequency; l – drill rod length	p_φ – torsional natural frequency	p – transverse oscillations natural frequency		
$X_u' = 0$	$X_\varphi' = 0$	$X_y = 0, \ X'_y = 0$	$x = 0$	Boundary conditions
$X_u' = 0$	$X_\varphi' = 0$	$X_y = 0, \ X'_y = 0$	$x = l$	
$\cos\left(\dfrac{p_u}{C_1} \cdot l\right) = 0$	$\sin\left(\dfrac{p_\varphi}{C_2} \cdot l\right) = 0$	$X_y = C_1 \cdot sh(S_1 \cdot x) + C_2 \cdot ch(S_1 \cdot x) +$ $+ C_3 \cdot \sin(S_2 \cdot x) + C_4 \cdot \cos(S_2 \cdot x),$	The equations for the determination of natural frequencies	

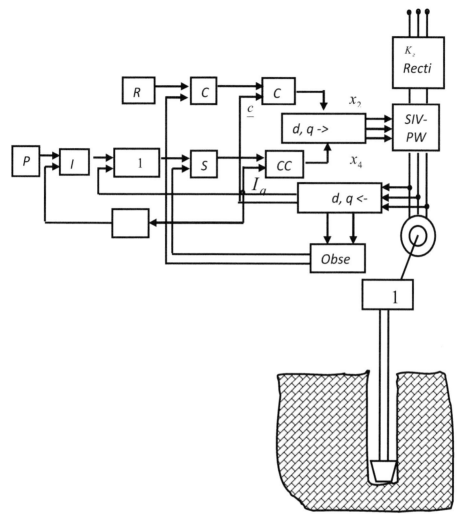

Figure 3. Roller cone bits rock failure power vector control system: *PA, IA, RLA* - setting devices of power, intensity and nominal linkage; *SC, CL, CC, CCI* - speed controller, linkage and stator currents; *SO* - vibration sensor; *MDD* - multiplier-divider unit; *SIV-PWM* - autonomous voltage inverter with *PWM*

The influence of non-linear power correction on the stability of the control system was researched. The characteristic equation of the external control loop has allocated, which is a third-degree polynomial. Therefore, the control quality is assessed by the Vyshnegradsky diagram (Fig. 4). The diagram used Vyshnegradsky coefficients A, B, which are quantitatively determined by the equations:

$$A = (a^2_{ш}b^2_{ш}T^2_c + a_{ш}b_{ш}T_cP_3T_M/M/K_{ш})/(a^2_{ш}b_{ш}T^3_c)^{2/3};$$

$$B = (a_{ш}b_{ш}T_c + a_{ш}b_{ш}T_cP_3T_M/M/K_{ш})/(a^2_{ш}b_{ш}T^3_c)^{1/3},$$

where $a_{ш}$, $b_{ш}$ – tuning coefficients; T_c– circuit current time constant; P_3– the reference signal; M – motor torque; T_M – mechanical constant; $K_{ш}$ – transfer coefficient of the speed sensor.

In Fig. 4 line *GH* – is the locus of the points belonging to the vertexes of the parabolas at the current values setting to the destroyed rock power and moment resistance at the roller cone bit. It characterizes the quality of the transition process. At

97

the minimum values of the Vyshnegradsky the coefficients $A_{min}=(a^2_{ul}b_{ul})^{1/3}$, $A_{min}=(a_{ul}b^2_{ul})^{1/3}$ and the coefficients of the velocity loop tuning $a_{ul}=b_{ul}=1$ line *GH* begin on the stability boundary, while $a_{ul}=b_{ul}=1$ lies in the sustainable transition. Line

Quality transient GH is in zones I and III, not falling in zone IV. This suggests that necessary and sufficient condition for preserving the control system stability is not disturbed after the introduction of non-linear correction (Fig. 3).

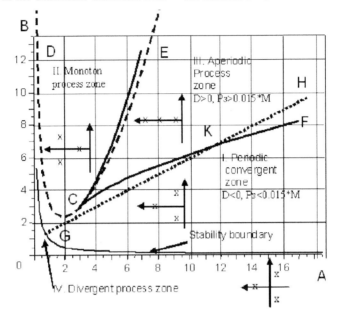

Figure 4. Zones of quality transient of control system with nonlinear power regulator

Table 3. Additional dynamic links in the control loops of the drive rotation

Loop	Loop additional dynamic link	Characteristic frequencies of the polynomial	
		Numerator (zeros of fractional-rational function)	Denumerator (zeros of fractional-rational function)
Current	$(T_I \cdot p +1)\cdot p \cdot T_{EM} \cdot \gamma \cdot$ $\cdot(p^2 \cdot T_y^2 + p \cdot 2 \cdot \xi_y \cdot T_y +1)$ $\overline{(T_I \cdot p +1)\cdot p \cdot T_{EM} \cdot \gamma \cdot}$ $\cdot(p^2 \cdot T_y^2 + p \cdot 2 \cdot \xi_y \cdot T_y +1)+$ $+ p^2 \cdot \gamma \cdot T_y^2 + p \cdot 2 \cdot \xi_y \cdot T_y +1$	$\omega_{KT1} =$ $= \dfrac{1}{\sqrt{T_I \cdot T_{EM}}} \cdot$ $\cdot \dfrac{1}{\sqrt{\gamma}},$ $\omega_{KT3} = \dfrac{1}{T_y},$	$\omega_{KT2,4} = \text{Im} \sqrt{\dfrac{1}{T_I \cdot T_{\ni M}}} \cdot \sqrt{\dfrac{T_I \cdot T_{EM} + T_y^2}{2}} \cdot$ $\cdot \dfrac{1}{T_y} \sqrt{\left(\pm \sqrt{1 - \dfrac{4 \cdot T_I \cdot T_{EM} \cdot T_y^2}{\gamma \cdot (T_I \cdot T_{EM} + T_y^2)^2}} -1 \right)}$
Rotation frequency of motor	$\dfrac{p^2 \cdot \gamma \cdot T_y^2 + p \cdot 2 \cdot \xi_y \cdot T_y +1}{p^2 \cdot T_y^2 + p \cdot 2 \cdot \xi_y \cdot T_y +1}$	$\omega_{ЧB1} = \dfrac{1}{\sqrt{\gamma \cdot T_y}}$	$\omega_{ЧB2} = \dfrac{1}{T_y}$
Linear speed of drill rod	$\dfrac{p \cdot 2 \cdot \xi_y \cdot T_y +1}{p^2 \cdot \gamma \cdot T_y^2 + p \cdot 2 \cdot \xi_y \cdot T_y +1}$	–	$\omega_{C2} = \dfrac{1}{\sqrt{\gamma \cdot T_y}}$

When there are more than three drill rods in the drill string (Section 2) the torsion vibrations of the drill string are low-frequency ones and are falling into the pass band of AC drive rotation control system of the drill rig СБШС-250Н, i.e. they affect the operation of the control system in drilling at completely assembled drill string that degrade the quality of governance indicators across electromechanical installation and increase the vibration load on the rig in general.

Only the most low frequency oscillation of the drill string was investigated, the one getting into the band pass, other attenuation band frequencies were not counted. Torsion vibration calculation scheme is a system with lumped masses of rotor induction motor and the weight of the drill string reduced to the frequency of the rotor rotation. These two masses are connected with elastic element stiffness determined by the lowest frequency torsion vibrations of the drill rod. In the selected calculation scheme there are additional dynamic links in the control loops (Table 3, where T_I, T_{EM}, T_y – time constants: electromagnetic, electromechanical, torsional vibrations composition; Γ, γ – coefficients ratio of inertial mass; ξ_y – damping coefficient of elastic vibrations drilling rod).

The dynamic characteristics of the control loops under the influence of the elastic properties placed on the regulatory processes in the drive system were analyzed. It is established that the application of classical PID controllers is possible only for locally optimal tuning of the regulation system.

If the number of rods attached to the drill string is changed from one to four, PID controllers classic tuning does not provide optimal dynamic parameters throughout the range. It is possible only with the use of adaptive controllers that must change not only its parameters, but also the management structure within wide limits, as well as provide the ability to find the derivative of the error signal of the fourth order in the presence of noise in the signal control. Therefore, the classical control system can function effectively only if the decrease in its rapidity of action to the level where its own characteristic frequencies are shifted to the zone of damping control loops. In the field of classical control systems the only possible construction of the control systems is the one with a low rapidity of action, which is contrary to the quality requirements of the electromechanical system of the rotation hocked.

The studies that have been performed allow to form the concept of automated rotation drive drill string of the drilling rigs and to create the system according to which, with the purpose of energy and resource saving, depending on the strength of the rock there exist supporting control mechanical characteristic on the roller cone bit with constant or variable stiffness and limitation of longitudinal vibrations of the drill rod. In the concept regulation the following principles are used: longitudinal vibrations are limited by acting onto the intensity setting of frequency rotated with simultaneously controlling and limiting the process parameters which leads to concentration of mechanical power flow to the bottom zone.

3.4 The EGM transmission properties research as the object of drill rod STF drive system control.

The drilling rig type СБШС-250Н by Closed Joint-Stock Company NKMZ significantly differs from machines СБШ-250МН-32 by EGM system with the mechanism of STF of the drill rod. In the kinematic link between the drive motor and the drilling string in rig type СБШС-250Н in rod translation displacement a hydraulic pump hydraulic motor (HM) with reducer gear and rope-tackle system (RTS) are included.

A mathematical model of the control object has been obtained based on the decomposition approach in the assumption that the motor and the gear have lumped parameters; mass of oil which circulates in the hydraulic system (HS) is not taken into account due to its small amount in comparison with other masses; the change in viscosity with temperature is neglected, i.e. isothermal processes are considered; the oil leakage from HS is directly proportional to the pressure in the pipes; the changes in the volume of lubricating oil in the pipeline are directly proportional to the volume of supply pipeline and inversely proportional to the bulk modulus of the pipeline material.

Block diagram of the control object of electro-drive STF is shown in Figure 4, where x_1 - speed $\omega 1$ of the drive motor; x_2 - the moment of M_c, that is developing by HS; x_3 - speed ω_2 HM; x_4 - M_y moment that corresponds to the force that is transmitted by rope; x_5 - linear speed of the drill, which is reduced to a rotational speed ω_3; J_1, J_2, J_3 - moment of inertia of the motor, HM, drill string; K_c - hydraulic transmission coefficient; T_c – time constant of the HS; δ - coefficient of leakage in the HS; c, b - coefficients of stiffness and dissipative forces CPS; M, M_C - torque and resistance moment.

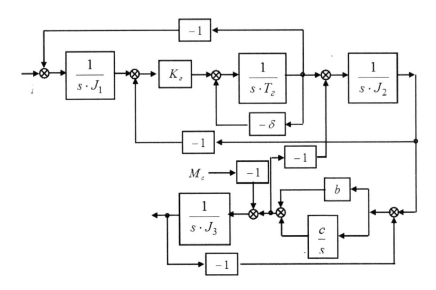

Figure 4. Block diagram of the control object of the drive tripping operations and the displacement of the drill rod

From the analysis of the resulting mathematical model of the control object of electric tripping and rod displacement for the implementation of the developed method for controlling the drilling process must be monitored, in addition to traditional coordinates - current and motor speed, and adding technological coordinates - the pressure in the hydraulic system, the frequency of hydraulic motor and line speed of the rod displacement, which needs appropriate connection of control loops.

The presence of elastic links in the transmission leads to the appearance of additional rational function, the zeros and poles of which are moved from the low to the high frequency region and cross the cutoff frequency of the circuit with a decrease in the number of rods in drill string from four to zero.

3.5 Constructing a system of motor control round-trip and rod displacement

According to the developed control method there has been proposed and investigated a drilling EGM drive control system with the distributed correction into the loops of the dynamic parameters. The system, except the EM parameters of the drive motor, controls technological parameters that have been defined in Section 3.4, allowing efficient optimization of the dynamic loads in the entire drive-train, automate processes of the round-trip and rod displacement.

At the beginning of the movement the drill string with the number of rods attached more than three the current loop includes not only traditional aperiodic link, but rational function of sixth order too. When the loop at the technical optimum is tuning up, the classical transfer function of the current regulator should have a retroactive transfer function of the control object, which is a fairly complex regulator to be implemented, because one needs to find the derivatives of higher order (up to sixth order inclusive) from the signal error. In addition, the regulator should have the properties of adaptation or the robustness from changing stiffness of the KPS and of the drill rod weight. During the decrease of the length of the cable rope and the reducing of rods number in the drill string, the characteristic oscillation frequency is shifted to higher values in the complex plane, crossing the cutoff frequency of the loop.

Traditional PI controller in the classical tuning suppresses none of the loop control object pole, and the presence of forcing element in the object control through a significant impact rotating electromotive force, leads to current loop astatism loss.

Table 4. The equation for calculating the characteristic frequencies of the rational function of STF and displacement drive

Loop	Numerator characteristic frequency (rational function zeros)	Denumerator characteristic frequency (rational function poles)
Current	$$\omega_{51} = \frac{1}{\sqrt{T_I \cdot T_{EM1}}};$$ $$\omega_{11,31} = \frac{1}{T_y} \cdot$$ $$\sqrt{\frac{1+T_y^2 \cdot \gamma_{12}/T_{\not e}/T_{M2}}{2}} \cdot$$ $$\cdot \sqrt{\left(\pm\sqrt{1-\frac{4 \cdot T_{\not e} \cdot T_{M2} \cdot T_y^2}{\gamma_{23} \cdot (T_{\not e} \cdot T_{M2}+T_y^2 \cdot \gamma_{12})^2}}-1\right)}$$	$$\omega_{21} = \frac{1}{T_y} \cdot \sqrt{\frac{1+T_y^2 \cdot \gamma_{12}/T_{\not e}/T_{M2}+}{+T_y^2/T_I/T_{EM1}}};$$ $$\omega'_{41,61} = \frac{1}{T_y} \cdot \frac{1}{\sqrt{T_I \cdot T_{EM1}}} \cdot \frac{1}{\sqrt{T_{M2} \cdot T_{\not e} \cdot \gamma_{23}}} \cdot$$ $$\sqrt{\frac{1-(1/T_y^2+\gamma_{12}/T_{\not e}/T_{M2}+1/T_I/T_{EM1}) \cdot}{\cdot (T_I \cdot T_{EM1} \cdot \Gamma+T_{M2} \cdot T_{\not e} \cdot \gamma_{23}+\gamma_{23} \cdot T_y^2)}{2 \cdot (1/T_y^2+\gamma_{12}/T_{M2}/T_{\not e}+1/T_I/T_{EM1})^2}}$$ $$\cdot \left(\pm\sqrt{1-\frac{4 \cdot T_I \cdot T_{EM1} \cdot T_{M2} \cdot T_{\not e} \cdot \gamma_{23} \cdot \left(1+T_y^2 \cdot \gamma_{12}/T_{M2} \cdot T_{\not e}+ +T_y^2/T_I/T_{EM1}\right)^3}{\left(T_y^2-(1+T_y^2 \cdot \gamma_{12}/T_{M2} \cdot T_{\not e}+ +T_y^2/T_I/T_{EM1}) \cdot \left(T_I \cdot T_{EM1} \cdot \Gamma+ T_{\not e} \cdot T_{M2} \cdot \gamma_{23}+T_y^2 \cdot \gamma_{23}\right)\right)^2}}+1\right)$$
Motor speed	$$\omega_{12,32} = \frac{1}{T_y} \cdot \sqrt{\frac{1+T_y^2/T_{\not e}/T_{M2}}{2} \cdot \left(\pm\sqrt{1-\frac{4 \cdot T_{\not e} \cdot T_{M2} \cdot T_y^2}{\gamma_{23} \cdot (T_{\not e} \cdot T_{M2}+T_y^2)^2}}-1\right)}$$	$$\omega_{22,42} = \frac{1}{T_y} \cdot$$ $$\sqrt{\frac{1+T_y^2 \cdot \gamma_{12}/T_{M2}/T_{\not e}}{2} \cdot \left(\pm\sqrt{1-\frac{4 \cdot \Gamma \cdot T_y^2 \cdot T_{\not e} \cdot T_{M2}}{(T_{\not e} \cdot T_{M2}+T_y^2 \cdot \gamma_{12})^2 \cdot \gamma_{23}}}-1\right)}$$
Pressure	$$\omega_{13} = \frac{1}{T_y}$$	$$\omega_{23,43} = \frac{1}{T_y} \cdot \sqrt{\frac{1+T_y^2/T_{\not e}/T_{M2}}{2} \cdot \left(\pm\sqrt{1-\frac{4 \cdot T_{\not e} \cdot T_{M2} \cdot T_y^2}{\gamma_{23} \cdot (T_{\not e} \cdot T_{M2}+T_y^2)^2}}-1\right)}$$
HM speed	$$\omega_{14} = 1/T_y/\sqrt{\gamma_{23}}$$	$$\omega_{24} = 1/T_y$$
Drill rod speed	–	$$\omega_{25} = 1/T_y/\sqrt{\gamma_{23}}$$

In the control object of the speed loop besides the induction motor stator current control closed loop, the mechanical link of the drive, an additional transfer link, is appearing: the elastic properties of the transmission have been taken into account. For electric round-trip and rod displacement of a medium type СБШС-250Н vibration characteristic frequency of the additional dynamic link while moving three or fewer drill rods lie in a region which is above the cutoff frequency of the speed loop, so the dynamic processes do not affect the loop. Moving four drill rods, which are guaranteed by the manufacturer as a regular mode of drill rig, the characteristic vibration frequencies fall within the bandwidth and significantly change the dynamics of the closed loop speed - with four or more rods adjusted not only to the dynamic processes in the current loop by the introduction of the first additional regulator, but also the frequency rotation by the introduction of the second additional regulator.

The characteristic frequencies of the numerator and denominator of additional dynamic elements (Table 4, the designations given in Table 3) that were defined, allow to determine the regularities at which the characteristic frequencies in adjacent loops of the numerators and denominators (zeros and poles) have periodic repetition. Only one characteristic frequency of the current in the circuit is determined by the actual parameters of the induction motor and is independent of changes in the parameters of the transmission. The loop characteristic frequencies of the current, the motor speed and the pressure are lateral relative to the frequency $1/T_y$ and by elastic vibrations are generated CPS considering the influence of the HS, nonlinearly varies depending on the amount the rods screwed onto the drill string and the cables length.

The regularities were determined between the parameters of the control object in which the dynamic links, due to the presence of flexibility in the transmission, will not be negatively denoted on the dynamic loads in the transmission. So for the STF electric drive of the average type, at three and fewer rods moving, the characteristic oscillation frequency of additional dynamic links lie outside the loop speed band pass area, so they do not affect.

On the basis of the studies we have developed a scientific basis for the construction of the automated electric drive and round-trip and rod displacement of open-pit drill rigs of new generation, according to which to support the optimal technology characteristics they are monitored and limited to an acceptable level of technological processes - linear speed of the drill rod, the speed of the hydraulic motor.

To implement the scientific concept of building control systems of the rotation electric drive (Section 3.3) and round-trip and rod displacement (Section 3.5) with the correction distributed into the loops, to the necessity of solving the problem are: getting normalized transients in the control loops in the presence of a control object is not only the transfer function of the second order, and the rational function including the sixth order with zeros and poles, which can move through the cut-off frequency of the contour. The solution is found in the field of control systems with fuzzy laws that allow to maintain quality transients when changing the quantitative characteristics of the control object in wide limits.

3.6 The method of active sequence correction using fuzzy control systems and regulation of complicated EM and EGM facilities including drive systems of rotation, tripping and drilling rod displacement .

Generalized block diagram of the control object complicated EM and EGM facilities is shown in Figure 5, where $W_{окп}(s)$ – transfer function of the control object compensated part; $\Sigma a_{mn}s^m / \Sigma b_{mn}s^m$ – additional transfer function that has defined by the transmission elastic properties and is represented as a rational function.

To compensate the action of the additional transfer function each basic classic controller must have the additional regulator connect in series with the generalized transfer function with the form $W'_n(s) = \Sigma b_{mn}s^m / \Sigma a_{mn}s^m$ or they must be in parallel with the transmission function $W_n(s) = 1/(sT_{OT}W_{окп}(s)) (\Sigma b_{mn}-a_{mn})s^m /(\Sigma a_{mn}s^m)$, which has the distribution parameters into the control loops.

If such compensation have been received, they are additionally distributed into control loops having the numerator and denominator in sixth degree - in the loop of the drive current round-trip and drilling rod displacement, and the fourth degree is in the loops of the STF drive motor speed and rotation drive in current loop, which creates difficulties when implementing them in the classical form.

If the original signals of all other regulators reduce to the output of the internal regulator, then instead of several connected in parallel to each contour the regulator comes to the single concentration regulator, which the classical regulators cover in parallel. In this case, we get one regulator that has concentrated in the inner-loop controller, and its work algorithm becomes more complex than the work algorithms of each distributed controllers.

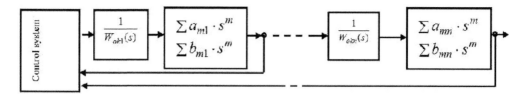

Figure 5. Generalized block diagram of the control object

The problem of compensation of elastic oscillations in the control loop is solved by applying the principles of fuzzy control of complex objects. Instead of classic corrective regulators in each circuit fuzzy loop regulators are introduced (Fig. 6), or the actions of all additional regulators are reduced to the output of the internal controls of the classic controller (Fig. 7).

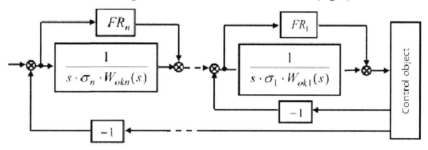

Figure 6. The control system with additional parallel fuzzy corrective regulators, which are distributed in loops

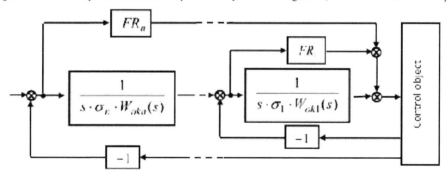

Figure 7. The control system with additional parallel fuzzy corrective regulators concentrated in the inner loop

In additional corrective fuzzy regulators the priori assumptions about the number and the arrangement of the membership functions are absent, so the synthesis of a fuzzy controller is carried out in two stages. At the first stage the number of membership functions is based on the cluster analysis of the vector input fuzzy controller, and at the second - to set up an artificial neural network dynamics of his work.

For a quantitative comparison of the quality of transients (Fig. 8) we have introduced integral estimates Iu, Ic, which take into consideration the damping rate and the deviation of the coordinates in the aggregate. The calculation of the deviation, not only from the given level of x, but the deviation of the current engine speed and deviation to the second and third derivative, respectively - Yager & Filev 1984, Zadeh 1976, Pivnyak 2004, Khilov 2011, Khilov & Glukhova 2004, Terekhov 2001, Terekhov 1996, Leonenkov 2003.

These estimate Iu, Ic characterize the approach of transient to an extremals, due to the solution of differential equations characteristic polynomials of optimized contours of the revolution frequency and motor current. The analysis of the values of integral estimates Ish, Ic of transition indicator (Figure 8), the largest deviation from the optimally tuned control system have been observed with PI fuzzy controller, and the minimal - PD-fuzzy controller. Intermediate indicators of quality control have a PID controller. Therefore, in the algorithms of fuzzy controller current and the speed is sufficient have limiting of two components - proportional and differential.

The use of classical systems with distributed along the loops of the correction one can effectively limit the current values and the limited ones of the coordinates controlled, and the use of fuzzy laws effectively compensates the action of the elastic properties of the transmission in the control loops. Thus, we obtain the adaptation of control systems of rotating electric drive, round-trip operations and the displacement of drill rigs operating modes and parameters of the control object, which allow the use of the proposed energy and resource saving method to be managed in the drilling process.

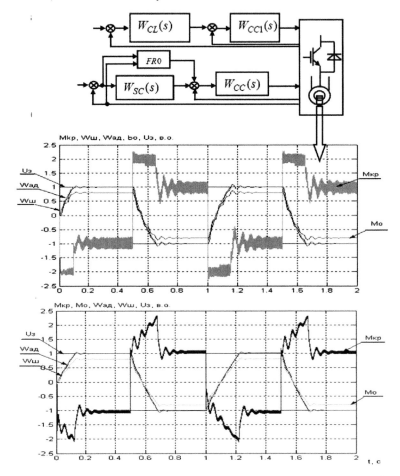

Figure 8. Vector control system of the rotation drive with hysteresis current regulator, PID speed controller $W_{SC}(s)$, flux linkage $W_{CL}(s)$, current $W_{CCI}(s)$, PD-fuzzy-correcting FR0 (a); transients in the system (a) without the fuzzy adaptation (b), and fuzzy-correcting (c) where M_{KP} - drive torque, which is turned by phase in angle in π radians, M_0 - reactive resistance moment at roller cone bits, W_{ad} - frequency shaft rotation of AD, W_{uu} - speed of roller cone bit, U_3 - reference signal to an external speed loop (all variables are given according to their rated values and in relative units are expressed)

Such approach to the suppression of additional rational functions of high order in the control loops allows the implementation of the proposed concept of object-oriented automated electric rotation STF rod of open-pit drill rigs of the new generation, in accordance with which it is possible to improve the quality of drill rod natural torsion vibrations regulation and elastic vibrations in hydro- and rope tackle systems are suppressed introducing the laws of the loop classical regulators and additional fuzzy control laws.

4 CONCLUSIONS

In the result of our studies we have arrived at the following conclusions:

1. The analysis of the dynamic properties of automated driving systems and frequency characteristics of transmission mechanisms of rotation, round-trip and drilling rod displacement of drilling rigs allowed to put forward the idea of building an automatic control system, which concentrates the flow of mechanical power in the area of the bottomhole suppressing forced transverse vibrations of rotating and natural torsional oscillations, minimizing the dynamic load on the nodes of the drilling rigs and providing energy and resource efficiency at the penetration of boreholes blasting.

2. Our analysis of rotating drilling automatic control methods, at the present level of technological achievements, showed that it is efficient to carry out the drilling on that basis of energy criterion, according to which the roller cone bits at the uniform energy load have the greatest durability, i.e. supporting the linear growth of the mechanical energy of rock destruction or keeping constant mechanical power, which is released in the contact zone chisel - work face. In this case, depending on the strength of the rock it is necessary to form the mechanical characteristic at the bit by the means of variable or permanent stiffness.

3. The following regularities of blastholes drilling are found to be of maximum value for the development of the theory of drill string automated rotation drives:

• while drilling with a fully assembled drill string with the length of 33 m in drill rigs СБШС-250Н there are generated their own transverse low-frequency oscillations with the cyclic frequency of 1.05 Hz (for machine СБШ-250МН- 32 -2,48 Hz with drill string 24 m long), which do not directly affect the dynamics of the drive system, but lead to additional loads on the components of the drilling rigs, the suppression of which is possible only by reducing the drill string frequency rotation, which leads to poor performance of mining operations;

• while drilling with a drill string having three or more angular drill rods the frequency of natural torsion vibrations in the drill rigs СБШС-250Н have values less than 186 1/s and get into the pass band of AC transistor drive system with a cutoff frequency of the external control loop 200 1/s, which significantly impairs the quality of indicators control of electromechanical systems. Such fluctuations are advisable to be compensated by the means of the drive system controls;

• low-frequency torsion oscillations of the drill string give rise to additional rational functions of higher order in the speed loop as well as in the current loop of the electric drive of rotation, which while using classical PID controller leads to the need to reduce the speed of the entire control system;

• zeros and poles of additional transmission functions in the control loops, which are caused by elastic vibrations in the drill string in the adjacent control loops, are repeated and, while changing the added mass in the drill string, are shifted relatively to the cutoff frequency of the loop.

4. On the basis of the established regularities in the drilling of subversive wells the scientific basis of automated electric rotation has received further development, according to which the proposed and studied object-oriented drive systems of the rotation drill rod of the drilling rigs with power flow controller in the fracture zone of the rock with a nonlinear correction in the feedback channel. These systems implement the control method of drilling based on the energy criterion which provides the formation of rigid mechanical characteristics of the actuator while drilling rocks having their strength coefficient less than 10-13 score according to the scale by M.M. Protodyakonov and transition to the soft mechanical characteristics while drilling in harder rock, which will provide the resource-saving of roller cone bit.

5. It was found that the development of the theory of automated electric drilling rod ripping and displacement on the work face had provided important regularities:

• for the implementation of energy and resource saving method of drilling it is necessary to control not only the shaft speed and motor current but also the pressure in the hydraulic system, hydraulic motor shaft speed and linear velocity of the drill rod;

• elastic properties of hydraulic and rope-polyspast systems in the loop objects control make oscillations, which are quantitatively defined by the additional rational function of the sixth order in the current loop and the fourth - in the frequency rotation loop;

• in additional dynamic links in the adjacent loops there appear repetitive zeros and poles, which are excited by elastic vibrations in hydro and cable-polyspast systems. The zeros and poles, that are repeated while changing the added mass in the drill string, are displaced relatively to the cutoff frequency control loop.

6. On the basis of the analysis of the regularities in the rotation system, drilling rod tripping and displacement on the work face, there was a further

development of the scientific concept of automated electric drives for the drilling rigs, according to which the purpose of energy and resource saving depends on the strength of the drilled rock: at the roller cone bit there are formed mechanical characteristic with constant or variable stiffness and the limitation of transverse and torsional vibrations of the drill string, as well as fluctuations in hydraulic and rope-polyspast systems. This concept is characterized by transverse vibrations induced by the drill rod, these vibrations are limited through the correction on the reference signal ramp-function of speed rotation machine in the rotation mechanism, and its own torsional vibrations and elastic vibrations in hydraulic and rope-polyspast systems - an introduction into the control laws of classical loop regulators the additional fuzzy control laws with simultaneous control and limit at the allowable level of technological parameters: the rotation drive - mechanical power flow in the area of the work face; in the drilling rod tripping and displacement on the work face drive - the linear speed of the drill string, and the rotation rate of the hydraulic drive and pressure in the hydraulic system.

7. Within the frameworks of the developed scientific concept of automated electric drives:

• the behavior of the characteristic frequencies of closed control loops are researched; it has been found that they cross the cut-off frequency and get into the pass band from the band attenuation in frequency-controlled induction motors in the new generation drill rigs СБШС-250Н in the rotation drive - in the current loop and frequency rotation, and in drilling rod tripping and displacement drive - the current loop at the maximum number of rods attached to the drilling string. It was found that in the most high-speed inner current loop the lowest low-frequency vibrations are got, they are caused by the sideband frequencies that are generated by fluctuations composition, hydro and cable-polyspast systems;

• for effective suppression of elastic vibrations in control loops of the rotation drive and those of the drilling rod tripping and displacement on the work face drive there has been established the need for the use of additional fuzzy controllers, distributed in loops or lumped in the internal loop. It is proved that fuzzy controllers, which are concentrated or distributed into the loops, should apply in the loops band pass getting rational functions caused by the elastic properties of the transmission, the characteristic frequencies of which are lower than the cutoff frequency of the loop while the maximum number of rods attached to the drill string;

• there has been developed the method for the calculation of the training set for the neuro-fuzzy controllers: to find the training set it is necessary to use transfer functions in the form of the rational function of higher order, caused by the properties of the control object.

REFERENCES

Quang N.P., Dittrich J.-A. 2010. *Vector Control of Three-Phase AC Machines: System Development in the practice (Power Systems)*. New York: Springer: 340.

Novotny D.V., Lipo T.A. 1996. *Vector control and dynamics of AC drives*. New York: Oxford science publications: 450.

Boldea I., Nasar S.A. 1992. *Vector Control of AC Drives (Handcover)*. New York: Taylor & Francis Group: CRC Press: 240.

Boldea I., Nasar S.A. 2006. *Electric drives*. New York: Taylor & Francis Group: CRC Press: 522.

Khilov V.S., Beshta A.S., Zaika V.T. 2004. *Experience of using frequency-controlled drive drill rigs in quarries in Ukraine* (in Russian). Moscow: Moscow State Mining University: Mountain information-analytical bulletin, Issue 10: 285-289.

Pivnyak G.G., Beshta A.S., Khilov V.S. 2005. *AC drive system for actuator's power control*. Lviv: XIII International Symposium on Theoretical Electrical Engineering ISTET'05: 368-370.

Khilov V.S., Plakhotnik V.V. 2004. *Estimate the natural frequencies of the drill rod under unsteady conditions* (in Russian). Proc. Scientific Transactions NGU, Issue 19: 145-150.

Pivnyak G.G., Beshta O.S., Khlov V.S. 2003. *Building Principles of Control systems for electric rotating machine cutting drilling pool* (in Ukrainian). Kharkov: NTU "KhPI": Bulletin of the National Technical University "Kharkiv Polytechnic Institute". Scientific Transactions KhPI, Issue 10, Vol.1: 141-143.

Khilov V.S. 2006. *The changing dynamics driving the spinner rig in the application of the AC drive system* (in Russian). Krivoy Rog: Development of ore deposits. Bulletin of KTU, Issue 1 (90): 180-184.

Pivnyak G.G., Beshta O.S., Khlov V.S. 2005. *Drive system the round-trip operations drills* (in Ukrainian). Kharkov: NTU "KhPI": Scientific Transactions of the National Technical University "Kharkiv Polytechnic Institute", Issue 45: 223-225.

Khilov V.S. 2006. *Application of computer-aided drives in new generation boring rigs for open pit's in Ukraine*. Dnipropetrovsk: Scientific Bulletin of NMU, Issue 5: 72-76.

Pivnyak G.G., Beshta O.S., Hilov V.S. 2004. *Adaptive fuzzy power controller to manage the drilling* (in Ukrainian). Kyiv: Technical electrodynamics, Issue 6: 47-52.

Khilov V.S. 2011. *Drill spinner drive dynamic performances correction of blast hole boring rig* Dnipropetrovsk: NGU: The materials of the international conference of "Miners Forum 2011": 90-95.

Khilov V.S. 2012. *The information-analytical characteristics of the busbar field parameters* (in Russian). Dnipropetrovsk: NGU: The materials of the international conference of "Miners Forum 2012": 90-95.

Terekhov V.M. 2001. *Algorithms for fuzzy controllers in electrical systems* (in Russian). Electricity, Issue 12: 55-63.

Terekhov V.M., Baryshnikov A.S. 1996. *Stabilization of slow-moving traffic drives on the basis of Fuzzy-logic* (in Russian). Electricity, Issue 8: 61-64.

Zadeh L.A. 1976. *The concept of a linguistic variable and its application to the adoption of approximate solutions* (in Russian). Academic Press: 165.

Yager R., Filev D. 1984. *Essentials of Fuzzy Modelling and Control*. USA: John Wiley & Sons: 387.

Leonenkov A.V. 2003. *Fuzzy Modeling with MATLAB environment and fuzzyTECH* (in Russian). St. Petersburg: BHV-Petersburg: 736.

Medvedev V.S., Potemkin V.G. 2002. *Neural Network MATLAB 6* (in Russian). Moscow: Dialog-mepi: 496.

Power Engineering and Information Technologies In Technical Objects Control – Pivnyak, Beshta & Alekseyev (eds)
© 2016 Taylor & Francis Group, London, ISBN 978-1-138-71479-3

Information technology of multi-criteria quality evaluation and increase in clustering results sustainability

O. Baybuz, M. Sidorova & A. Polonska
Dnipropetrovsk National University named after Oles Honchar, Dnipro, Ukraine

ABSTRACT: Clustering, as an important area of data mining, solves the problem of partitioning a set of objects into homogeneous groups based on the similarity of the analyzed features. Currently, despite the large number of clustering methods, there are no universal ones. The result depends mostly on the structure of the researched data as well as the ways of formalizing the knowledge of objects and clusters similarity. It is important to evaluate the quality of the results in order to select the partition which is most similar to the cluster structure relevant to the researched data. For this reason, the article presents the information technology of multi-criteria quality evaluation and increase in clustering results sustainability on the basis of the relative quality criteria, decision making theory methods and a set of algorithms.

1 INTRODUCTION

Sophisticated technology of data recording and storage explains a tendency of accumulating large amounts of information. Thus, there is a need for processing large sets of data in order to detect hidden data, such as laws, properties, tendencies and trends, which lead to a better understanding of the structure. This results in necessity of developing information systems and software to ensure solution of such problems.

Cluster analysis is one of the most important tasks of data mining. It allows us to understand the structure of multidimensional data, simplify further processing, reduce the initial sample, identify unusual objects or formulate and check the hypotheses based on the results. Requiring no basic knowledge of the nature of the researched phenomena, processes or situations and using minimal information, cluster analysis methods allow to find practically valuable knowledge about the nature of the research object (such as clusters of objects, hierarchical structures of data, standard objects and "emissions") and are useful in various application areas, such as the following: geology, ecology, economics, marketing, sociology, psychology, biology, medicine, computer science and others.

Currently there are no universal methods of solving the problem of clustering, despite the large number of various approaches. Every approach has its advantages and disadvantages. The result depends on the structure of investigated data, the choice of attributes, measures of proximity, ways of formalizing notions of objects and clusters similarity. Random unreasonable method choice may lead to the situation, where a partition will be absolutely different from the natural, relative to the researched data, cluster structure (Fig.1).

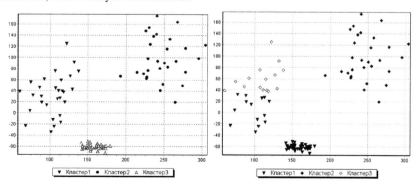

Figure 1. Results of data sets clustering using different methods

Therefore, the topical problems of cluster analysis are: 1) evaluation of the results quality (Halkidi 2001, 2002; Kovacs 2004; Wu 2009; Liu 2010; Pascual 2010) in order to find the partition which is most suitable to the data structure and 2) development of the collective methods (construction of a set of algorithms) (Kuncheva 2006; Reza 2009; Sarumathi 2013), which are used to get the most consistent and stable solutions.

Analyzing the results, it is important to take into consideration the properties of clusters: density, dispersion, size, shape, isolation, etc. [83-86]. The unambiguous definition of qualitative and quantitative characteristics of cluster structure are currently missing in the related literature. There is only intuitive concept that objects within clusters must be located as close to each other as possible, and clusters themselves must be significantly separated.

There are three approaches to the research into the quality of the precise clustering results. The first is based on external criteria (Halkidi 2001, 2002; Gusarova 2003), which means the results of clustering algorithm are being evaluated, based on pre-determined structure, that applies to the data set and reflects our assumptions. The second approach is based on internal criteria (Halkidi 2001, 2002; Gusarova 2003; Bezdek 1998; Pascual 2010; Liu 2010; Wu 2009). It allows us to evaluate the results of clustering algorithm in terms of values, which are associated with vectors and matrices of data vicinity. The third approach, which is the most popular in practice, uses the relative criteria (Mirkin 2011; Halkidi 2001, 2002; Bezdek 1998; Kovacs 2004; Maulik 2002). Its basic idea is to assess patterns of clustering, comparing it with other clustering schemes as the results of using other algorithms or different values of method parameters. The comparative analysis in other studies (Bandyopadhyay 2004; Halkidi 2001; Milligan 1985) suggests a large number of quality criteria for data sets with different structure and different number of clusters.

However, defining of the quality criteria is mainly based on such concepts as compactness and separation of clusters, yet each of them carries different ideas about clusters and their homogeneity, that is why they often demonstrate totally different results, choosing most suitable partitions. The promising area of future research is to develop new methods and technologies that will allow multi-criteria quality assessment of a cluster structure to obtain a more precise definition of the cluster. In addition, an important area of research in cluster analysis is the ensemble approach, that allows us to get compromising partition results, using various methods.

2 FORMULATION OF THE PROBLEM

Suppose, there is a set of objects $X = \{x_1, x_2, \ldots, x_N\}$, where each object $i = \overline{1, N}$ $x_i = \{x_{i1}, x_{i2}, \ldots, x_{ip}\}$ is characterized by a set of p features, x_{ij} – is a value of j $j = \overline{1, p}$ features of object i, is a real number. According to the results of cluster analysis, conducted using different methods or different parameter values, we need to define a partition $G^* = \{g_1^*, g_2^*, \ldots, g_K^*\}$ of the initial set into clusters, which are most similar to the cluster structure appropriate to the investigated data.

For this purpose, it is necessary to develop information technology which takes into account different criteria in assessing the quality as well as to aggregate the best partitioning to improve sustainability of the results obtained.

3 MATERIALS FOR RESEARCH

We propose information technology of multi-criteria quality evaluation and clustering results sustainability increase in order to determine the best partitioning based on the relative quality criteria using methods of decision-making theory and building an ensemble of algorithms, which allow us to rank results by quality criteria and determine a stable cluster structure. The developed technology comprises the following steps:

1. Clustering of the initial set of objects using different methods or different parameter values. Thus, we get a set of groups $G = \{G_1, G_2, \ldots, G_n\}$, among which we need to find the one with the highest quality, where $G_i = \{g_1^{(i)}, g_2^{(i)}, \ldots, g_{K_i}^{(i)}\}, i = \overline{1, n}, g_j^{(i)} = \{x_l\},$

$j = \overline{1, K_i}, \ l = \overline{1, N_j^{(i)}}, \ x_l = \{x_{lh}\}, \ h = \overline{1, p},$

K_i – number of clusters in group i, $N_j^{(i)}$ – number of objects in cluster j of group i,

$\sum_{j=1}^{K_i} N_j^{(i)} = N, \quad \bigcup_{j=1}^{K_i} g_j^{(i)} = X, \quad g_j^{(i)} \cap g_t^{(i)} = \varnothing,$

$j, t = \overline{1, K_i}, j \neq t$.

2. Selecting quality criteria, which will help evaluate and compare partitions obtained. In case there is no information about the genuine cluster

110

structure, we use the relative quality criteria which are based on such concepts as compactness and separation of clusters, in order to compare and rank different clustering results. Let us consider the most popular ones:

- Calinski-Harabasz index

$$Q_{CH}(G) = \frac{B/(K-1)}{W/(N-K)} \to \max,$$

$$B = \sum_{i=1}^{K} N_i \left\| C_i - C \right\|^2, \; W = \sum_{i=1}^{K} \sum_{j=1}^{N_i} \left\| u_j^{(i)} - C_i \right\|^2,$$

$C_i = (\overline{u}_1^{(i)}, \overline{u}_2^{(i)}, \ldots, \overline{u}_T^{(i)})$ – the center of cluster i, $C = (\overline{u}_1, \overline{u}_2, \ldots, \overline{u}_T)$ – the center of the entire set of objects.

- Dunn index

$$Q_D(G) = \min_{\substack{i,j \in 1,K, \, i \neq j \\ l \in 1, N_i, \, m \in 1, N_j}} \left\{ \frac{d(u_l^{(i)}, u_m^{(j)})}{\max_{h \in 1,K} \{diam(g_h)\}} \right\},$$

$$diam(g_h) = \max_{i,j = 1, N_h, i \neq j} \{d(u_i^{(h)}, u_j^{(h)})\}$$

- Bejdek-Dunn index

$$Q_{BD}(G) = \frac{\min_{i,j \in 1,K, \, i \neq j} \{\delta(g_i, g_j)\}}{\max_{h \in 1,K} \{\Delta(g_h)\}} \to \max,$$

$$\delta(g_i, g_j) = \frac{1}{N_i \cdot N_j} \sum_{l=1}^{N_i} \sum_{m=1}^{N_j} d(u_l^{(i)}, u_m^{(j)}),$$

$$\Delta(g_h) = \frac{2}{N_h} \sum_{i=1}^{N_h} d(u_i^{(h)}, C_h).$$

- Quality functionals, such as: the amount of internal cluster dispersions by all indications; the sum of squared distances to the centers of the clusters; the ratio of internal and average external cluster distances; the sum of internal cluster distances.

3. Calculating evaluation of each criterion for each alternative clustering. We present the results obtained in matrix $Q = \{q_{ij}; i = \overline{1,n}, j = \overline{1,m}\}$, where n – number of compared partitions, m – number of quality evaluation criteria, q_{ij} – quality evaluation of partition i using criterion j.

4. Determining the best option of splitting the initial set of objects into clusters by several criteria can be formulated in terms of decision-making theory. As alternative options, we will consider clustering results obtained by different methods or for different parameter values, as expert assessments – the value of quality criteria calculated for each alternative. Thus, we use the matrix obtained in the previous step $Q = \{q_{ij}; i = \overline{1,n}, j = \overline{1,m}\}$. The methods of collective choice allow us to move from individual (one criterion) to generalized (all criteria) assessments for comparable alternatives. Thus, the alternatives are ranked by multi-criteria evaluations, which allows to choose the highest quality solutions. It is proposed to use the following collective methods of decision making (Emelianenko 2005):

- the Borda Method. For each criterion we evaluate the alternatives in terms of quality depreciation and assign rank places r_i^j, $i = \overline{1,n}, \; j = \overline{1,m}$. It is possible to ascribe the same numbers to equivalent options. Then, we build the utility collective scale as follows: we attribute the number r_i^* to each alternative, which is equal to the sum of ranked places in individual orderings by each criterion: $r_i^* = \sum_{j=1}^{m} r_i^j$, i.e., the utility scale is built - "the sum of rank places". We consider the option that will have a minimum value r_i^* to be the best.

- Plurality Method. The first step of this procedure is the same as in the Borda count, i.e. we organize the alternatives and rank them by each criterion. During the construction of supporting collective scale for each alternative, we compute an assessment equal to the number of criteria, which makes it the best quality one, i.e. ranked first $r_i^j = 1$ in individual orderings. Thus we construct a scale "sum of the first places" and complete maximization based on this scale. The best option is considered to be the one with the highest collective assessment.

- the Copeland Method. We construct an auxiliary collective structure – majoritary graph for a profile M of individual orderings by each criterion. That would be a direct graph, the vertices of which are alternatives, and focused curve from vertex i to vertex j can only be possible if the number of criteria, indicating ordering i to be better than ordering j, will make up more than a half of the total number of criteria. According to the Copeland method, each possible option will be ascribed with the number s_i, which is equal to the subtraction of the number of curves, that come out of vertex i and take place in the majority graph M:

$$s_i = \left|\{j : iMj\}\right| - \left|\{j : jMi\}\right|, \; j = \overline{1,n}, i \neq j$$

The number s_i can be interpreted as the number of i-option's "wins" over other options minus the number of its "loses". Taking into account the majority graph M, this method builds the additional auxiliary structure – the collective scale $S = \{s_i\}, i = \overline{1,n}$. The best alternatives are considered to be the ones that have the maximum values by this scale.

– Method of Plural Analysis. We transform the evaluations using the following formula:

$$q_{ij} = q_{ij} / \sum_{i=1}^{n} q_{ij}$$. If among the selected criteria of quality there are minimizing as well as maximizing ones, we need to take this fact into account and convert appropriate assessments $q_{ij} = q_{ij}^{-1}$.

Quality evaluation of alternatives can be conducted using the recurrent procedure. On every step i, we calculate $a_l^i = \sum_{j=1}^{m} q_{lj} k_j^{i-1}, l = \overline{1,n}$,

where $k_j^0 = 1/m$, $k_j^i = \dfrac{1}{\lambda^i} \sum_{l=1}^{n} q_{lj} a_l^i, j = \overline{1,m}$,

$\lambda^i = \sum_{l=1}^{n}\sum_{j=1}^{m} q_{lj} a_l^i \; \sum_{l=1}^{n} k_l^i = 1$, until the process will not lie with the certain accuracy ε. It is considered, that a_l – collective assessment of l-alternative's quality, k_j – adequacy of assessments by criterion j. It is proved, that this process is convergent. The best result is considered to be the one with minimum or maximum assessment according to minimization or maximization of quality functionals. This method also allows us to evaluate the consistency of assessments by different criteria based on dispersion coefficient of concordance. For this, the matrix $Q = \{q_{ij}; i = \overline{1,n}, j = \overline{1,m}\}$ is converted into the matrix of ranks $\{R_{ij}\}$, based on which the characteristics are calculated:

$$D = \frac{1}{n-1}\sum_{i=1}^{n}(r_i - \overline{r}), \text{ where } r_i = \sum_{j=1}^{m} R_{ij}, i = \overline{1,n};$$

$\overline{r} = \dfrac{1}{n}\sum_{i=1}^{n} r_i$. Then we get the concordance dispersion coefficient $W = D / D_{max}$,

$0 < W < 1$, $D_{max} = \dfrac{m^2(n^3 - n)}{12(n-1)}$. The value W, and thus, the hypothesis check H_0: $W = 0$ if $n > 7$ is based on statistical characteristics for unrelated ranks $\chi^2 = W \cdot m(n-1)$, or connected ranks

$$\chi^2 = \frac{12\sum_{i=1}^{n}\left(\sum_{j=1}^{m} r_{ij} - \overline{r}\right)^2}{mn(n+1) - \dfrac{1}{n-1}\sum_{j=1}^{m}\sum_{k=1}^{H_j}(h_k^3 - h_k)},$$

where H_j – the number of groups of equal ranks j, h_k – number of equal ranks in group k connected with ranking of j. Both statistical characteristics are distributed as χ^2 - Pearson with $\nu = n - 1$ number of fluency degrees.

The partition $G^* = \{g_1^*, g_2^*, \ldots, g_K^*\}$, which is the one with the highest quality, considering the results of implementing collective methods of decision making can be regarded as the most similar to the cluster structure, inherent in the investigated data.

5. Sometimes several clustering results may represent different groups, though equivalent in quality. In this case, instead of selecting one of these solutions it is suggested using ensemble approach in order to be able to obtain a collective resultant solution. (Reza 2009; Kuncheva 2006; Sarumathi 2013; Fern 2003; Fred 2005).

The algorithm of determining a collective solution is based on the ensemble of individual clusterings.

Suppose, we are given L individual groupings, equivalent in quality $G_l, l = \overline{1,L}$. To identify a collective solution G^* on the basis of individual clusterings, we are going to use the following steps:

1. We create generalized similarity matrix $S = \{s_{ij}\}; \; i, j = \overline{1,N}$ and initialize it with zeros:

$s_{ij} = 0;\ i, j = \overline{1, N}$. The values s_{ij} determine the similarity of objects i and j by clustering ensemble.

2. One by one, we consider the groupings from the set of individual clusterings $G_l : l = \overline{1, L}$. If objects i and j in the 1st grouping belong to the same cluster, we increase s_{ij} per unit: $s_{ij} = s_{ij} + 1$, otherwise – value s_{ij} remains the same.

3. We build the matrix elements of similarity to the unit scale: $s_{ij} = \dfrac{s_{ij}}{L}$, $i, j = \overline{1, N}$. After this transformation, s_{ij} acquire values on a segment from 0 to 1.

4. We complete transition from the similarity matrix $S = \{s_{ij}\}$, $i, j = \overline{1, N}$ into the distances matrix $D = \{d_{ij}\}$, $i, j = \overline{1, N}$ as it follows:

$d_{ij} = 1 - s_{ij}$, $i, j = \overline{1, N}$, i.e., the more similar the objects i and j in matrix S, the smaller the distance between them in matrix D.

5. We get a final partitioning

$$G^* = \{g_1^*, g_2^*, \dots, g_K^*\}, \quad \bigcup_{i=1}^{K} g_i = X,\ g_i \cap g_j = \varnothing,$$

$i, j = \overline{1, K}, i \neq j$ which can be obtained by applying the algorithms of cluster analysis to matrix D. These algorithms use the matrix of distances between objects as the initial information (such as hierarchical, graph, some optimization, fuzzy methods, etc.).

The flow chart of the proposed technology is shown in Fig. 2.

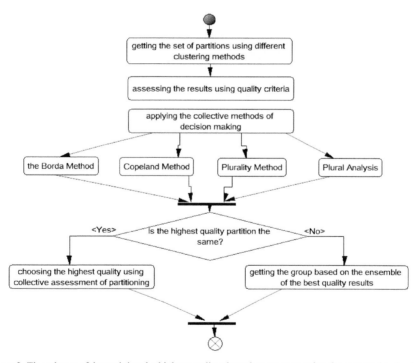

Figure 2. The scheme of determining the highest quality clustering outcome using the proposed technology

To check the adequacy of the proposed multi-criteria quality evaluation technology we completed a series of experiments comparing the quality of the results obtained with the known clustering data structure. For example, let's consider the sets "Irises" and "Wheat Grains", which are often used to illustrate methods of cluster analysis and classification (Fisher 1936; Charytanowicz 2010).

In order to have a set of individual solutions containing the most complete information about groupings inherent to the investigated data, it is suggested to apply various algorithms of cluster analysis. The analysis in this study had been conducted using the following methods: hierarchical agglomerative single connection (SC), complete connection (CompC), average connection (AC), central connection (CentC), Ward connection; K-average in variations Ball-Hall (BH) and McKean (MK), genetic algorithm (GA), Forel, graph algorithm of the shortest non closed lane (MST). Quality evaluation was based on the following criteria: the sum of inside cluster dispersions by all indicators (Q1); the sum of squared distances to the centers of the clusters (Q2); the ratio of the average inside cluster and average distances between clusters (Q3); the sum of inside cluster distances

(Q4); Calinski-Harabasz index (QCH); Dunn index (QD); Bejdek-Dunn index (QBD).

First, we consider the clustering results of a classic set "Irises", which contain the data (the length and width of a sepal, the length and width of petals) about 150 copies of iris flowers, which is 50 copies of three species –iris setosa, iris virginica and iris versicolor.

Table 1 presents the individual quality assessments of the obtained results by every criteria. Table 2 shows the results of collective multi-criteria quality assessment.

Table 1. Quality evaluation of clustering results regarding the data set "Irises" by relative criteria

Clustering Methods	Quality Criteria						
	Q1	Q2	Q3	Q4	QCH	QD	QBD
Hierarchy SC	0,74	5,78	0,62	1541,67	1241,67	0,20	1,08
Hierarchy CompC	0,80	2,83	4,38	645,95	1366,28	0,18	1,33
Hierarchy AC	0,81	2,89	4,42	649,83	1354,03	0,15	1,29
Hierarchy CentC	4,13	5,94	0,32	1584,67	1250,76	0,15	1,09
Hierarchy Ward	0,80	2,83	4,38	645,95	1366,28	0,18	1,33
Ball-Hall	0,98	4,15	5,11	751,49	1130,98	0,04	1,18
McKean	0,81	2,92	4,46	655,55	1355,43	0,07	1,27
MST	0,74	5,78	0,62	1541,67	1241,67	0,20	1,08
GA	0,80	2,83	4,38	645,95	1366,28	0,18	1,33
Forel	0,93	5,62	0,89	1495,51	1190,36	0,12	1,13

Table 2. The results of collective multi-criteria quality evaluation of the clustering results of the data set "Irises"

Methods of Collective Choice	Clustering Methods									
	Hierarchy SC	Hierarchy CompC	Hierarchy AC	Hierarchy CentC	Hierarchy Ward	Ball-Hall	McKean	MST	GA	Forel
Borda Method	28	12	20	34	12	37	25	28	12	32
Plurality Method	2	4	0	1	4	0	0	2	4	0
Copeland Method	-2	7	3	-9	7	-7	1	-2	7	-5
Method of Plural Analysis	0,09	0,084	0,087	0,14	0,084	0,13	0,1	0,09	0,084	0,1

Analyzing the obtained results, we can come to the conclusion that the genetic method and hierarchical methods of complete connection and Ward demonstrate the same partition, which is the highest quality one. In addition, the fact that some results match, underlines the significance of their firmness. Since the real cluster structure in the investigated data set is known, we conducted an additional quality evaluation based on external criteria:

– Rand Index:

$$R(G,\tilde{G}) = \frac{SS+DD}{SS+SD+DS+DD} \to \max,$$

– Jaccard Index:

$$J(G,\tilde{G}) = \frac{SS}{SS+SD+DS} \to \max,$$

– Fowlkes-Mallows Index:

$$FM(G,\tilde{G}) = \sqrt{\frac{SS}{SS+SD}\frac{SS}{SS+DS}} \to max,$$

where SS – the number of pairs of objects that belong to the same cluster as G, as well as $\tilde{G}$, SD – the number of pairs of objects that belong to the same cluster in G and different clusters in $\tilde{G}$, DS –

the number of pairs of objects that belong to different clusters in G and the same cluster in $\tilde{G}$, DD – the number of pairs of objects that belong to different clusters in G, as well as in $\tilde{G}$.

The results of quality evaluation based on external criteria are shown in Table 3. As we can see, the same partitioning was considered to be the highest quality one.

Table 3. The quality assessments of the clustering results of the data set "Irises" based on external criteria

External Criteria	Clustering Methods									
	Hierarchy SC	Hierarchy CompC	Hierarchy AC	Hierarchy CentC	Hierarchy Ward	Ball-Hall	McKean	MST	GA	Forel
Rand	0,78	0,96	0,94	0,78	0,96	0,84	0,93	0,78	0,96	0,78
Jaccard	0,59	0,86	0,85	0,59	0,86	0,62	0,80	0,59	0,86	0,59
Fowlkes-Mallows	0,76	0,92	0,91	0,77	0,92	0,76	0,89	0,76	0,92	0,76

Next, we consider the results of clustering in the data set "Wheat Grains", which contains such information as measurement of geometric properties (area, perimeter, compactness, length, width, asymmetry coefficient, gutter length) for 210 wheat grains of three species: Kama (70 grains), Ross (70 grains) and Canadian (70 grains). The data was prepared by the Institute of Agrophysics, Polish Academy of Sciences in Lublin.

Table 4 shows the individual quality assessments of the obtained results for each of the relative criterion. Table 5 represents the results of collective multi-criteria quality evaluation.

Table 4. Quality assessments of the clustering results for the data set "Wheat Grains" by the relative criteria

Clustering Method	Quality Criteria						
	Q1	Q2	Q3	Q4	QCH	QD	QBD
Hierarchy SC	6,45	15,10	0,44	7358,89	72,32	0,14	0,69
Hierarchy CompC	1,12	7,75	4,16	2831,04	923,14	0,09	1,16
Hierarchy AC	1,05	5,30	11,09	1461,66	1200,81	0,09	1,19
Hierarchy CentC	7,26	14,69	0,44	7209,14	105,17	0,12	1,01
Hierarchy Ward	1,04	5,32	11,11	1463,04	1197,76	0,09	1,17
Ball-Hall	1,23	8,17	9,06	2558,88	887,48	0,03	0,69
McKean	1,08	5,68	11,01	1653,02	1176,43	0,05	1,15
MST	6,45	15,10	0,44	7358,89	72,32	0,14	0,69
GA	1,07	5,27	11,15	1467,80	1214,16	0,02	1,19
Forel	1,49	10,56	5,10	4230,86	493,45	0,04	1,04

Table 5. The results of the collective multi-criteria quality evaluation of the clustering results for the data set "Wheat Grains"

Methods of Collective Choice	Clustering Method									
	Hierarchy SC	Hierarchy CompC	Hierarchy AC	Hierarchy CentC	Hierarchy Ward	Ball-Hall	McKean	MST	GA	Forel
Borda Method	44	29	17	42	21	40	29	44	24	41
Plurality Method	2	0	2	1	1	0	0	2	3	0
Copeland Method	-8	1	9	-5	7	-1	3	-8	5	-3
Plural Analysis	0,177	0,059	0,057	0,157	0,057	0,086	0,06	0,18	0,08	0,09

Analyzing the results, we can make a conclusion that the best partitioning is demonstrated by the hierarchical method of average connection. We also conducted a quality evaluation based on external criteria and known cluster structure, inherent to the investigated data (Table 6).

Table 6. The quality assessments of the clustering results for the data set "Wheat Grains" by external criteria

External Criteria	Clustering Method									
	Hierarchy SC	Hierarchy CompC	Hierarchy AC	Hierarchy CentC	Hierarchy Ward	Ball-Hall	McKean	MST	GA	Forel
Rand	0,34	0,73	0,89	0,34	0,88	0,67	0,80	0,34	0,84	0,47
Jaccard	0,33	0,52	0,72	0,33	0,72	0,41	0,54	0,33	0,61	0,32
Fowlkes-Mallows	0,57	0,70	0,84	0,56	0,83	0,59	0,70	0,57	0,76	0,52

Thus, the results obtained considering the external criteria coincide with the results obtained by the relative criteria without any information about a cluster structure.

The analysis of the results indicates that the proposed information technology of multi-criteria quality evaluation and sustainability increase in clustering results selects the solutions most adequate to the cluster structure and inherent to the investigated data.

4 CONCLUSIONS

We developed the information technology of multi-criteria quality evaluation and the sustainability increase in clustering results, which provides support for decision making when determining the cluster structure that fits best to the investigated data. The technology is based on the process of applying the relative quality criteria, methods of decision-making theory and building the algorithms ensemble; allows to rank the obtained results by multi-criteria quality evaluation and determine a stable cluster structure.

The proposed technology became a part of the authors' data mining system «Medisa» which implements clustering, classification, prediction, probabilistic and statistical analysis, visualization, processing and analysis of information, including data monitoring, and provides support for decision making under uncertainty.

We have conducted a series of successful experiments comparing the quality of clustering results, obtained on the basis of the proposed technology, knowing the cluster structure of the studied data. The technology has also been used in the analysis of the real data of medical and hydrochemical monitoring.

REFERENCES

Halkidi M., Batistakis Y., Vazirgiannis M. 2002. *Cluster Validity Methods: Part I.* CM SIGMOD Record, Vol. 31, Issue. 2: 40–45.

Halkidi M., Vazirgiannis M. 2001. *Clustering validity assessment: Finding the optimal partitioning of a data set.* Proceedings of ICDM: 187–194.

Gusarova L., Yatckiv I. 2003. *Checking the validity of the cluster solution.* Proceedings of International Conference RelStat'03, Vol. 2: 34.

Bezdek J. C., Pal N. R. 1998. *Some new indexes of cluster validity.* IEEE Trans. Systems Man Cybernet. – Part B: Cybernetics 28 (3): 301–315.

Liu Y. 2010. *Understanding of Internal Clustering Validation Measures.* IEEE Int. Conf. Data Mining. – Sydney, NSW (Australia): 911–916.

Pascual D., Pla F., Sanchez J. S. 2010. *Cluster validation using information stability measures.* Pattern Recognition Letters, Vol. 31: 454–461.

Wu K.-L., Yang M.-S., Hsieh J.-N. 2009. *Robust cluster validity indexes.* Pattern Recognition, Vol. 42: 2541–2550.

Mirkin B. G. 2011. *Methods of cluster analysis to support decision making: a survey.* Ed. House NRU "Higher School of Economics": 88.

Halkidi M., Batistakis Y., Vazirgiannis M. 2002. *Clustering Validity Checking Methods: Part II.* ACM SIGMOD Record, Vol. 31, Issue 3: 19-27.

Kovacs F. 2004. *Cluster validity measurement techniques.* London: Academic Press: 388–393.

Maulik U., Bandyopadhyay S. 2002. *Performance evaluation of some clustering algorithms and validity indices.* Journal IEEE Transactions Pattern Analysis and Machine Intelligence, Vol. 24, Issue 12: 1650–1654.

Bandyopadhyay S., Pakhira M., Maulik U. 2004. *Validity index for crisp and fuzzy clusters.* Pattern Recognition, Vol. 37, Issue 3: 487–501.

Halkidi M., Batistakis Y., Vazirgiannis M. 2001. *On Clustering Validation Techniques.* Journal of Intelligent Information Systems, Issue 17:2/3: 107–145.

Milligan G., Cooper M. 1985. *An examination of procedures for determining the number of clusters in a data set.* Psychometrika, Vol. 50, Issue 2: 159–179.

Emelianenko T. G., Zberovsky A. V., Pristavka A.P., Sobko B. E. 2005. *Decision-making in monitoring systems.* Dnipropetrovsk: 224.

Reza G., Nasir S. Md., Hamidah I., Norwati M. 2009. *A Survey: Clustering Ensembles Techniques.* Proceedings Of World Academy Of Science, Engineering And Technology, Vol. 26: 636–645.

Kuncheva L. I., Hadjitodorov S. T., Todorova L. P., Sofia B. 2006. *Experimental comparison of cluster ensemble methods.* Proc. 9th International Conference on Information Fusion: 1–7.

Sarumathi S., Shanthi N., Santhiya G. 2013. *A Survey of Cluster Ensemble.* International Journal of Computer Applications (0975 – 8887), Vol. 65, Issue 9: 8–11.

Fern X.Z., Brodley C.E. 2003. *Clustering ensembles for high dimensional data clustering.* Proc. International Conference on Machine Learning: 186–193.

Fred A., Jain A. 2005. *Combining Multiple Clusterings Using Evidence Accumulation.* Transaction on Pattern Analysis and Machine Intelligence, Vol. 27, Issue 6: 835–850.

Fisher R.A. 1936. *The Use of Multiple Measurements in Taxonomic Problems.* Annals of Eugenics: 179–188.

Charytanowicz M., Niewczas J., Kulczycki P. [et al]. 2010. *A Complete Gradient Clustering Algorithm for Features Analysis of X-ray Images, in: Information Technologies in Biomedicine.* Springer-Verlag Berlin-Heidelberg: 15–24.

Power Engineering and Information Technologies In Technical Objects Control – Pivnyak, Beshta & Alekseyev (eds)
© 2016 Taylor & Francis Group, London, ISBN 978-1-138-71479-3

Preprocessing of digital image for compression in JPEG

O. Narimanova, K. Tryfonova, A. Agadzhanyan
Odesa National Polytechnic University, Odesa, Ukraine

V. Hnatushenko
Dnipropetrovsk National University named after Oles Honchar, Dnipro, Ukraine

ABSTRACT: The work discusses some properties of singular vectors of normal singular value decomposition of the blocks of digital image matrix (matrices). The results of computational experiments and analysis of detected properties allowed to develop a new method for preprocessing of DI with subsequent compression in JPEG. Results of computational experiment confirm the efficiency of preprocessing before JPEG compression, which allows to reduce image file size by 2 — 6% compared with conventional JPEG compression at the same level of visual perception.

1 INTRODUCTION

Digital images (DI) are widely used in almost all areas of the industry. At the same time real images used in industrial machine vision, inspection, monitoring processes, etc. may have a fairly large size, which leads to the necessity of solving the problem of storage of these images. In connection with this processing, including compression, of images is an important task in many technical systems for analyzing digital data (Gnatushenko 2003; Hnatushenko 2015).

To reduce the size of digital images files, different compression algorithms are used. The main requirement is that they have to satisfy two contradictory requirements: small size of image file and high quality or high level of perception reliability (Narimanova 2014). Today, compression algorithms are so sophisticated, that even a slight reduction in the size of compressed digital image file while maintaining the perception reliability is a significant result.

The most widely used format for storing DI, which provides high compression with acceptable levels of perception reliability is JPEG (Salomon 2002; Gonzalez 2002; Narimanova 2011). In connection with this it is expedient to develop a method of digital image preprocessing to reduce the DI file size while compression DI exactly in JPEG.

Currently, the authors have obtained some results concerning the decrease of size of the image file by replacing the basis vectors in the normal singular value decomposition of the matrix blocks before saving DI in JPEG (Narimanova 2015). However, it is necessary to carry out the research of the statistical properties of parameters of the normal singular value decomposition to make informed choice about replacing of its basis vectors. In addition, a convincing comparative analysis of the JPEG compression with and without preprocessing of DI is carried out in (Narimanova 2015).

Current research in some aspects intersects with the work (Narimanova 2015), but in general, it is its complement and continuation, which gives a complete picture about the development process and the developed method of DI preprocessing itself.

2 THE FORMULATION OF ARTICLE PURPOSES

It is necessary to develop a method for digital image preprocessing in order to reduce the size of its file with retaining the perception reliability while saving it in JPEG.

To achieve the aim, it is necessary to:

1. Conduct a study of the statistical properties of the parameters of normal singular value decomposition of blocks of DI matrix (matrices);

2. Select and justify the replacement of the basis vectors of normal singular value decomposition of the DI matrix (matrices) blocks;

3. Carry out the computational experiments for the analysis of JPEG compression with and without the proposed DI preprocessing.

3 THE BASIC MATERIAL

As a complete set of DI parameters, it is proposed to use the singular values and singular vectors of the DI matrix (matrices) obtained using the normal singular value decomposition (Demmel 1997; Horn 2012). However, during the DI analysis or processing singular value decomposition is not accepted to apply to the entire DI matrix (matrices) virtue of its (their) high dimension. So to start with, the original image is divided into non-overlapping blocks B of size $n \times n$ pixels and singular value decomposition is applied to each block of the image (Narimanova 2015).

Let B be $n \times n$ -matrix with elements $b_{ij}, i, j = \overline{1, n}$. The singular value decomposition of B is true:

$$B = U \Sigma V^T, \tag{1}$$

where U, V are the orthogonal matrices of dimension $n \times n$ containing left $u_1, ..., u_n$ and right $v_1, ..., v_n$ singular vectors of matrix B accordingly;

$\Sigma = diag(\sigma_1, ..., \sigma_n)$ is a diagonal matrix containing singular values $\sigma_i, i = \overline{1, n}$, $\sigma_1 \geq ... \geq \sigma_n \geq 0$.

Further let's assume that blocks B of DI matrices are non-degenerate and have pairwise distinct singular values $\sigma_i, i = \overline{1, n}$. Then it is possible to build a single normal singular value decomposition (1), in which the left (right) singular vector are lexicographically positive (Parlett 1980).

In the work (Narimanova 2015), development of method for DI preprocessing with following saving in JPEG is based on the replacement of the first (left and right) singular vectors u_1, v_1 of the DI matrix blocks corresponding to the largest singular value σ_1 on n -optimal vector of corresponding dimension, since, according to (Kobozeva 2013), these vectors are close.

Denote n -optimal in R_n is vector of the form

$$n\_optimal = \underbrace{\left(\frac{1}{\sqrt{n}}, ..., \frac{1}{\sqrt{n}} \right)}_{n}.$$

In this work, the deflection angles of the singular vectors u_1, v_1 from n -optimal vector for blocks of different sizes in DI of different compression ratios are studied.

To illustrate the results, Figure 1 shows the histograms of deviation angles for a single image in TIFF: in partitioning, for the block of 2×2 pixels (Fig. 1, a) the most frequent deviation angle of singular vectors u_1 from n -optimal vector is $0°$; for blocks of 4×4, 8×8, 16×16 and 32×32 pixels (Fig. 1, $b — e$) it is $1°$; for blocks of 64×64 pixels (Fig. 1, f) it is $3°$.

Some of the most representative results of the research for the images in TIFF file format are shown in Table. 1. The most frequent deviation angle in the Table is denoted α and the number of vectors equal to the n -optimal vector in percentage is indicated $\overline{\alpha_0}$

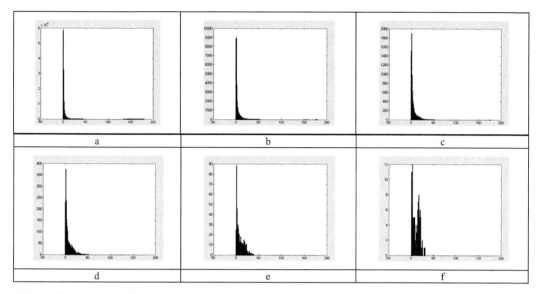

Figure 1. The histograms of deviation angles of vectors u_1 from n-optimal vector for digital image in file format TIFF in partitioning into blocks of: 2×2 pixels (a); 4×4 pixels (b); 8×8 pixels (c); 16×16 pixels (d); 32×32 pixels (e); 64×64 pixels (f).

Table 1. The deviation angles of vectors u_1 from n-optimal vector for digital images in file format TIFF

Block's size	2x2		4x4		8x8		16x16		32x32		64x64	
Image №	α, °	$\overline{\alpha_0}$, %	α, °	$\overline{\alpha_0}$, %	α, °	$\overline{\alpha_0}$, %	α, °	$\overline{\alpha_0}$, %	α, °	$\overline{\alpha_0}$, %	α, °	$\overline{\alpha_0}$, %
1	0	77,11	0	80,27	0	86,58	0	83,49	0	62,14	1	21,37
2	0	60,34	0	47,43	0	39,16	0	24,74	1	9,47	2	4,27
3	1	19,18	1	7,67	2	5,34	2	3,03	2	1,85	3	0,85
4	0	62,40	0	52,14	0	43,07	0	29,06	1	10,49	2	0,85
5	0	63,00	0	51,57	0	43,95	0	33,95	0	22,63	2	4,27
6	0	24,52	1	12,59	3	7,68	3	3,14	3	0,62	8	0,00
7	0	48,98	1	26,18	1	13,14	1	8,23	1	1,85	1	0,00
8	0	57,05	0	50,35	0	45,99	0	38,63	0	27,37	1	9,40
9	0	48,55	0	35,85	1	24,59	1	14,20	1	5,56	3	2,56
10	0	46,10	1	28,04	1	19,20	1	11,88	1	5,14	3	0,00

Since interesting are not only images stored in a lossless format, the study is carried out for d in JPEG format with different compression ratios.

Figures 2 and 3 show histograms of deviation angles for a single image in JPEG format with quality coefficient of 100% and 60%, respectively. In partitioning of DI in JPEG with quality coefficient of 100% into the blocks of 2×2 pixels (Fig. 2, a) the most frequent deviation angle of singular vectors u_1 from n-optimal vector is 0°; for blocks of 4×4, 8×8, 16×16 and 32×32 pixels (Fig. 2, b — e) it is 1°; for blocks of 64×64 pixels (Fig. 2, f) it is 3°.

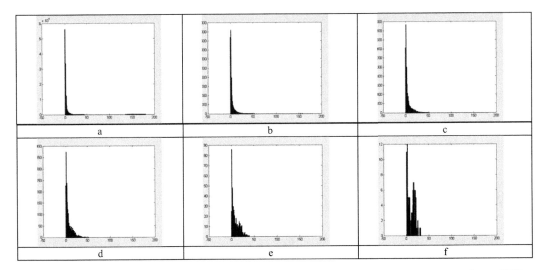

Figure 2. The histograms of deviation angles of vectors u_1 from n-optimal vector for digital image in file format JPEG with quality coefficient 100% in partitioning into blocks: 2×2 pixels (a); 4×4 pixels (b); 8×8 pixels (c); 16×16 pixels (d); 32×32 pixels (e); 64×64 pixels (f).

At high compression ratios, i.e. at the lower value of quality coefficient the most frequent deviation angles of the first singular vectors form n-optimal slightly change. Thus, for DI in JPEG with a quality coefficient of 60% in partitioning into the blocks of 2×2, 4×4, 8×8 and 16×16 pixels (Fig. 3, a —

d) the most frequent deviation angle of singular vectors u_1 from n-optimal vector is 0°; for blocks of 32×32 pixels (Fig. 3, e) it is 1°; and for blocks of 64×64 pixels (Fig. 3, f) it is 3°.

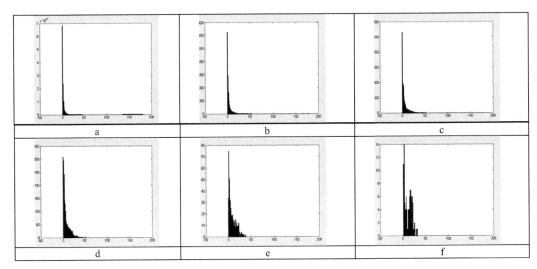

Figure 3. The histograms of deviation angles of vectors u_1 from n-optimal vector for digital image in file format JPEG with quality coefficient 60% in partitioning into blocks of: 2×2 pixels (a); 4×4 pixels (b); 8×8 pixels (c); 16×16 pixels (d); 32×32 pixels (e); 64×64 pixels (f).

Some of the most representative results of the research for the images in JPEG file format with

quality coefficient 60% are shown in Table 2. The most frequent deviation angle in the Table is

122

denoted α and the number of vectors equal to the n-optimal vector in percentage is indicated $\overline{\alpha_0}$.

Table 2. The deviation angles of vectors u_1 from n-optimal vector for digital images in file format JPEG with quality coefficient 60%

Block's size	2x2		4x4		8x8		16x16		32x32		64x64	
Image №	α ,°	$\overline{\alpha_0}$,%	α ,°	$\overline{\alpha_0}$,%	α ,°	$\overline{\alpha_0}$,%	α ,°	$\overline{\alpha_0}$,%	α ,°	$\overline{\alpha_0}$,%	α ,°	$\overline{\alpha_0}$,%
1	0	54,41	0	47,59	0	45,62	0	22,74	1	7,82	3	1,71
2	0	19,91	0	12,22	0	10,88	8	2,57	7	0,82	6	0,00
3	0	25,07	0	16,96	0	15,52	9	5,09	8	1,65	11	0,00
4	0	42,10	0	32,46	0	30,57	0	20,01	1	10,08	1	4,27
5	0	38,05	0	31,67	0	30,42	0	19,39	1	7,20	2	2,56
6	0	67,37	0	57,55	0	52,05	0	20,73	1	3,70	3	0,00
7	0	53,08	0	47,05	0	43,07	0	28,29	1	12,76	1	3,42
8	0	76,22	0	60,03	0	51,99	1	21,60	1	6,79	1	1,71
9	0	64,42	0	54,58	0	40,66	1	18,06	1	6,17	3	0,85
10	0	64,63	0	53,95	0	48,22	0	23,92	1	6,58	3	1,71

The results of experiments demonstrate a small deviation of first singular vectors from n-optimal vector and their frequent coincidence. In this regard, carried out is a computational experiment with full replacement of the first singular vectors of normal singular value decomposition of the DI matrix blocks of size 2×2 by n-optimal vector of corresponding dimension with following save in JPEG with different quality coefficient according to Algorithm 1.

Algorithm 1.

Step 1. Split the DI matrix I_0 on blocks B of size $n\times n$ pixels;

Step 2. Apply the normal singular value decomposition (1) to each block B;

Step 3. In each block B replace the left u_1 and the right v_1 vector corresponding to the largest singular value by the n-optimal vector of corresponding dimension:

$$u_{11},...,u_{1n} = \frac{1}{\sqrt{n}},...,\frac{1}{\sqrt{n}},$$

$$v_{11},...,v_{1n} = \frac{1}{\sqrt{n}},...,\frac{1}{\sqrt{n}}.$$

Carry out the orthogonalization of the vectors composing the matrices U,V according to the modified Gram-Schmidt algorithm (Golub 2013). The resulting matrices are $\overline{U},\overline{V}$;

Step 4. Recover the image matrix I_1 by recovering blocks $\overline{B}$:

$$\overline{B} = \overline{U}\Sigma\overline{V}^T;$$

Step 5. Save the images with matrices I_0 and I_1 in JPEG with different values of quality coefficient Q. Fix the portions of the original image size in % which are I_0's and I_1's sizes.

Shown in Figure 4 are the portions of the original image size in % which are I_0's and I_1's sizes after saving images with matrices I_0 and I_1 in JPEG with values of quality coefficient from 95 to 70% with step 5%.

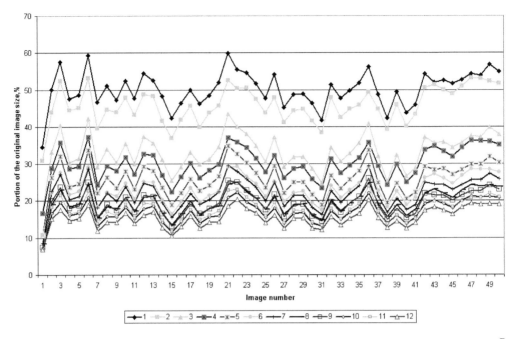

Figure 4. The portions of the original image size in % which are sizes of original image with or without preprocessing P with saving in JPEG with quality coefficient Q:

	1	2	3	4	5	6	7	8	9	0	11	12
P	-	+	-	+	-	+	-	+	-	+	-	+
Q,%	95	95	90	90	85	85	80	80	75	75	70	70

In Figure 4, one can see that the portion of original image size for images after preprocessing by Algorithm 1 is lower than for images without one. These results correlate with results obtained in (Narimanova 2015).

However, preprocessing according to Algorithm 1 through the replacement of the first singular vectors by the n-optimal one (where it really happens) leads to a significant disturbance of the first singular vectors u_1 and v_1, as well as the matrices U,V. Such disturbance of basic vectors of singular value decomposition gives the decrease of DI file size after saving in JPEG, but it may still lead to significant visual distortions of the image, which was confirmed by experiments in (Narimanova 2015).

For the possibility of qualitative DI processing (maintaining a high level of perception reliability), it is necessary to reduce the disturbance of the first singular vectors u_1 and v_1 during DI preprocessing.

For this purpose it is offered to select those first singular vectors that deviate from n-optimal vector (let us denote such a set by T) and carry out the research of this set.

At first glance, it seems that from the set T we can select one vector that is repeated more frequently than the others (let us denote it $\bar{t}$) and replace all vectors by this vector. However, even by the example of the two-dimensional space, it is clear that, since, obviously, $\bar{t}$ is not equal to n-optimal vector, $\bar{t}$ is deflected in the direction of the vector $e_1 = (1,0)$ or the vector $e_2 = (0,1)$. For simplicity, let us assume that $\bar{t}$ deviates from n-optimal vector in the direction of the vector e_1. Then those vectors that are deflected in the direction of the vector e_2 during the replacement by $\bar{t}$ apparently undergo larger perturbations than during the replacement with n-optimal vector. So, such a direct selection of the most frequently repeated vector $\bar{t}$ and replacement of all vectors can only reduce the level of DI visual perception.

Provided these arguments we come to the necessity of clustering vectors of the set T,

124

followed by the allocation of the most frequently repeated vectors $\bar{t}_i$ for each cluster i.

Clustering algorithm.

Step 1. Split the DI matrix I_0 into blocks B by $n \times n$ pixels;

Step 2. Apply the normal singular value decomposition (1) to each block B;

Step 3. Find the angle between the first singular vectors u_1 and all the basis vectors $e_1,...,e_n$ in R_n. Form a set T excluded from consideration of the vector equal to n-optimal.

Step 4. Determine

$$[z,k] = \min_{i=1}^{n}(\arccos(u_1,e_i)),$$

where z is the smallest angle of deviation u_1 from the basis vector e_k; k is a number of basis vector that defines the accessory to a cluster.

Clustering can be done for any n. To illustrate, Figure. 5 shows an example of set T clustering for $n = 3$.

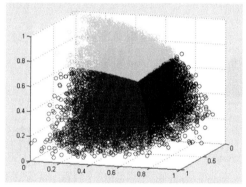

Figure 5. The clusterization of first singular vectors for blocks of 3×3 pixels.

After the clustering, all of the first singular vectors different from n-optimal are divided into n clusters. Now it is necessary to find in each cluster such a vector $\bar{t}_i, i = 1, n$, which is the most frequently repeated one. Let us recall, that it is necessary to replace all of the vectors of the cluster.

To find vectors $\bar{t}_i, i = 1, n$ let us construct the histograms of deviation angles of the first singular vectors from the n-optimal one. The histograms

can help to find the most frequent deviation angle $\bar{\alpha}_i, i = 1, n$, as well as vector $\bar{t}_i, i = 1, n$, which can be obtained by deflection of n-optimal vector in the direction of e_i on $\bar{\alpha}_i$.

To find the vector $\bar{t}_k$, let us denote

$$ON = \left(\frac{1}{\sqrt{n}},...,\frac{1}{\sqrt{n}}\right), \quad e_k = OE = \begin{cases} e_{k_i} = 0, i \neq k; \\ e_{k_i} = 1, i = k. \end{cases}$$

Then vectors ON and OE belong to the same plane, forming a triangle ΔNOE. Construct vector OK in the plane between the vectors ON and OE. Point K is on the side NE of ΔNOE:

$$\angle NOE = \arccos\left(\frac{1}{\sqrt{n}}\right),$$

$$\angle NOK = \bar{\alpha}_i, i = 1, n,$$

$$\angle KOE = \angle NOE - \angle NOK.$$

Thus, the following relation holds:

$$\lambda = \frac{NK}{KE} = \frac{\dfrac{\angle NOK \cdot 100}{\angle NOE}}{\dfrac{\angle KOE \cdot 100}{\angle NOE}} = \frac{\angle NOK}{\angle KOE} =$$

$$= \frac{\angle NOK}{\angle NOE - \angle NOK} = \frac{\angle NOK}{\arccos\left(\dfrac{1}{\sqrt{n}}\right) - \angle NOK}.$$

Therefore, vector $\bar{t}_k$ can be defined as follows:

$$\bar{t}_k = OK = \frac{ON + \lambda OE}{1 + \lambda}. \tag{2}$$

To illustrate the obtained results, Figure 6 shows histograms of the deviation angles of the first singular vectors from the n-optimal for the partition by blocks 4×4. Histograms are shown with different detailing from 0 to $180°$ with increments of $1°$ [0: 1: 180], from 0 to $5°$ with increments of $0.1°$ [0: 0.1: 5], from 0 to $1°$ with increments of $0.01°$ [0 0.01: 1].

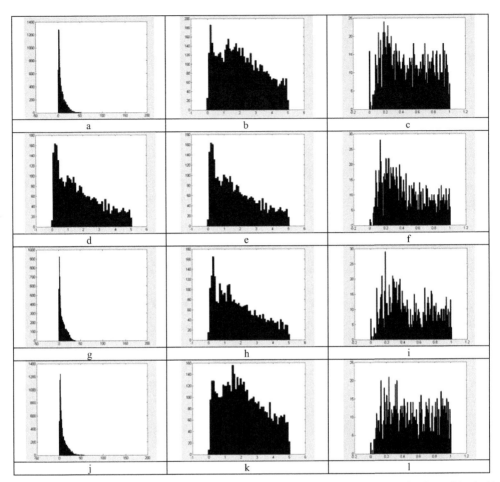

Figure 6. histograms of the deviation angles of the first singular vectors from the n -optimal for the partition by blocks 4×4 : for cluster 1 [0:1:180] (a), [0:0.1:5] (b), [0:0.01:1] (c); for cluster 2 [0:1:180] (d), [0:0.1:5] (e), [0:0.01:1] (f); for cluster 3 [0:1:180] (g), [0:0.1:5] (h), [0:0.01:1] (i); for cluster 4 [0:1:180] (j), [0:0.1:5] (k), [0:0.01:1] (l).

Computational experiments are conducted with the participation of 550 DI with the partition into blocks of different sizes. The results of clustering and analysis of deviation angles for different values n shows that the most frequent deviation angles are equal for all clusters (but still different for different n). Thus, in splitting DI matrix into blocks of 4×4 pixels (or which is the same in the construction of 4 clusters) the replacement of the first singular vectors should be done by the vector which deviates from n -optimal vector by 0.2°. For splitting into blocks of 8×8 pixels this value is 0.4°, and for 16×16 pixels it is 0.8°.

Based on the retrieved theoretical and experimental results it is possible to formulate the basic steps of the modified Algorithm 1.

Algorithm 2.

Step 1. Split the DI matrix I into blocks B by $n \times n$ pixels;

Step 2. Apply the normal singular value decomposition (1) to each block B ;

Step 3. Consider the left singular basis U ;

Step 4. Consider the first singular vector u_1 :

Step 4.1. Determine the deviation angle α_1 of u_1 from the n -optimal vector;

Step 4.2. Determine the deviation angles β_i of u_1 from $e_1,...,e_n$ in R_n ;

Step 4.3. If $\alpha_1 = 0$, i.e. vector u_1 coincides with n -optimal, then the u_1 is not subjected to modification;

Step 4.4. If $\alpha_1 \neq 0$, i.e. vector u_1 differs from n - optimal, then u_1 has to be modified:

126

Step 4.4.1. Determine

$$[z,k] = \min_{i=1}^{n}\left(\arccos\left(u_{1i}\right)\right),$$

where z is the smallest angle of deviation u_1 from the basis vector e_k; k is a number of basis vector that defines the accessory to a cluster;

Step 4.4.2. Construct a new vector $\overline{u}_1$ according to (2);

Step 5. Replace the first singular vector u_1 by $\overline{u}_1$;

Step 6. Use the modified Gram-Schmidt algorithm for the orthogonalization and get U_2;

Step 7. Consider the right singular basis V;

Step 8. Consider the first singular vector v_1;

Step 9. Replace the first singular vector v_1 by $\overline{v}_1$ (repeat step 4, instead u_1 consider v_1);

Step 10. Use the modified Gram-Schmidt algorithm for the orthogonalization and get V_2;

Step 11. Recover the image block $B_2 = U_2 S V_2^T$;

Step 12. Recover the image I_2 from blocks B_2;

Step 13. Save I_2 in JPEG with quality coefficient Q.

Based on the developed DI preprocessing algorithm for compression in JPEG (Algorithm 2), a computational experiment is carried out. The results are shown in Fig. 7 — 10. The original DI is saved in JPEG with a variety of quality coefficient values (from 90 to 60%) both without any preprocessing and with preprocessing according to the algorithms 1 and 2. At the same time, the percentage of the DI file size after use of Algorithms 1 and 2 from the DI file size without performing preprocessing is calculated and fixed.

It is noteworthy that the size of DI file without performing preprocessing for any value of quality coefficient is considered as 100%. Thus, the results presented in Figures 7 — 10 show the ratio of DI file size not to the original one (without compression), but to the DI file size after compression in JPEG with different compression ratios. So, if the value of portion of JPEG image size with some quality coefficient Q after preprocessing is lower than 100%, it means that such preprocessing gives better compression (since the file size is lower) then JPEG with this quality coefficient. Hence, to obtain such small image file size without using preprocessing, this DI should be compressed in JPEG with lower value of quality coefficient. Such a recompression leads to degradation of image visual perception.

The results in Figures 7 — 10 show excellence of both proposed algorithms, but Algorithm 2 shows better compression results compared to Algorithm 1 and JPEG without preprocessing.

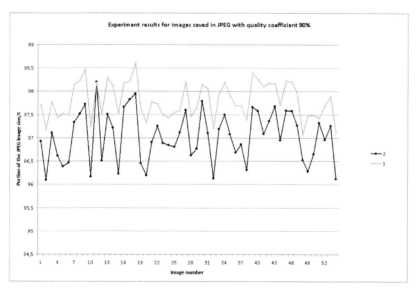

Figure 7. Portion of JPEG image sizes, quality coefficient is 90%: 1 — after preprocessing using Algorithm 1; 2 — after preprocessing using Algorithm 2.

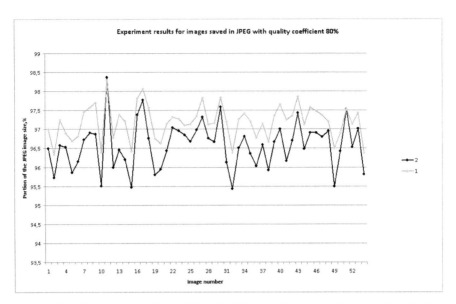

Figure 8. Portion of JPEG image sizes, quality coefficient is 80%: 1 — after preprocessing using Algorithm 1; 2 — after preprocessing using Algorithm 2.

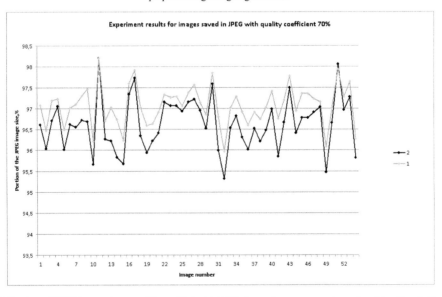

Figure 9. Portion of JPEG image sizes, quality coefficient is 70%: 1 — after preprocessing using Algorithm 1; 2 — after preprocessing using Algorithm 2.

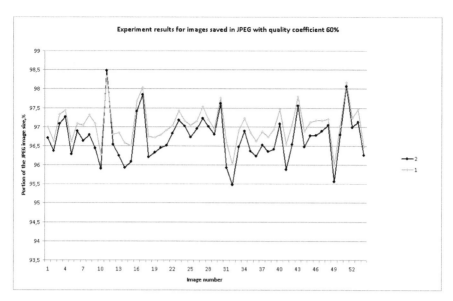

Figure 10. Portion of JPEG image sizes, quality coefficient is 60%: 1 — after preprocessing using Algorithm 1; 2 — after preprocessing using Algorithm 2.

The results of computational experiment in Fig. 7 — 10 show the effectiveness of using Algorithm 2 in order to reduce the file size of DI stored in JPEG. So when using the preprocessing according to the Algorithm 2 the DI file size can be reduced up to 2 — 6% compared to JPEG format with the same quality without preprocessing.

4 CONCLUSIONS

The paper studies the statistical properties of the parameters of the normal singular value decomposition of the DI matrix blocks, in particular — the analysis of the deviation angle of the first singular vectors of the DI matrix blocks from n - optimal vector is carried out. The results of computational experiments and analysis of detected properties allowed to develop a new method of DI preprocessing for compression in JPEG.

Comparative analysis of the work of JPEG compression algorithm with usage of the developed algorithms demonstrates the feasibility of using Algorithm 2, which allows to achieve lower (from 98 to 94% of the DI file size after compression JPEG) DI file sizes at high quality coefficients in JPEG.

REFERENCES

Gnatushenko V.V. 2003. *The use of geometrical methods in multispectral image processing.* Journal of Automation and Information Sciences, Vol. 35, Issue 12.

Hnatushenko V.V., Hnatushenko Vik.V., Kavats O.O., Shevchenko V.Yu. 2015. *Pansharpening technology of high resolution multispectral and panchromatic satellite images.* Scientific Bulletin of National Mining University, Issue 4: 91-98.

Narimanova O, Trifonova K., Kilin A., Kuchma M. 2014. *Technique of Quantitative Assessment of Digital Image Perception Reliability* (in Russian). Informatics and Mathematical Methods in Simulation, Issue 4: 332-336.

Salomon D. 2002. *A Guide to Data Compression Methods.* Springer New York: 295.

Gonzalez R.C., Woods R.E. 2002. *Digital Image Processing, 2nd Edition.* Prentice Hall: 793.

Narimanova O.V. 2011. *Verification of digital signal integrity* (in Russian). Donetsk: Tsifrovaya tipografiya: 180.

Narimanova O.V., Agadzhanyan A.R., Kuchma M.S. 2015. *Preprocessing of digital image for compression.* Informatics and Mathematical Methods in Simulation, Issue 3: 209-216.

Demmel J.W. 1997. *Applied numerical Linear Algebra.* SIAM: 424.

Horn R.A., Johnson C.R. 2012. *Matrix Analysis, 2nd Edition.* Cambridge University Press: 662.

Parlett B.N. 1980. *The symmetric eigenvalue problem.* Prentice-Hall Inc., Englewood Cliffs: 348.

Kobozeva A., Melnik M. 2013. *Sensitivity Analysis of Singular Vectors of Cover Matrix as Basis of Steganography Algorithm, That Steady to Lossy Compression* (in Russian). Information Security, Issue 2: 49-58.

Golub G.H., Van Loan C.F. 2013. *Matrix Computations, 4th Edition.* Baltimore: Johns Hopkins University Press: 756.

Power Engineering and Information Technologies In Technical Objects Control – Pivnyak, Beshta
& Alekseyev (eds)
© 2016 Taylor & Francis Group, London, ISBN 978-1-138-71479-3

The new computational schemes for time series change-point risk estimation and their usage in the foreign exchange market trading

O. Lutsenko, O. Baybuz
Dnipropetrovsk National University named after Oles Honchar, Dnipro, Ukraine

ABSTRACT: The computational schemes of change-point risk functions for the time series of foreign exchange market quotations has been proposed. The way of using the proposed computational schemes in financial trading has been described. The simulation models of trading systems has been built to determine the impact of proposed schemes on the trading process efficiency. The simulation on the history of market quotations has been conducted.

1 INTRODUCTION

The recent tendencies of financial transactions volume growing, increasing of the number of market players, increasing of an information distribution speed and reducing of a time window for decision making on the foreign exchange market lead to a considerable increase of the demand for automated technology for market technical analysis that will allow monitoring and minimizing trading risks and provide a decrease of financial losses.

However, the study of the current state of mathematical methods and software of decision making support in the financial trading sphere shows some problems that, for today, could not find its full solution and are not reflected in the current software. Probabilistic analysis of processes with variable trend and assessment of time series change-point risk in an interval of time in the future can be examples of such problems.

Due to these, developing of the information technology for probabilistic forecasting of the foreign exchange market state, which will be based on the latest scientific developments in the field of statistical and probabilistic analysis of time series, is an actual problem.

2 FORMULATION OF THE PROBLEM

In their previous works, (especially in Lutsenko & Baybuz 2013) the authors of this article described the probabilistic model of the foreign exchange market, which was based on the assumption that the time of a change-point appearance on market quotations time series and the difference of quotations price in neighboring change-points is a vector random value, the density function of which is slowly changing over time. In Lutsenko & Baybuz 2014, the computational schemes for assessing the risk of change-point appearance on some time interval in the future, as well as for assessing the risk of non exceeding a certain value by the quotation price in the moment of next change-point.

This work contains the description of above mentioned computational schemes usage in the process of foreign exchange market trading and the analysis of their impact on the trading effectiveness.

3 MATERIALS FOR RESEARCH

Let us write the risk functions obtained in Lutsenko & Baybuz 2014, as follows.

Let the probability of change-point appearance during time interval Δt in the future be:

$$P_{cpa} = \frac{P_1}{P_2} \tag{1}$$

$$P_1 = \int_t^{t+\Delta t} p_1(x_s \mid y_s = \int_{-\infty}^{\infty} y_s(\Delta y_s, t) f(\Delta y) d\Delta y) \, dx_s$$

$$P_2 = \int_{t+\Delta t}^{\infty} p_1(x_s \mid y_s = \int_{-\infty}^{\infty} y_s(\Delta y, t) f(\Delta y) d\Delta y) \, dx_s$$

where t – the current moment of time, x_s – the axis of time passed since the last change-point observation, y_s – the axis of price increments since the last change-point observation, Δy is an increment of price during one time interval and is a random value. Since y_s can be written as a function of its own random increments Δy and time t, the part

$$y_s = \int_{-\infty}^{\infty} y_s(\Delta y, t) f(\Delta y) d\Delta y$$ is the math estimation of

the resulting random function y_s.

The probability of non-exceeding a certain value by the price during a trend of known direction is:

$$
P_{ne} = \begin{cases} \dfrac{\displaystyle\int_{t'y_{cp}}^{\infty}\displaystyle\int^{\infty} p(x_s,y_s)dydx}{\displaystyle\int_{t'y_{st}}^{\infty}\displaystyle\int^{\infty} p(x_s,y_s)dydx}, & y_{st} < y_{cp}, \\[20pt] \dfrac{\displaystyle\int_{t'-\infty}^{\infty y_{cp}}\displaystyle\int p(x_s,y_s)dydx}{\displaystyle\int_{t'-\infty}^{\infty y_{st}}\displaystyle\int p(x_s,y_s)dydx}, & y_{st} > y_{cp}, \end{cases}
\tag{2}
$$

where y_{st} – the current value of price increment since the last change-point observation, y_{cp} – the price level of the estimated change-point. The condition $y_{st} < y_{cp}$ corresponds to a buy deal, since opening a buy deal implies that the price moves up and the estimated change-point will appear above the current price level; the condition $y_{st} > y_{cp}$ corresponds to a sell deal.

Using of the scheme (1) in trading is quite obvious: the value of risk can serve as an additional condition when closing transactions. Setting the threshold value, beyond which further trading is too risky is required from a trader. The scheme (2) has a more complex application, which will be described in details below.

While trading indicators define points of opening and closing, the choice of quantitative properties of a deal, such as values of take profit (the price level at which the deal will be closed with profit) and stop loss (the price level at which the deal will be closed with loss to prevent further losses), usually remains to the trader. Exceptions are some systems based on the Japanese candlestick charts or bars charts (numerical representations of market quotations time series where each point contains additional information of time interval: the minimum and the maximum prices, as well as the prices at the beginning and the end of period), but not all the systems work this way.

For the vast majority of trade systems the methodology of take profit and stop loss selection is of recommendation nature, unclear or absent, reducing to usage of common rules, which are non-specific to a particular trading system.

To formalize the choice of take profit value from probabilistic point of view the following functions which follow from (2) are proposed:

1) maximizing the mathematical estimation of profit from the deal:

$$takeprofit =$$
$$= \underset{y_{cp} \in (y_{st};\infty)}{\arg\max} \begin{cases} (y_{cp} - y_{st})P_{ne}, y_{st} < y_{cp} \\ (y_{st} - y_{cp})P_{ne}, y_{st} > y_{cp} \end{cases}.
\tag{3}$$

The function (3) is better for getting big profits, but from more rare deals.

2) maximizing the mathematical estimation of the next change-point price level:

$$takeprofit =$$
$$= \underset{y_{cp} \in (y_{st};y_{max})}{\arg\max} \begin{cases} (y_{cp})P_{ne}, y_{st} < y_{cp} \\ (-y_{cp})P_{ne}, y_{st} > y_{cp} \end{cases}.
\tag{4}$$

The function (4) improves win/loss ratio, but with less maximum profit from deals then in the previous case.

To incorporate the previously developed model and the computational schemes into the process of foreign exchange market trading, the information system for market analysis was developed. The system contains tools for change-point detection on the time series of market quotes, probability of density estimation and computing risk functions. The system allows to assess risk functions given above either in real time or in the simulation mode, on market historical data. The simulation can take place with the use of the proposed methods and/or in control experiment mode, with constant values of take profit and stop loss.

The series of tests on the four different trading systems (system based on MACD indicator; Diver system based on the combination of MACD, Stochastic and commodity channel index; a system based on the RSI indicator; Bagovino system based on RSI combined with exponential average) was conducted to study the change of their effectiveness before and after integrating the proposed functions of risk to their structure.

As the criteria of trading system efficiency for simulation the following values were used:

1) maximum intraday drawdown – the maximum difference between local maximum and minimum of the profit time series;

2) profit factor:

$$PF = \frac{S_p}{S_l},$$

where S_p – sum of all profits from profitable deals, S_l – sum of all losses from non profitable deals.

3) average profit and average loss from single operation;

4) quantity of operations with profit and with loss.

The simulation was conducted on the stock market historical data from the last 10 years, on time sets of 1 hour and 6 hours. For the time set of 2 minutes the time retrospective size of 3 months was considered sufficient due to transition to small scale trends.

The functions (3) and (4) were used for optimal setting of take profit and stop loss during testing (the (4) function was used for MACD, Diver and RSI, while the function (3) shown better results for Bagovino), and the function (2) was used as an additional condition for closing deals.

Constant values of take profit and stop loss for the control experiments were obtained by averaging the values of take profit and stop loss used in probabilistic optimization mode testing. Thus,

testing was conducted on values of the same order, and the only difference is the presence of their dynamic adjustment in the case of probabilistic models.

Given that the goal of this study is to check the efficiency of the probabilistic model for the integration to existing trading systems and not checking the effectiveness of trading systems themselves, no fine-tune of the indicators parameters (periods of moving averages, etc) was conducted. These parameters were set to recommended values listed by the authors of indicators or recommended in literature, particularly (Welles Wilder J. 1978, DeMark T.R. 1994, Kahn. M.N. 2010).

Table 1 shows the results of simulation in the form of relative changes of efficiency criteria for each system averaged by tested currency pairs. Absolute values in Table 1 are omitted: different currency pairs are values of different order, so averaging of their absolute values have no sense. The summary profit relative growth is not included in the table because of its tendency of taking negative values if the sign of resulting profit changes. The increase or decrease in the profit can be measured by the change of profit factor.

Table 1. The results of trading systems testing with trailing stop enabled – relative change in %

	Total number of deals	Profit factor	MIDD	Average profit	Average loss	Number of profitable deals	Number of deals with loss
2 minutes							
MACD	-9,0950	14,3434	-0,1340	6,7055	52,4787	19,1974	-25,1510
Diver	-7,7821	8,1515	-14,0780	21,2806	11,1383	-0,8929	0,0000
Bagovino	2,3707	18,6175	-31,5410	-25,5600	39,3342	41,5954	-36,1440
RSI	-8,8170	6,2566	21,6740	15,3201	20,9140	-3,7743	-13,5840
1 hour							
MACD	-0,9360	24,7560	-32,5800	5,6450	-25,1300	-7,2590	5,1587
Diver	-16,3200	10,3230	-24,5000	74,3990	-2,6680	-36,7300	2,9390
Bagovino	1,7516	7,4120	-29,3700	-27,1300	26,5600	28,2660	-29,4400
RSI	-20,2100	0,4222	24,0070	27,8200	43,6220	-16,9100	-23,7100
6 hours							
MACD	-3,0451	25,4483	-19,0668	5,7140	22,8304	19,2577	-17,8425
Diver	-2,8503	47,3533	-39,2438	27,9338	-9,5116	1,6082	-3,1915
Bagovino	3,5901	14,3854	0,5360	-29,2095	25,2552	32,5475	-34,4444
RSI	-8,4511	3,9604	48,0511	6,1441	25,7958	1,1062	-17,6025

Table 1 shows that the use of schemes (3) and (4) for the take profit and stop loss values optimization leads to higher profits, particularly because of improving the profitable to unprofitable operations ratio. Depending on the method and the time scale it occurs either in the form of an increase in the number of profitable deals with a reduction of number of unprofitable, or in a reduction of the number of both profitable and unprofitable deals, but with a more significant reduction of the latter. The average loss of one transaction in some cases are increased, which, however, did not outweighs the positive effects of other factors. The result of it is an increase of profit, which is reflected in the profit factor increase.

As an accompanying effect in most cases there is a slight but regular decrease in the number of concluded deals. This is due to an increase of the deals duration as with the probabilistic optimization some optimized deals get high take profit which takes more time to reach.

The value of the maximum intraday drawdown (MIDD) can deviate in both directions: on the one hand, minimizing losses quantity leads to a decrease in drawdowns; on the other hand, in some cases the value of the maximum drawdown can be larger than in control test, even when the resulting profit is higher, due to the increase of peak profits.

A typical example of the chart of profit dynamic obtained from simulation is shown in the Figure 1.

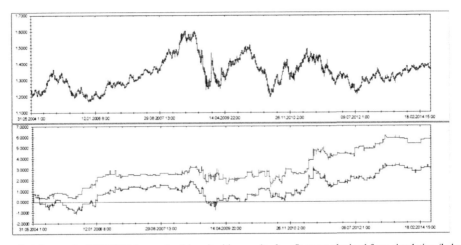

Figure 1. A fragment of EURUSD time series (above) with a result of profit curve obtained from simulation (below)

In the tests above the trading systems operate in a mode which is close to real trading, using the indicator signals, as well as the exceeding of take profit and stop loss levels, for closing deals. The method of trailing stop (the technique of trade when stop loss adjusts itself when market trend moves away from it) is also used.

Unlike the first series of tests, in the second series of tests, the results of which are shown in the Table 2, deals are closed only when the price exceeds take profit and stop loss levels which were chosen when opening deal. Trailing stop in this case is disabled, and closing signals of indicators are ignored.

Table 2. The results of trading systems testing with trailing stop disabled – relative change in %

	Total number of deals	Profit factor	MIDD	Average profit	Average loss	Number of profitable deals	Number of deals with loss
2 minutes							
MACD	-18,3770	61,6616	-17,8450	9,7736	-3,3217	5,3922	-26,2780
Diver	-16,1020	43,4769	-39,0240	2,6215	9,7447	13,3333	-26,1360
Bagovino	-16,8160	7,8185	-22,8750	-12,6020	5,8155	-1,5126	-24,2370
RSI	-28,1610	23,6682	1,5630	-17,0200	2,1665	-4,3883	-38,0660
1 hour							
MACD	-32,9300	20,8460	45,9430	257,0600	115,4700	-34,0300	-30,7100

Continuation of Table 2.

	Total number of deals	Profit factor	MIDD	Average profit	Average loss	Number of profitable deals	Number of deals with loss
Diver	-11,9300	35,7910	-45,1900	-1,5560	-0,1970	8,5659	-20,3100
Bagovino	-13,9000	9,7619	-20,5400	-19,7800	12,0950	14,4230	-25,0900
RSI	-6,8270	11,3450	-28,4100	-7,5710	0,8098	6,6524	-12,1800
6 hours							
MACD	6,9804	19,9499	6,9685	-4,6417	7,1334	29,9242	-3,8816
Diver	-6,7805	63,2319	-41,3767	0,0255	-0,8622	25,6494	-21,5933
Bagovino	-3,7823	10,4241	-1,9845	-8,7351	11,1781	13,3776	-16,0256
RSI	-31,3514	30,0889	1477,0440	-5,1102	-2,8919	-16,6667	-37,4046

Table 2 shows that the results of the methods which use probabilistic optimization of take profit and stop loss show a significant advantage over the control methods results. Results of individual criteria are similar to the first group of tests, with the efficiency difference from the control methods even more significant. For example, the reduction of the number of deals reaches 30%.

Trading without trailing stop, unlike the methods simulated in the first group of tests, can not be recommended for actual use in market due to high risk and the total absence of insuring techniques, but the results obtained from this simulation show the positive impact of the proposed computational schemes on the effectiveness of trade in conditions of absence of all other settings and factors which could have an impact on simulated systems efficiency.

4 CONCLUSIONS

The efficiency of integrating the earlier proposed computational schemes of risk assessment to the automated trading systems was analyzed.

The simulation of 4 different trading systems, based on systems MACD, Diver, RSI level close and Bagovino on three time scales: 2 minutes, 1 hour and 6 hours was conducted.

Data obtained in the series of tests, in which the logic of trading systems was identical to that used in the actual real life trading, shown an increase in the efficiency of trading systems obtained compared with the same systems, but without the use of probabilistic optimization with the proposed computational schemes.

To ascertain the impact of the proposed computational schemes on the trading efficiency in the condition of absence of influence of all settings

except that which where optimized, the second series of tests was conducted, in which the mechanisms of trailing stop, as well as ability of systems to use signals of economical indicators to close a deal were disabled, so that the effectiveness of tested trading systems was dependent only on optimized values. The results obtained during this series of tests confirmed the previously obtained results.

The experiments have shown that the proposed probabilistic model is adequate, and the computational schemes of risk functions obtained on its basis can be effectively utilized in the process of trade in the financial market.

An information decision support system based on the proposed methods of solving the problem of risk assessment when trading on the financial market was developed. The system allows its user to calculate the optimal take profit and stop loss values for newly opened trades for both the current time and at any time of history loaded.

REFERENCES

Lutsenko O. & Baybuz O. 2013. *Model of probabilistic assessment of trend stability at financial market.* Kharkiv: Eastern European Journal of Enterprise Technologies, Vol 6, Issue 3(66): 50-54

Lutsenko O. & Baybuz O. 2014. *Information system of analysing risk functions during the financial trade* (in Ukrainian). Dnipropetrovsk: Actual Problems of Automation and Information Technology, Vol. 18: 42-51

DeMark T.R. 1994. *The New Science of Technical Analysis.* John Wiley: 247.

Kahn. M.N. 2010. *Technical Analysis Plain and Simple: Charting the Markets in Your Language (3rd Edition).* FT Press: 352.

Welles Wilder J. 1978. *New Concepts in Technical Trading Systems.* Trend Research: 142.

Improved algorithm for detecting and removing shadows in multichannel satellite images with high information content

V. Hnatushenko
Dnipropetrovsk National University named after Oles Honchar, Dnipro, Ukraine

O. Kavats & Yu. Kavats
National metallurgical academy of Ukraine, Dnipro, Ukraine,

ABSTRACT: The main goal of the immediate monitoring based on the remote sensing data is detection of changes in different natural and anthropogenic processes and objects. Monitoring bases on remote sensing data is irreplaceable in observation of dangerous natural phenomena and catastrophes, investigating territories, which are difficult to access, determination of unauthorized construction works, etc. Immediacy and quality of monitoring in the conditions of town infrastructure directly depends on spatial resolution of the satellite data, as well as on the presence of additional interferences, such as shadows of tall buildings and others. However, to extract shadows, the automated detection of shadows from images must be accurate. In this paper, an improved algorithm for detecting and compensating shadows with the use of the infrared channel is proposed. The results after the evaluation on the satellite images exhibit that the recovered shadow regions are consistent with their neighboring nonshadow region.

1 INTRODUCTION

Images of modern remote-sensing systems allow us to solve various problems such as carrying out operational monitoring of land resources, city building, monitoring state of the environment and influence of anthropogenic factors, detecting contaminated territories, unauthorized buildings, estimating the state of forest plantation and other (Gnatushenko 2003; Schowengerdt 2007). Remote sensing data can be used in circumstances where it is impossible terrestrial research methods, such as fires and floods. Development of this problematics is stipulated by the advantages intrinsic to the digital multichannel images obtained by scanning systems of modern space vehicles:

1) observability of the scene allowing us to analyze big territories;
2) possibility of investigating territories that are difficult to access;
3) fast obtaining of the information;
4) obtaining of information of high spatial resolution;
5) high credibility of the obtained information, and so on.

The problem of monitoring is closely related to solving of the following problems:

1. Improvement of quality of the source multichannel image with the use of technologies of improvement of the information capacity.
2. Application of algorithms for detection and compensation of shadows in multichannel images.
3. Classification of objects of the town infrastructure.

Let us consider the above mentioned stages in more detail. Source multichannel satellite images are characterized by low spatial resolution, for this reason, the task of the first stage is to increase their information capacity.

In satellite images, the lower spatial resolution multispectral images are fused with higher spatial resolution panchromatic images. To receive a synthesized image with the best value of information content it is necessary to choose the most effective fusion method because it affects the thematic problem solving. The quality of the fusion method is determined by the increase of spatial resolution and the existence (absence) of color distortion. Numerous works related to preprocessing of multichannel digital images focus on improving their visual quality excluding physical mechanisms of fixing information, especially, correlation between channels, making it impossible to determine the image information content from the standpoint of analysis and interpretation. In paper

(Hnatushenko 2013) a new information technology based on ICA- and wavelet transformations is proposed, which can significantly increase the information content of the primary data and does not lead to color distortion.

Space images of the town infrastructure, as a rule, contain many shadows cast by tall objects (buildings, bridges, towers, etc.) when the scene is lit by the sun. Shadows are a substantial obstacle in recognition of objects because they make it difficult to determine boundaries of those objects. On the other hand, a shadow is a deciphering feature making it possible to learn a lot about the shape, location and other properties of an object casting it. In object recognition, an important step is detection and compensation of shadows, as well as reconstruction of the scene in the shadowy area.

At the moment, there are classical algorithms of image processing for detection and selection of shadows, namely, contour, cluster, and frequency algorithms (Huang 2004; Chung 2009; Chen 2014; Azevedo 2015). A variety of image enhancement methods have been proposed for shadow removal, such as histogram matching, gamma correction, linear correlation correction and restoration of the colour invariance model (Ma 2008; Singh 2012; Tiwari 2015). The main shortcoming of these algorithms is formation of false outlines, which is caused by the substantial difference of brightness of the shadow and surrounding background and, subsequently, by a sharp change of brightness at the boundary of the shadow with the background and the object, which substantially complicates the stage of reconstruction of objects in the shadowy area (Tsai 2006; Kulkarni 2015).

2 FORMULATION THE PROBLEM

There is a necessity to improve the algorithm of detection and compensation of shadows in the multichannel satellite images.

3 THE BASIC MATERIAL

In this work, an improvement of the algorithm for detection and compensation of shadows, the foundations of which include:

1. The preliminary processing stage, increase of the information capability on the base of ICA and Wavelet transforms (Hnatushenko 2013).

2. Selection of the vegetation component.

3. Determination of the optimal binarization threshold.

4. Selection of the outlines of the objects.

5. Segmentation of the shadow area.

6. Compensation of the shadow area.

The structure chart of the improved algorithm is represented in Figure 1. The proposed method has been examined with variety of high resolution satellite images obtained under dissimilar illumination conditions in urban areas. The paper investigates satellite images from Worldview-2, the original size of the panchromatic image is 4600x4604 and 1150x1151 multichannel. The multichannel bands (Band1 = Coastal, Band2 = Blue, Band3 = Green, Band4 = Yellow, Band5 = Red, Band6 = Red Edge, Band7 = Near-Infrared 1, Band8 = Near-Infrared 2) cover the spectral range from 400 nm - 1050 nm at a spatial resolution of 1.84 m, while the panchromatic band covers the spectrum from 450 nm – 800 nm with spatial resolution 0.46 m. Figure 2 represents images obtained after each stage of work of the algorithm.

We proposed to select the vegetation component with the use of the infrared channel. As the R channel, we select the IR, second and third (G and B, respectively). This operation is represented by this expression:

$$NDI = \frac{IR - R}{IR + R} \qquad (1)$$

where IR - infrared channel, R - red channel.

Suppose R, G, and B are the red, green, and blue values of a color. The HSI intensity is given by the equation:

$$I = (R + G + B)/3. \qquad (2)$$

Now let m be the minimum value among R, G, and B. The HSI saturation value of a color is given by the equation:

$$S = 1 - m / I \text{ if } I > 0, \text{ or}$$
$$S = 0 \text{ if } I = 0 \qquad (3)$$

To convert a color's overall hue, H, to an angle measure, use the following equations:

$$H = cos^{-1}\left[\frac{(R - \frac{1}{2}G - \frac{1}{2}B)}{\sqrt{R^2 + G^2 + B^2} - G - RB - GB}\right],$$
$$\text{if } G \geq B, \text{ or} \qquad (4)$$

$$H = 360 - cos^{-1}\left[\frac{(R - \frac{1}{2}G - \frac{1}{2}B)}{\sqrt{R^2 + G^2 + B^2} - RG - RB - GB}\right],$$
$$\text{if } B > G, \qquad (5)$$

where the inverse cosine output is in degrees.

The task of segmentation of the image is implemented on the base of the two-level hierarchical pyramidal algorithm, which allows us to use various color and texture difference of areas.

As the criterion of homogeneity, the evaluation of closeness of the elements and areas of the image in the combined texture and color space of features. In the process of segmenting, the image transformed into HSI color model underwent binarization with selection of the optimal threshold. As a result of binarization, shadow areas were set to 1, and areas without shadows were set to 0 (Figure 2 (e)). Then the mask is applied to the obtained image, and we get segmented of shadow areas.

On the next stage of the algorithm, we carry out the operation of subtraction of the vegetation component out of the shadow one, which allows us to leave out only shadowy areas in the image.

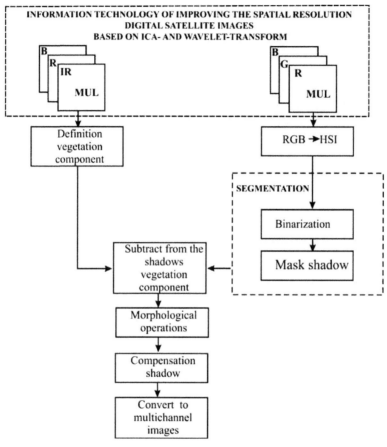

Figure 1. Algorithm scheme

Compensation of shadow pixels is done owing to the increase of the brightness component for those pixels, the brightness component of which is below the threshold value (*Tiwari 2015*).

Then we convert the obtained image from the HSI color space back into RGB. To convert hue, saturation, and intensity to a set of red, green, and blue values, you must first note the value of H. If $H = 0$, then R, G, and B are given by

$$R = I + 2IS$$
$$G = I - IS \qquad (6)$$
$$B = I - IS.$$

If $0 < H < 120$, then

$$R = I + IS * \cos(H) / \cos(60 - H)$$
$$G = I + IS * \left[1 - \cos(H) / \cos(60 - H)\right] \qquad (7)$$
$$B = I - IS.$$

If $H = 120$ then the red, green, and blue values are

$$R = I - IS$$
$$G = I + 2IS \qquad (8)$$
$$B = I - IS.$$

139

If $120 < H < 240$, then

$R = I - IS$
$G = I + IS * cos(H - 120) / cos(180 - H)$ 　　(9)
$B = I + IS * \left[1 - cos(H - 120) / cos(180 - H) \right]$.

If $H = 240$ then

$R = I - IS$
$G = I - IS$ 　　　　(10)
$B = I + 2IS$

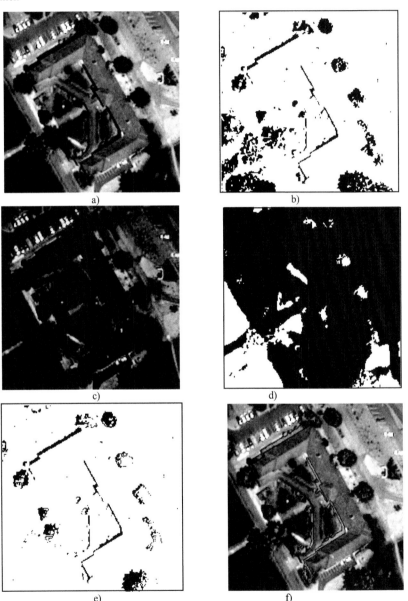

Figure 2. Algorithm scheme: a) original image (R G B); b) binary image showing shadows;
c) original image (IR R B); d) binary image showing vegetation; e) subtraction of vegetation component from shadow;
f) result image

And if $240 < H < 360$, we have

$$R = I + IS*\left[1 - cos(H - 240)/cos(300 - H)\right]$$
$$G = I - IS \qquad\qquad (11)$$
$$B = I + IS*cos(H - 240)/cos(300 - H).$$

4 CONCLUSIONS

In the paper, an improved algorithm is proposed for detection and compensation of shadows in multichannel satellite images with increased information capacity on the base of ICA- and wavelet transforms. As the source, a multichannel image with an IR channel is used, which makes it possible to reliably determine the vegetation component, as well as shadows cast by the vegetation. In order to get a shadow detection result, image segmentation considering shadows is applied. The shadow removal method based on brightness component matching can effectively restore the information in a shadow area. The parameters calculated by using the radiation difference between inner and outer homogeneous sections can retrieve a shadow very effectively. The results after the evaluation on the satellite images exhibit that the recovered shadow regions are consistent with their neighboring nonshadow region. In the future, we will explore more feature information to estimate more accurate the shadow coefficient, and obtain better result of shadow removal.

REFERENCES

Gnatushenko V.V. 2003. *The use of geometrical methods in multispectral image processing.* Journal of Automation and Information Sciences, Vol. 35, Issue 12.

Schowengerdt R. 2007. *Remote sensing: models and methods for image processing.* New York: Academic Press: 560.

Hnatushenko V., Kavats A. 2013. *Information technology increase spatial fragmentation of digital satellite images based on wavelet transformation and ICA* (in Ukrainian). Lviv: Proceedings of the National University "Lviv Polytechnic" series "Computer Science and Information Technology": 28-32.

Tsai V.J.D. 2006. *A comparative study on shadow compensation of color aerial images in invariant color models.* IEEE Trans. Geosci. Remote Sens., Vol. 44, Issue 6: 1661–1671.

Huang J., Xie W., & Tang L. 2004. *Detection of and compensation for shadows in colored urban aerial images.* Proc. 5th World Congr. Intell.

Control Autom., Hangzhou, China, Jun. 15–19: 3098–3100.

Chung K-L., Lin Y-R., Huang Y-H. 2009. *Efficient shadow detection of color aerial images based on successive thresholding scheme.* IEEE Trans Geosci Remote Sens, Issue 47: 671–682.

Azevedo S.C., Silva E.A., Pedrosa M.M. 2015. *Shadow detection improvement using spectral indices and morphological operators in urban areas in high resolution images.* The International Archives of the Photogrammetry, Remote Sensing and Spatial Information Sciences, Volume XL-7/W3: 587-592.

Ma H., Qin Q. & Shen X. 2008. *Shadow segmentation and compensation in high resolution satellite images.* Proc. IEEE IGARSS, Jul., Vol. 2: 1036–1039.

Hnatushenko V.V., Kavats O.O., Kibukevych I.O. 2015. *Efficiency Determination of Scanner Data Fusion Methods of Space Multispectral Images.* International Young Scientists Forum on Applied Physics «YSF-2015», September 29 - October 2, 2015/Dnipropetrovsk, Ukraine.

Singh K.K., Pal K, Nigam M.J. 2012. *Shadow Detection and Removal from Remote Sensing Images Using NDI and Morphological Operators.* International Journal of Computer Applications (0975 – 8887), Vol. 42, Issue 10: 37-40.

Tiwari S., Chauhan K. & Kurmi Y. 2015. *Shadow detection and compensation in aerial images using Matlab.* International Journal of Computer Applications (0975 – 8887), Vol. 119, Issue 20: 5-9.

Chen D., Shang S., & Wu C. 2014. *Shadow-based Building Detection and Segmentation in High-resolution Remote Sensing Image.* Journal of Multimedia, Vol. 9, Issue 1: 181-188.

Kulkarni S.S., Hingmire K., Kute P., Kusalkar S., Pethe S. 2015. *Survey on Shadow Detection and Reconstruction in VHR Images.* International Journal of Advanced Research in Computer Science and Software Engineering, Vol. 5, Issue 3: 318-320.

Power Engineering and Information Technologies In Technical Objects Control – Pivnyak, Beshta
& Alekseyev (eds)
© 2016 Taylor & Francis Group, London, ISBN 978-1-138-71479-3

The statistical method of effective data structures designing for particular software and hardware environments

V. Shynkarenko, H. Zabula & D. Petin
Dnipropetrovsk National University of Railway Transport, Dnipro, Ukraine

ABSTRACT: The parameters of time efficiency of data structures and the methods of their assessment with the confidence intervals have been developed. The authors have shown the approaches to forming scenarios (the sequences of the data processing operations). The authors have developed the order of parameters determining on the basis of computer experiment. The results of computer experiments compared with a number of the data structure implementations have been given. The obtained objective parameters provide a basis for the automated data structure adaptation.

1 INTRODUCTION

Simple and structured data are differentiated at all the levels of data designing, i.e. at the logical and hardware ones as well as at the level of language implementation (Ziegler 1982).

The indexed array, record, list, tree, stack, deque and queue are the most widely used classical structures (Cherkassky 1999; Knuth 1997; Perevozchikova 2007). At present the associative array, collection and other structures are widespread. These structures are predefined and trivial in programming languages.

The design and development of data structures consist in choosing them from predefined structures or designing more complex ones (Bentley 1982; Cormen 2009). Designing is carried out by putting some structures into other ones, so the elements of structures are structures (i.e. Hoare 1975).

Designing is performed if it is necessary for the solution of the problem, with the solution being done algorithmically. The array of arrays, the list tree, the stack of records, the collection of array trees and so on are the examples of designed structures.

In programming practice, the development of data structures is done intuitively but if there are specific requirements to the time characteristics of the software, the objective efficiency parameters are to be taken into account.

The basis of the design of data structures at the physical level is partially specified at the logical one (Shynkarenko 2009, 2012, 2014).

The data structures are not functional therefore the time efficiency of the data structure is determined by the time efficiency of the processing operations that are implemented in the form of algorithms (Shynkarenko 2001) and can consist of the operations of access as well as changing the order and location of structured elements.

The memory location of the data structures can considerably influence the time efficiency of programs and program systems using structured data. It has been shown (Kaspersky 2003; Shynkarenko 2006, 2009) that the time span of the operation of access to a location can vary by two orders of magnitude depending on the memory location of the data. There are many different approaches and methods of analyzing the algorithm time execution using computational complexity (Chase 1990; Cormen 2009; Galil 1983; Kozen 1992).

2 FORMULATION OF THE PROBLEM

The problem of developing of time efficiency parameters of the data structures and the methods of assessing these parameters experimentally is solved in this paper.

3 THE TIME EFFICIENCY OF THE DATA STRUCTURES

We define the time efficiency of the data structures as the time efficiency of the set of operations (algorithms) of data processing (not including calculations).

Let us specify what is meant by the operations of data processing. Data processing is carried out on the basis of atomic or primitive processing operations. All primitive operations of processing data structures can be divided into:

- operations that do not change the structure and the values of its elements. Such operations are to be strictly abridged with selecting and searching for values and exclude other functions such as determining mean values in a selection and so on;
- operations that are the data structure conversion including:
  - changing the logical structure (for example, substituting the array for the list, adjusting the tree);
  - adding or removing some element(s);
  - operations that change the values of its elements without changing the structure. Such operations are to be restricted entirely to searching for the position (if necessary) and changing the value without any data conversion;
- operations that change the values of the structure elements with the ensuing structure changing, for example, introducing an element into the ordered list.

Let the use of the data structures by a particular application be represented in the form of algorithm $A\vert_X^Y$ in the notation according to (Shynkarenko 2009):

$$A\vert_X^Y = \prod_i (B_i\vert_{X_i}^{Y_i} \cdot C_i\vert_{Z_i}^{Q_i}),$$

where X and Y are the domain and range of algorithm $A\vert_X^Y$, respectively; $B \cdot C$ is the operation of the consecutive execution of algorithm C after algorithm B; B_i are the algorithms of processing the data structure (all of which corresponding to the above mentioned classification); C_i are the algorithms of control, data conversion as well as processing other data and structures. Some of algorithms C_i can be empty.

Then the time efficiency of the data structure can be defined as the time efficiency of algorithm $\overline{A}\vert_X^Y$:

$$\overline{A}\vert_X^Y = \prod_i B_i\vert_{X_i}^{Y_i} \tag{1}$$

4 THE PARAMETERS OF THE TIME EFFICIENCY OF THE DATA STRUCTURES

The time efficiency of two algorithms can be compared by means of the parameter of the degree of superiority of one algorithm A_i over the other

algorithm A_j in bounded subsets

$$\Omega = ((\overline{V} \subseteq V^*),(\overline{U} \subseteq U^*),(\overline{X} \subseteq X^*),$$
$$(\overline{\Psi} \subseteq \Psi^*),(\overline{\mathfrak{R}} \subseteq \mathfrak{R}^*) \qquad \text{(Shynkarenko 2001)}:$$

$$SUP_{ij}\vert \overline{V},\overline{U},\overline{X},\overline{\Psi},\overline{\mathfrak{R}} =$$

$$\frac{1}{N}\sum_{v\in\overline{V}}\sum_{u\in\overline{U}}\sum_{x\in\overline{X}}\sum_{\psi\in\overline{\Psi}}\sum_{r\in\overline{\mathfrak{R}}} \frac{t_j - t_i}{\max(t_j,t_i)} \cdot 100\%,$$

and that of the region of superiority:

$$REG_{ij}\vert \overline{V},\overline{U},\overline{X},\overline{\Psi},\overline{\mathfrak{R}} =$$

$$\frac{1}{N}\sum_{v\in\overline{V}}\sum_{u\in\overline{U}}\sum_{x\in\overline{X}}\sum_{\psi\in\overline{\Psi}}\sum_{r\in\overline{\mathfrak{R}}} sign(t_j - t_i) \cdot 100\%$$

where V^* is a set of possible values of the data capacity, U^* is a set of possible data types, X^* is a set of possible data values, Ψ^* is a set of possible software environments, $\mathfrak{R}^*$ is a set of computer architectures in which the algorithms are expected to be executed, t_i (function $t_i(v,u,x,\psi,r)$) is the time needed to execute algorithm A_i with the definite values $v \in \overline{V}$, $u \in \overline{U}$, $x \in \overline{X}$, $\psi \in \overline{\Psi}$, $r \in \overline{\mathfrak{R}}$,

$$sign(a) = \begin{cases} 1, & \text{if } a \geq 0 \\ 0, & \text{if } a < 0 \end{cases}.$$

Taking into account the fixed data types of the data structure elements, the parameters of time efficiency of the data structure are:

$$SUP_{ij}\vert \overline{V},\overline{X},\overline{\Psi},\overline{\mathfrak{R}} = \tag{2}$$

$$\frac{1}{N}\sum_{v\in\overline{V}}\sum_{u\in\overline{U}}\sum_{x\in\overline{X}}\sum_{\psi\in\overline{\Psi}}\sum_{r\in\overline{\mathfrak{R}}} g_{ij}(v,x,\psi,r) \cdot 100\%$$

$$REG_{ij}\vert \overline{V},\overline{X},\overline{\Psi},\overline{\mathfrak{R}} = \tag{3}$$

$$\frac{1}{N}\sum_{v\in\overline{V}}\sum_{u\in\overline{U}}\sum_{x\in\overline{X}}\sum_{\psi\in\overline{\Psi}}\sum_{r\in\overline{\mathfrak{R}}} sign(g_{ij}) \cdot 100\%,$$

where N is the total number of terms, g_{ij} is the degree of superiority of the ith algorithm over the jth algorithm at the point, which is determined as

144

$$g_{ij} = \frac{t_j(v,x,\psi,r) - t_i(v,x,\psi,r)}{\max(t_j(v,x,\psi,r), t_i(v,x,\psi,r))} \quad (4)$$

where t_i is the time needed to execute the ith algorithm.

5 ASSESSMENT OF THE PARAMETERS OF THE TIME EFFICIENCY OF THE DATA STRUCTURE

Assuming that sets $\overline{V}, \overline{X}, \overline{\Psi}, \overline{\Re}$ are discrete and bounded as they are represented in discrete memory, they can be mapped into the set of integers

$$\overline{V} \to \overline{\overline{V}} = \{1,2, \dots N_1\}, \ \overline{X} \to \overline{\overline{X}} = \{1,2, \dots N_2\},$$

$$\overline{\Psi} \to \overline{\overline{\Psi}} = \{1,2, \dots N_3\}, \ \overline{\Re} \to \overline{\overline{\Re}} = \{1,2, \dots N_4\}$$

} respectively with:

$$\hat{g}_{ij}(\overline{\overline{V}}, \overline{\overline{x}}, \overline{\overline{\psi}}, \overline{\overline{r}}) = g_{ij}(v,x,\psi,r). \quad (5)$$

The assessment of the parameters is done according to the Monte Carlo method. To do that, we proceed from summing to integrating for the step function.

In the range $\widetilde{V} \times \widetilde{X} \times \widetilde{\Psi} \times \widetilde{\Re} = [1 \dots N_1 + 1] \times$

$[1 \dots N_2 + 1] \times [1 \dots N_3 + 1] \times [1 \dots N_4 + 1]$ let us define:

$$\widetilde{g}_{ij}(\widetilde{v}, \widetilde{x}, \widetilde{\psi}, \widetilde{r}) = \hat{g}_{ij}(\overline{\overline{v}}, \overline{\overline{x}}, \overline{\overline{\psi}}, \overline{\overline{r}}), \quad (6)$$

where $\widetilde{v} \in [\overline{\overline{v}}, \overline{\overline{v}} + 1], \ \widetilde{x} \in [\overline{\overline{x}}, \overline{\overline{x}} + 1],$

$\widetilde{\psi} = \overline{\overline{\psi}}, \widetilde{r} = \overline{\overline{r}}$, and substitute the sum for the integral:

$$SUP_{ij} \mid \overline{V}, \overline{X}, \overline{\Psi}, \overline{\Re} =$$

$$\frac{1}{N} \sum_{v \in \overline{V}} \sum_{u \in \overline{U}} \sum_{x \in \overline{X}} \sum_{\psi \in \overline{\Psi}} \sum_{r \in \overline{\Re}} g_{ij}(v,x,\psi,r) = I =$$

$$\frac{1}{N_1 N_2 N_3 N_4} \int\int\int\int \widetilde{g}_{ij} d\widetilde{v} d\widetilde{x} d\widetilde{\psi} d\widetilde{r} \quad (7)$$

In the set Ω let us randomly select $\overline{N}$ points $p = (\widetilde{v}, \widetilde{x}, \widetilde{\psi}, \widetilde{r})$ and define the value of the random variable $\xi = \widetilde{g}_{ij}(\widetilde{v}, \widetilde{x}, \widetilde{\psi}, \widetilde{r})$, with the integral (7) being the mathematical expectation of the random variable ξ, i.e. $M[\xi] = I$.

As we know, the assessment of the mathematical expectation is the value

$$\theta_S = \frac{1}{\overline{N}} \sum_{k=1}^{\overline{N}} \xi_k = \sum_{k=1}^{\overline{N}} g_{ijk}(v,x,\psi,r), \quad (8)$$

where ξ_k is the obtained experimentally k^{th} value of the random variable ξ, $g_{ijk} = (g_{ij})_k$

Then, let us define the S assessment of the degree of superiority of the ith algorithm over the jth algorithm (2) as

$$S_{ij} \mid \overline{V}, \overline{X}, \overline{\Psi}, \overline{\Re} = \theta_S \cdot 100\% = \sum_{k=1}^{\overline{N}} g_{ijk}(v,x,\psi,r). (9)$$

Let us define the assessment of the confidence intervals in the following way (Shynkarenko 2009):

$$\theta_S - \tau_{2,\beta} \sqrt{\frac{1}{6} \sum_{k=1}^{3} (\zeta_k - \theta_S)^2} \cdot 100 <$$

$$S_{ij} \mid \overline{V}, \overline{X}, \overline{\Psi}, \overline{\Re}, \beta <$$

$$\theta_S + \tau_{2,\beta} \sqrt{\frac{1}{6} \sum_{k=1}^{3} (\zeta_k - \theta_S)^2} \cdot 100, \quad (10)$$

where $\zeta_k = \frac{1}{\overline{N}/3} \sum_{i=1+(k-1)\overline{N}/3}^{k\overline{N}/3} \xi_i$.

Thus, we obtain the S- and R- assessments of the time efficiency of the data structures:

$$S_{ij} \mid \overline{V}, \overline{X}, \overline{\Psi}, \overline{\Re} = \sum_{k=1}^{\overline{N}} \frac{\tau_{jk} - \tau_{ik}}{\max(\tau_{jk}, \tau_{ik})} \cdot 100\% \quad (11)$$

$$R_{ij} \mid \overline{V}, \overline{X}, \overline{\Psi}, \overline{\Re} = \sum_{k=1}^{\overline{N}} sign(\tau_{jk} - \tau_{ik}) \cdot 100\%, (12)$$

where τ_{ik} is the time needed to execute the algorithm of processing the ith data structure during the kth testing, with v,x,ψ,r selected randomly,

τ_{jk} is the time needed to execute the algorithm of processing the jth data structure with the same v,x,ψ,r.

6 ASSESSMENT OF THE PARAMETERS OF THE TIME EFFICIENCY OF THE DATA STRUCTURE

To make the measurement of τ_{ik} and τ_{jk}, the modeling of the environment in which the data structure is operated needs to be performed. The best way of doing that is certainly taking the measurements in the soft- and hardware environment in which the structure is operated as part of the software (here the soft- and hardware environment is understood as the computer, operating system and concurrent application programs). There are various approaches to modeling the process of executing the structure, i.e. algorithm $A\,|_X^Y$.

The first approach. Let the quantitative proportion of the operations of data structure processing given by vector

$$N^T = [n_1, n_2, \ldots, n_m], \qquad (14)$$

where n_i is the approximate number of performing the ith basic operation of the data structure processing in a particular operational environment, m is the number of the data structure processing operations of various types be known beforehand. Then, $A\,|_X^Y$ can be represented as:

$$\overline{A}\,|_X^Y = \prod_{p=1}^{m} n_p B_p\,|_{X_p}^{Y_p} \qquad (15)$$

and on the basis of (14) and (16), τ_{ik} can be defined as:

$$\tau_{ik} = \frac{1}{m} \sum_{p=1}^{m} n_p \widetilde{\tau}_{ikp}, \qquad (16)$$

where $\widetilde{\tau}_{ikp}$ is the mean time period needed to perform the pth basic operation of processing the ith structure in the kth testing. The assessment of the efficiency parameters is done according to (11) and (12).

This approach requires keeping the record of the periods needed for executing the basic algorithms of the data structure processing.

To get a more precise S- R- assessment, it is necessary to take into account the dependence of $\widetilde{\tau}_{ikp}$ not only on V and X, but on the order of performing operations $B_p\,|_{X_p}^{Y_p}$ as well.

The other approaches are based on the algorithm modeling according to (1).

Let us define the modeled sequence of the data processing operations as the scenario of operating the data structure (hereinafter the scenario).

We offer the following way of defining the scenario. The approximate number range of the operations of the data structure processing is given ($\underline{n}_i$ is the bottom bound of the approximate number of performances of the ith basic operation of the data structure processing, $\overline{n}_i$ is the upper one):

$$\overline{N}^T = [(\underline{n}_1, \overline{n}_1), (\underline{n}_2, \overline{n}_2), \ldots, (\underline{n}_m, \overline{n}_m)]\,.$$

The number of repetitions of each basic operation within the given bounds is randomly determined for a particular scenario. The sequence of operations is ordered randomly in accordance with the frequency with which they are used $n_i / \sum_{k=1}^{m} n_k$:

$$A\,|_X^Y = \prod_{p=1}^{k} \widetilde{B}_p\,|_{X_p}^{Y_p}\,,$$

where $k = \sum_{p=1}^{m} \widetilde{n}_p$, $\widetilde{n}_p = P(\underline{n}_p, \overline{n}_p)$, $\widetilde{B}_p = B_i$ so that $j = P(1,k)$, $\sum_{p=1}^{i-1} \widetilde{n}_p \le j \le \sum_{p=1}^{i} \widetilde{n}_p$, $P(a,b)$ is a random variable on segment $[a,b]$ and

$$(\sum_{l=1}^{p-1} (\widetilde{B}_l == B_i)) < n_i\,.$$

Besides

$$\tau_{ik} = \tau_{ik}^e - \tau_{ik}^b, \qquad (17)$$

where τ_{ik}^b, τ_{ik}^e is the start and termination time of executing the kth scenario while processing the ith structure, «==» is the operation of comparing implemented in C++. The assessment of the efficiency parameters is performed according to (11) and (12).

The authors have implemented one of the approaches while adapting the data structures (Shynkarenko 2012). Arbitrary scenarios are performed by the user by means of the corresponding interface. The user aware of the usage patterns of data structure of a particular program can set the scenario. Forming the scenario consists in making a list of the data structure operations. The elements of this list correspond to one of the operations of the data structure processing.

It is possible to automatically form scenarios, which requires built in software tools of collecting the information on the data structure processing in the software that is operated. These tools allow recording the sequence of data structure operations in particular software in order to further adapt the data structure.

These approaches allow forming the scenario more realistically and achieving more accurate time efficiency assessments.

7 ASSESSMENT OF THE PARAMETERS OF THE TIME EFFICIENCY OF THE DATA STRUCTURE

To assess the parameters of the time efficiency of the algorithms of data structure processing experimentally, we are to:
- select and develop competitive data structures;
- prepare the computer experiment;
- perform the experiment;
- develop the time efficiency parameters;
- analyze the results.

Let us demonstrate the method developed by the authors using the implementation in C# and the widely spread data structures from the library of the .NET framework classes.

The following data structures have been analyzed:
- the authors' implementation
  - of a single-linked list (DS_1);
  - a double-linked list (DS_2);
- .NET framework implementation (Richter 2012) of:
  - a double-linked list (DS_3);
  - a dynamic vector (DS_4);
  - a hash set (DS_5).

The experiments were performed on PCs whose characteristics are given in table 1.

Table 1. Hardware design of the computers used in the experiment

	Processor	Frequency of the processor bus, MHz	Processor caches (L1 Data / L1 Instruction / L2 / L3), Kbytes	Mother-board	Memory	Frequency of the memory bus, MHz
1	Intel(R) Pentium(R) CPU P6000 @1.87GHz	133	2 x 32 / 2 x 32 / 2 x 256 / 3072	chipset Intel HM55 rev. 05	DDR3 Transcend (800 MHz) 4 Gb; DD3 Kingstone (667 MHz) 2 Gb	532.2
2	Intel(R) Core(TM) i5-3330 CPU @3.00GHz	100	4 x 32 /4 x 32 /4 x 256 / 6144	Intel Z77 rev. 04	DDR3 Kingston (667 MHz) 4GB, DD3 NO-NAME (667 MHz) 4GB, DDR3 Kingston (667 MHz) 4GB	668.5

Two scenarios of the data structure processing were set. Scenario SC_1 consisted in adding 10 000 000 elements. Scenario SC_2 included 40 000 alternate operations of data adding and data searching. The assessments of the parameters of the time efficiency are given in tables 2 and 3. The top parameter is the S-assessment while the bottom one is the R-assessment. The indexes show the deviations of the confidence intervals with the reliability level of 0.95.

While getting ready for the experiment, we prepared the software that implemented the data processing operations, the tools of assessing the time needed for performing the data processing, as well as the software for conducting the experiment,

collecting the assessment results and developing the time efficiency parameters.

The conditions under which the experiment was made approached the real ones. The threads, processes, operating system, antivirus software, computer characteristics, nature and capacity of the data were taken into account.

The results given below show the dynamic vector has the best time efficiency in the structure forming (adding elements). However when the forming alternates with the searching, the hash set shows the best performance.

Table 2. S-/R- assessments of the data structures $DS_1...DS_5$ with scenario SC_1, %

	DS_1	DS_2	DS_3	DS_4	DS_5
DS_1	X	$12{,}52^{+1{,}36}_{-1{,}36}$ $100{,}00^{+0{,}00}_{-0{,}00}$	$12{,}05^{+0{,}84}_{-0{,}84}$ $99{,}92^{+0{,}08}_{-0{,}36}$	$-25{,}00^{+0{,}28}_{-0{,}28}$ $0{,}00^{+0{,}00}_{-0{,}00}$	$14{,}51^{+1{,}50}_{-1{,}50}$ $99{,}92^{+0{,}08}_{-0{,}36}$
DS_2	$-12{,}52^{+1{,}36}_{-1{,}36}$ $0{,}00^{+0{,}00}_{-0{,}00}$	X	$-0{,}54^{+2{,}46}_{-2{,}46}$ $32{,}58^{+35{,}25}_{-32{,}58}$	$-34{,}41^{+0{,}77}_{-0{,}77}$ $0{,}00^{+0{,}00}_{-0{,}00}$	$2{,}25^{+3{,}23}_{-3{,}23}$ $68{,}42^{+31{,}58}_{-37{,}68}$
DS_3	$-12{,}05^{+0{,}84}_{-0{,}84}$ $0{,}08^{+0{,}36}_{-0{,}08}$	$0{,}54^{+2{,}46}_{-2{,}46}$ $67{,}42^{+32{,}58}_{-35{,}25}$	X	$-34{,}04^{+0{,}88}_{-0{,}88}$ $0{,}00^{+0{,}00}_{-0{,}00}$	$2{,}78^{+0{,}82}_{-0{,}82}$ $90{,}67^{+2{,}80}_{-2{,}80}$
DS_4	$25{,}00^{+0{,}28}_{-0{,}28}$ $100{,}00^{+0{,}00}_{-0{,}00}$	$34{,}41^{+0{,}77}_{-0{,}77}$ $100{,}00^{+0{,}00}_{-0{,}00}$	$34{,}04^{+0{,}88}_{-0{,}88}$ $100{,}00^{+0{,}00}_{-0{,}00}$	X	$35{,}91^{+1{,}35}_{-1{,}35}$ $100{,}00^{+0{,}00}_{-0{,}00}$
DS_5	$-14{,}51^{+1{,}50}_{-1{,}50}$ $0{,}08^{+0{,}36}_{-0{,}08}$	$-2{,}25^{+3{,}23}_{-3{,}23}$ $31{,}58^{+37{,}68}_{-31{,}58}$	$-2{,}78^{+0{,}82}_{-0{,}82}$ $9{,}33^{+2{,}80}_{-2{,}80}$	$-35{,}91^{+1{,}35}_{-1{,}35}$ $0{,}00^{+0{,}00}_{-0{,}00}$	X

Table 3. S-/R- assessments of the data structures $DS_1...DS_5$ with scenario SC_2, %

	DS_1	DS_2	DS_3	DS_4	DS_5
DS_1	X	$1{,}81^{+0{,}63}_{-0{,}63}$ $99{,}58^{+0{,}36}_{-0{,}36}$	$46{,}98^{+2{,}57}_{-2{,}57}$ $100{,}00^{+0{,}00}_{-0{,}00}$	$43{,}57^{+4{,}53}_{-4{,}53}$ $100{,}00^{+0{,}00}_{-0{,}00}$	$-99{,}13^{+0{,}05}_{-0{,}05}$ $0{,}00^{+0{,}00}_{-0{,}00}$
DS_2	$-1{,}81^{+0{,}63}_{-0{,}63}$ $0{,}42^{+0{,}36}_{-0{,}36}$	X	$46{,}02^{+2{,}29}_{-2{,}29}$ $100{,}00^{+0{,}00}_{-0{,}00}$	$42{,}57^{+4{,}27}_{-4{,}27}$ $100{,}00^{+0{,}00}_{-0{,}00}$	$-99{,}15^{+0{,}05}_{-0{,}05}$ $0{,}00^{+0{,}00}_{-0{,}00}$
DS_3	$-46{,}98^{+2{,}57}_{-2{,}57}$ $0{,}00^{+0{,}00}_{-0{,}00}$	$-46{,}02^{+2{,}29}_{-2{,}29}$ $0{,}00^{+0{,}00}_{-0{,}00}$	X	$-5{,}67^{+3{,}23}_{-3{,}23}$ $6{,}25^{+7{,}05}_{-6{,}25}$	$-99{,}54^{+0{,}01}_{-0{,}01}$ $0{,}00^{+0{,}00}_{-0{,}00}$
DS_4	$-43{,}57^{+4{,}53}_{-4{,}53}$ $0{,}00^{+0{,}00}_{-0{,}00}$	$-42{,}57^{+4{,}27}_{-4{,}27}$ $0{,}00^{+0{,}00}_{-0{,}00}$	$5{,}67^{+3{,}23}_{-3{,}23}$ $93{,}75^{+7{,}05}_{-7{,}05}$	X	$-99{,}51^{+0{,}01}_{-0{,}01}$ $0{,}00^{+0{,}00}_{-0{,}00}$
DS_5	$99{,}13^{+0{,}05}_{-0{,}05}$ $100{,}00^{+0{,}00}_{-0{,}00}$	$99{,}15^{+0{,}05}_{-0{,}05}$ $100{,}00^{+0{,}00}_{-0{,}00}$	$99{,}54^{+0{,}01}_{-0{,}01}$ $100{,}00^{+0{,}00}_{-0{,}00}$	$99{,}51^{+0{,}01}_{-0{,}01}$ $100{,}00^{+0{,}00}_{-0{,}00}$	X

4 CONCLUSIONS

The parameters of the data structure time efficiency have been developed on the basis of the characteristics and parameters of the algorithm time efficiency. The obtained parameters allow the objective comparing of the data structures based on the time characteristics of their operation.

The assessments of the data structures time efficiency allow automated processes of adapting the data structures to their operational environment. The operational environment is determined by the technical characteristics of the computing facilities, operating system and the set of application programs running concurrently with the data structure processing.

The authors have described the methods of the experimental assessment of the parameters of the data structure time efficiency developed by them. The theoretical hypothesis stating that the time efficiency of the data structures is predetermined by the order of executing the data processing operations has been proven.

REFERENCES

Bentley. J. L. 1982. *Writing Efficient Programs.* New York: Prentice Hall: 186.

Chase D. R., Wegman M., Zadeck K. F. 1990. *Analysis of pointers and structures.* SIGPLAN Notices, Vol. 25. Issue 6: 296-310.

Cherkassky B.V., Goldberg A.V., Silverstein C. 1999. *Buckets, heaps, lists, and monotone priority queues.* SIAM Journal on Computing, Vol. 28, Issue 4: 1326-1346.

Kaspersky K. 2003. *Code optimization: effective memory usage.* A-List Publishing: 240.

Cormen T.H., Leiserson C.E., Rivest R.L., Stein C. 2009. *Introduction to algorithms (3rd ed.).* Cambridge: MIT press: 1312.

Galil Z., Seiferas J. 1983. *Time-space-optimal string matching.* Journal of Computer and System Sciences, Vol. 26, Issue 3: 280–294.

Hoare C. A. R. 1975. *Recursive data structures.* International Journal of Computer & Information Sciences, Vol. 4, Issue 2:105-132.

Knuth D. E. 1997. *The art of programming, Volume 1 (3rd ed.). Fundamental algorithms.* Redwood City: Wesley: 650.

Kozen. D. C. 1992. *The design and analysis of algorithms.* New York: Springer: 322.

Perevozchikova O. L. 2007. *Information systems and data structures* (in Russian). Kyiv: KMA: 287.

Shynkarenko V. I. 2009. *Experimental research of algorithms in hardware and software environments* (in Russian). Dnipropetrovsk: DNURT: 279.

Shynkarenko V. I. 2006. *The time estimation of structured data processing operations with command piping and data caching taken into account* (in Russian). Problems in Programming, Issue 3-4: 43-52.

Shynkarenko V. I., Ilman V. M., Skalozub V. V. 2009. *Structural models of algorithms in problems of applied programming. I. Formal algorithmic structures.* Cybernetics and systems analysis, Vol. 45, Issue 3: 329-339.

Shynkarenko V. I. 2001. *Comparative analysis of time efficiency of functionally equivalent algorithms* (in Russian). Problems in Programming, Issue 3-4: 31-39.

Richter J. 2012. *CLR via C# (4rd ed.).* Pearson Education: 896.

Shynkarenko V. I., Ilman V. M., Zabula H. V. 2014. *The constructive- synthesizing model of the data structures at logical level* (in Russian). Problems in Programming, Issue 2-3: 10-16.

Shynkarenko V. I., Zabula H. V. 2012. *Application of genetic algorithms to the data structures adaptation* (in Russian). Artificial intelligence, Issue 3: 323-331.

Shynkarenko V. I., Zabula H. V. 2012. *Improvement of the time efficiency of data structures in RAM on the basis of adaptation* (in Russian). Problems in Programming, Issue 2-3: 211-218.

Shynkarenko V. I., Ilman V. M. 2014. *Constructive-synthesizing structures and their grammatical interpretations. Part I. Generalized formal constructive-synthesizing structure.* Cybernetics and systems analysis, Vol. 50, Issue 5: 655-662.

Ziegler C. 1982. *Programming system methodologies.* N. York: Prentice Hall PTR: 304.

Power Engineering and Information Technologies In Technical Objects Control – Pivnyak, Beshta
& Alekseyev (eds)
© 2016 Taylor & Francis Group, London, ISBN 978-1-138-71479-3

The representative of national problems
in the field of cybersecurity

Y. Kovalova, T. Babenko
State Higher Educational Institution "National Mining University", Dnipro, Ukraine

ABSTRACT: Scientific and technological progress of recent years has become a significant basis for the development of the information society. Technological breakthroughs in information and telecommunications has become the most important aspect for the formation of new directions in implementing cyber security measures. Today countries are in the stage of information warfare. This fact sets the basic trend for the development of cybersecurity at the national level in Ukraine. The present paper reveals the characteristic difficulties in the existing legal framework of cybersecurity in Ukraine in compavison with countries possessing high experience in this area of activity.

1 INTRODUCTION

Recently, the problem of security has become topical in global distribution of information sources and occupied the central place on the world governmental portals, because this priority area is the foundation of a developed, independent and democratic state.

Technological advances of the last decades played the key role in the formation of the global policy of the international community. In the era of the information environment and ultimate integration into innovation process, protection of information resources including the whole range of information technologies is becoming a cutting-edge issue.

By combining information and communication technology (ICT) and the rapid development of information and telecommunication systems (ITS) for today takes effect large-scale distribution of factors Information Society, which plays a significant role in the economic and social development of countries.

Genesis of the Information Society, at first glance, makes it possible to build an effective model of modern global community society, but on the other hand it provides new security threats and great vulnerabilities of individual information sphere from outside interventions.

In such circumstances, it is vital to find new opportunities to ensure the security of the state and creation of a reliable system of cybersecurity.

However, it should be noted that in the information space of our country the concept of "cyber-security" has appeared only recently, so this area of information security is only at the stage of formation.

Considering today's unstable geopolitical situation the question of information resources security from outside interventions and cyber security is critical importance in Ukraine.

Absence and disregard for implementation of cybersecurity system may result in loss of leverage in the sphere of influence of national interests and political independence of any country in the world.

2 FORMULATION OF THE PROBLEM

The main purpose is to determine the basic provisions of the legislation in the field of cybersecurity in Ukraine compare them with the ones in other countries and identify problematic issues in the existing state system.

3 BASICS OF ANALYSIS

The modern concept of cyberspace, due to its relative novelty, does not have a clear and balanced legislation base that meets the general principles of international law. Cyberspace undoubtedly is an urgent problem on the global level. Cyberspace opens possibilities not only for hacking various types of intellectual, governmental and strategic property, but also for the special operations carried out by military or intelligence units at the state level. In this case, any military intervention into the country through cyberspace becomes possible without any further formal sanctions both from the part of the country which suffered from the intervention and from the international community.

Therefore, governments of many countries started to implement measures for controlling their own cyberspaces.

In today's world the Internet is increasingly used for military purposes which has prompted countries to focus on the problem of security.

To a large extent, it is a threat of outside interventions and cyber attacks on governmental infrastructure projects, which are of strategic importance for the countries of the international community, led to the adoption at the NATO's new strategic concept of the Alliance for 2011-2020, which clearly spelled out the facts of the threat of cyber-terrorism, which is equivalent to the military threat and are entitled to use the armed forces for the benefit of members of the alliance (Strategic Concept for the Defence and Security of the Members of the North Atlantic Treaty Organization).

In February 2013, the European Commission and the External Action Service issued a joint document which present a vision of an open, safe and secure cyberspace. European Strategy in the field of cyber security has become a foundation for the secure, free and open cyberspace. The Strategy priorities are to increase stability, reduce cybercrime, develop defense policy, promote of EU industry and fundamental European values at international level.

One of the principal aspects of the Strategy is to support research and innovation at the state level, which in turn is a fundamental basis for the development of the security sphere the whole. During the two years since the adoption of the Strategy, Europe has made breakthrough in the development of cyber security.

A comprehensive approach to Security Strategy implies activities across the entire spectrum of political sectors: from justice and internal affairs, to the customs, finance, transport and the environment. And, above all, the most important factor of the global activities in the segment of cyber security is the basis of European values - such as freedom and security of European citizens, regardless of religion, ethnicity and place of residence.

The basis for a European approach is proposed in the Directive Network and Information Security, which provides for the establishment of a high overall level of cyber security within the EU. Also this normative act coordinates cooperation between EU Members in the field of cyber-security, which means organizing of the necessary national structures in the same field (The European Cyber Security Strategy 2013).

Directive of the European Union could help to provide a higher level of fundamental cybersecurity and increase resistance to cyber threats if it was aimed at improving the readiness of the most important elements of the European Union infrastructure for such threats for the establishment of harmonized processes of reporting and publishing of information across the whole single market.

The European Union provides for the security on the Internet as a matter of vital importance for the development of the information society and establishes certain actions to improve cybersecurity, which include the creation of a well-functioning network group CERT (Computer Emergency Response Team Emergency) at the national level, covering the whole Europe; model of cyber attacks and the EU's readiness to support greater cooperation in the field of cybersecurity. In addition, the policy of protection of information infrastructure (CIIP) is aimed at strengthening the security and stability of vital ICT infrastructure by stimulating and supporting the development of high-level capacities for preparedness, security and resiliency, both at national and at EU levels (Proceedings of conference on the EU Cybersecurity Strategy 2015).

Special attention of countries is placed on ensuring the high level of risk management practices at the sites of infrastructures on the prediction of cyber attacks, which is the traditional approach to the security of critical infrastructures of countries such as energy, transport, economy, industry. Implementation of these measures is the key to national security and technological sovereignty.

All the efforts of the leading countries in the field of cybersecurity are based on the principles of strong and effective legislation and expansion of the operational capabilities to combat cybercrime, support the combat capability of the member countries and improve coordination at EU level.

At present, the EU-countries have no clear and effective policy measures to ensure cybersecurity. According to the analysis made by the Association of software producers "BSA | The Software Alliance", legislation and policies of the EU in the area of cybersecurity testify to discrepancies in normative documents and acts, as well as insufficient preparedness of EU-countries to the spread of virtual threats.

For the purposes of this analysis, BSA carried out a comprehensive assessment of national laws, regulations and policies in 28 member countries of the European Union on a number of criteria, that are considered the most important to ensure effective protection against threats of cyber attacks represented in the table below (EU Cybersecurity Dashboard: A Path to a Secure European Cyberspace).

Table 1. European Union cybersecurity maturity dashboard (2015)

- YES - NO - PARTIAL	What year was the national cybersecurity strategy adopted?	Is there legislation/policy that requires (at least) an annual cybersecurity audit?	What year was the computer emergency response team (CERT) established?	Does legislation/policy include an appropriate definition for "critical infrastructure protection" (CIP)?	Is there legislation/policy that requires an inventory of "systems" and the classification of data?	Is there legislation/policy that requires security practices/ requirements to be mapped to risk levels?	Is there legislation/policy that requires a public report on cybersecurity capacity for the government?
Austria	2013		2008				
Belgium	2012		2008				
Bulgaria	-		2008				
United Kingdom	2011		2014				
Greece	-		2009				
Denmark	-		2009				
Estonia	2014		2008				
Ireland	-		-				
Spain	2013		2008				
Italy	2014		2014				
Cyprus	2013		-				
Latvia	2014		2006				
Lithuania	2011		2006				
Luxembourg	2013		2011				
Malta	-		2002				
Netherlands	2013		2012				
Germany	2011	draft	2012				draft
Poland	2013		2008				
Portugal	-		2008				
Romania	2013		2011				
Slovakia	2008		2009				
Slovenia	-		2010				
Hungary	2013		2013				
Finland	2013		2014				
France	2011		2008				
Croatia	-		2009				
Czech Republic	2011		2011				
Sweden	-		2003				

As a result of this analysis the following conclusions were made:

• most EU-member states consider the issue of cyber security a key national priority, especially for critical infrastructure;

• presence of significant differences in state policies in the field of cyber security, as well as in legal mechanisms and operational capabilities of various EU member states determines the existence of significant gaps in the overall system to ensure cyber security in Europe as a whole;

• almost in all EU-member states rapid response teams were established to resolve cybersecurity incidents, but missions and expertise of the teams in each country differ considerably;

• there are some worries about absence of a systematic approach to public-private partnerships and cooperation on cyber security between the governments of the EU countries, non-governmental organizations and international partners.

Rapid tendency to strengthen the national security system of the world leading countries explains the need to streamline and strengthen the national policy of cybersecurity. First of all, events of 2013-2015 years require that analysis of threats to national security include not only forces and tools to defeat the enemy, but also consider action of non-military methods of warfare (Gorbulin 2015).

It is necessary to pay special attention to the development of a specific regulatory framework that meets norms and standards of international documents, where basic concepts in the field of cybersecurity will be clearly defined.

The National Security Strategy of Ukraine determines new topical threats to the national security of the country. In terms of global security instability and a new foreign policy strategy of Ukraine the National Security Strategy determines the threats of cyber security and security of information resources, including the vulnerability of critical infrastructure, government information resources to cyber attacks. Priorities of ensuring cyber security and security of information resources comprise:

• development of information infrastructure of the state;

• establishing a system to ensure cyber security, development of networks for reacting to computer emergencies;

• monitoring of cyberspace in order to timely detect, prevent and neutralize cyber threats;

• developing the capacity of law enforcement agencies to investigate cybercrimes;

• providing protection of critical infrastructure objects, government information resources from cyber attacks, denial from software such as anti-virus, developed in the Russian Federation;

• reforming the system of protection of state secrets and other classified information, protection of state information resources, systems of e-governance, technical and cryptographic protection of information using the practice of states - members of NATO and the EU;

• creating system of training employees in the field of cyber security for the needs of security and defense sector;

• developing of international cooperation in the field of cyber security, intensification of cooperation between Ukraine and NATO, in particular within NATO Trust Fund to strengthen Ukraine's capabilities in cyber security ("About the Strategy of National Security of Ukraine" 2015).

Recently, the issue of cyber security systems has been studied in more detail. The countries of the world use variety of approaches to the definition of cybersecurity and cyberspace but still there is no consensus on the definition of this concept. The same is true for Ukraine.

In Ukraine cybersecurity is governed by laws and legal instruments, including: the Law of Ukraine "About Protection of Information in Telecommunication Systems", " About National Security", "About Information", "About the General Principles of Information Society Development in Ukraine for 2007-2015 years " and others, as well as the Council of Europe Convention on Cybercrime, which was ratified by Parliament in 2005 (The Law of Ukraine № 80/94-ВР, № 964-15, № 2657-12, № 537-16, № 2824-IV).

Detail analysis of Ukraine legislation on cyber security testifies that the main problem in providing legal cybersecurity is absence of single regulatory system in this area.

Today the Cabinet of Ministers of Ukraine has developed and submitted to the Verkhovna Rada of Ukraine the draft Law of Ukraine "About Cybersecurity in Ukraine" that is the first regulatory document with narrow specifications. The aim of the Law is to create a national system of cybersecurity as a combination of political, social, economic and information measures related to the organizational, administrative and technical-technological activities through an integrated approach in close collaboration between public, private sectors and civil society. The project identified the basic principles of the state policy aimed at protecting vital interests of individuals, society and the state, implementation of which depends on the proper functioning of information, telecommunication, information and telecommunication systems (Draft Law of Ukraine registration number 2126 of 06/19/2015). It should be noted that the project revealed the concept of the terms "cyber security" and "cyberspace". According to the project, "cyber security (cybersecurity) - a state of protecting vital interests of man and a citizen, society and the country in cyberspace; cyber space (cyberspace) - is the environment that is the result of the operation based on common principles and general rules of information, telecommunication and information-telecommunication systems. "

The project identified the main threats to cyber security of Ukraine, such as:

• use of cyberspace for military purposes, the establishment cyber troops of other states , cyber units in traditional military branches;

• development of new cyber weapons in foreign states;

• existence of plans and intelligence offensive military operations in cyberspace in other countries;

• development of foreign special services which employ methods of intelligence-subversive activity in cyberspace and methods of manipulating public opinion by the cyberspace;

• possibility of Ukraine's involvement in an armed conflict or in a conflict with other countries through the use of the national segment of cyberspace;

• attempts to interfere in the internal affairs of countries using social networks to spread through the national segment of the cyberspace cult of violence, cruelty, pornography;

• intensification of the manifestations of cyber terrorism;

• spread of cybercrime;

• critical dependence of the national information infrastructure on foreign manufacturers of high-tech products, inclusion into software and hardware of hidden malicious functions;

• increased risks of emergency situation due to low level of protecting objects of critical information in the infrastructure of the country (Draft Law of Ukraine registration number 2126 of 06/19/2015).

A significant flaw of the project is the lack of integrity model and comprehensive system of cybersecurity of the state. And most importantly, the draft reflected the new cybersecurity institution activities, but there is a risk of duplication of functions relevant bodies, which in turn causes lack of important institutions. The absence of a competent public discussion of this project with the involvement of prominent scientific experts and industry professionals is critical for implementation of the law that will move the issue of cybersecurity from the space.

4 CONCLUSIONS

Today the cybersecurity system of Ukraine take important steps toward to the formation, institutionalization and creating legal and juristic support of the National Cybersecurity system.

One of the significant steps in this direction should be the adoption of cybersecurity strategy which was presented during the expert consultations between NATO and Ukraine at the questions of cybersecurity.

Ukraine needs to review and make their own audit capabilities to ensure the cybersecurity sector and it would be started from National Security in the segment of critical infrastructure.

REFERENCES

Strategic Concept for the Defence and Security of the Members of the North Atlantic Treaty Organization. NATO HQ, 2010 [See www.nato.int/lisbon2010/strategic-concept-2010-eng.pdf]

The European Cyber Security Strategy. 2013.

Proceedings of conference on the EU Cybersecurity Strategy, 28 May 2015, Belgium.

EU Cybersecurity Dashboard: A Path to a Secure European Cyberspace [See http://cybersecurity.bsa.org/assets/PDFs/study_eucybersecurity_en.pdf]

Gorbulin V. 2015. *"Hybrid war" as a key instrument of Russian geostrategy revenge* (in Russian). Dzerkalo Tijdnja, Issue 2.

The decision of the National Security and Defense Council of Ukraine on May 6, 2015 "About the Strategy of National Security of Ukraine".

The Law of Ukraine "About protection of information in telecommunication systems", № 80/94-ВР of 05.07.1994.

The Law of Ukraine "About National Security", № 964-15 of 2003.

The Law of Ukraine "About Information", № 2657-12 of 1992 (amended on 21.05.2015).

The Law of Ukraine "About the Fundamentals of Information Society Development in Ukraine for 2007-2015", № 537-16 of 09.01.2007.

The Law of Ukraine "About ratification of the Convention on Cybercrime", № 2824-IV of 07.09.2005.

Draft Law of Ukraine "About basic principles of ensuring cybersecurity Ukraine", registration number 2126 of 06/19/2015.

Power Engineering and Information Technologies In Technical Objects Control – Pivnyak, Beshta
& Alekseyev (eds)
© 2016 Taylor & Francis Group, London, ISBN 978-1-138-71479-3

The estimation of traffic descriptions in the information telecommunication networks

V. Kornienko

State Higher Educational Institution "National Mining University", Dnipro, Ukraine

ABSTRACT: The efficiency of the estimation procedure of traffic descriptions in the information telecommunication networks, being the component part of the compound identification method of the complex nonlinear processes, is being studied in the article. It has been proved that the traffic of multiservice switching system is the fractal self-similar process, and the Internet traffic is a multi-fractional one.

1 INTRODUCTION

The constant traffic expansion in the information telecommunication networks (ITN) causes the necessity of its identification and predicting in order to prevent the overload in the networks and to avoid the deterioration of quality of the services provided.

2 FORMULATING THE PROBLEM

The traffic in ITNs is a nonlinear stochastic process with the self-similar properties, with chaotic and fractal dynamics (Crovella & Bestravos 1997). Besides, it is established (Riedi & V'ehel 1997; Sheluhin 2011) that the aggregated traffic from different sources on the small time scales displays the multi-fractal character.

The development of telecommunication networks is connected with their planning that allows to determine the required network resources and to provide the specified switching capacity. The designing of telecommunication facilities based on statistic data and the required mathematical apparatus using predicting allows to balance the load on routers, switchboards and communication channels and, thus, to improve reliability and efficiency of the network.

The estimation of the network traffic descriptions is required not only for classification of the parent processes, but also for construction of its adequate simulation model that allows to ensure the quality of service in the ITNs.

Therefore, it is relevant to construct and use the models of network traffic based on the estimation of its characteristics in order to control the information transfer.

The study of the self-similar property of the network traffic resulted in arising of the models based on the fractal stochastic processes. In the work (Kornienko, Gulina & Budkova 2013) it is offered the combined method of identification and estimation of characteristics of the complex nonlinear processes, which includes the following procedures:

1) estimation of process characteristics;
2) selection of the type (structure) of the process model;
3) determination of the model parameters (parametrical identification of the process).

The aim of the work is the study of the efficiency of the estimation procedure of traffic descriptions in the ITNs.

3 PROCEDURE OF CHARACTERISTICS ESTIMATION

The procedure of characteristics estimation includes the following stages:

a) time-frequency analysis:
– analysis of the time signal type, autocorrelation function (AF) and spectral density of the process capacity;

b) statistical analysis:
– analysis of the sampling average dispersion;
– determination of experimental distribution adequacy by theoretical one;
– determination of the 'tail burden' of distribution index;

c) fractal analysis:
– analysis of attractor phase portrait;
– computing of correlation entropy, predictability interval (depth of the correct prediction) and attractor dimension of the process;
– determination of the attractor embedding dimension (dimension of the phase space – memory depth);
– determination of Herst exponent;

– analysis of fluctuation and partial functions of the process.

Let us consider the stages mentioned above.

The qualitative characters of the system motion randomness are: irregularity of time signal, exponential decay of its AF and bandpass components at the low frequencies in its spectrum.

For self-similar processes, AF has the durable dependence, and spectral concentration is submitted to power law.

If the dispersion of the sampling average has a slower decay than the value reciprocal to sample length, then this process is a self-similar one.

Mostly, for approximation of histograms of the self-similar processes experimental data the functions of sub-exponential distribution laws are used, for verification of adequacy of which Kolmogorov and Pearson's goodness-of-fit tests are used.

The distribution with 'tail burden' is typical of self-similar processes with exponential decrease of AF.

According to the Takens theorem, at search by time steps, there is a discrete set of points in d-dimensional space, which is the attractor's phase portrait at the constant system mode. The analysis of the phase portrait may reveal the directions of motions at different initial conditions, as well as determine the qualitative properties of dynamic system, which originates the process.

To determine the parent process mode its Kolmogorov entropy K is estimated, which is equal to the sum of Lyapunov senior exponents and describes the speed of the data loss about the dynamic system state in time. K-entropy is equal to zero at a regular motion, infinite for random systems, positive and limited for chaotic systems.

The value of correlation entropy is the lower limit of K-entropy and allows to estimate the interval of the process accurate predictability.

The phase space d dimension, beginning with which the correlation dimension stops changing, is the attractor nesting minimum dimension (the smallest phase space integral dimension that holds the whole attractor). Thus, the d dimension determines the generating system order (its memory depth).

The Herst exponent H characterizes the level of process self-similarity. It indicates the presence of the trend or the the process randomness, as well as defines the system (process) evolution. If $0,5 < H < 1$, then the process is characterized by a long-term memory and is persistent, for which the retention of the system evolution tendency is probable in future. If $0 < H < 0,5$, then its time

realizations are changeable consisting of frequent reverses recession-raising. Notably, for them, the increase (decrease) of values in the past means their probable decrease (increase) in the future.

The analysis of fluctuation functions allows to reveal the properties of self-similarity and long-term dependences of nonstationary time range with the availability of probable trends, without knowledge of their origin and form. The scaling index for those functions corresponds to Herst exponent; besides, those functions in log-log scale for mono-fractal process have the form of straight lines, and for multi-fractal – the form of curved lines.

The values of partial functions (generalized statistical sums) in log-log scale for mono-fractal processes are constant, and for multi-fractal are variable. .

4 ANALYSIS OF NETWORK TRAFFIC

In order to check the methods developed, the experimental signals of traffic of automatic telephone exchange (ATE) on the base of switching systems: EWSD DNEPR (signal 1) and F-1500 (signal 2), and traffic transmitted through Internet (signal 3) have been used.

The EWSD DNEPR system renders the superset of telecommunication services to the subscribers, such as: integral service, data exchange and voice data exchange, while the switching system F-1500 renders only the last one. As well, ATE on the base of EWSD switching system is areal, and ATE on the base of F-1500 switching system is urban. These peculiarities of construction and use of the systems determine the difference of their traffics. The aggregation (discretization) time for signals 1 and 2 is 1 hour.

The data of Internet network traffic realization represent the dependence of the Ethernet frames size in bytes on the time. The aggregation procedure with the step of 5 s was carried out in order to set the initial data in equidistant scale by hour-angle axis.

The results of the signals analysis are showed on Fig. 1-3.

By the form of AF (Fig. 1.a) and spectrum (Fig. 2.a) of signal 1, we can say about irregularity of the parent process. It arises from the fact that its spectral power is approximately inversely proportional to frequency. At the same time, from the form of its AF, we cannot affirm that the parent process has the property of self-similarity (AF approximation in the form of slowly (SDD) and quickly decreasing dependences (QDD) are showed on Fig. 1).

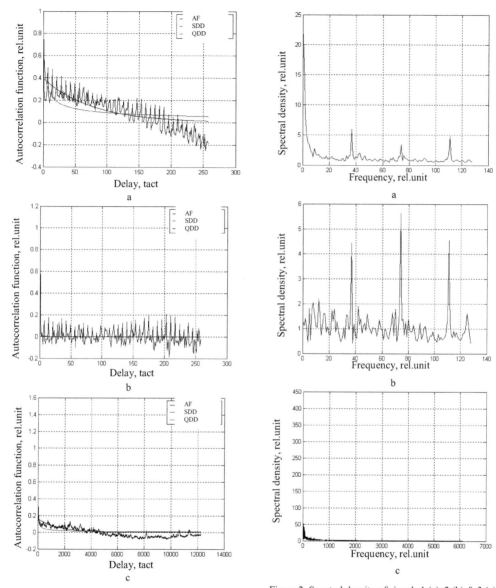

Figure 1. AF of signals 1 (a), 2 (b) & 3 (c)

Figure 2. Spectral density of signals 1 (a), 2 (b) & 3 (c)

The value of Herst exponent of signal 1 is made up depending on the methods used (0,814-0,923). For such signals, the events of recent past produce a much stronger effect upon the process than the events of the distant past. The slowly decreasing dispersion of the process (Fig. 3.a) confirms the availability of such property in the process under study, which causes signal 1.

The construction of the experimental distribution histograms and determination of their adequacy to theoretical distribution using Kolmogorov and Pearson's goodness-of-fit tests gives the opportunity to establish that the logarithmically normal distribution law corresponds to signal 1.

The quick AF decay of experimental signal 2 (Fig. 1.b) indicates the process stochasticity. The energy of signal spectrum (Fig. 2.b) is shown both in the low-frequency and high-frequency region. The value of Herst exponent changes in the range

159

(0,671-0.909). The graph of signal 2 dispersion (Fig. 3.b) indicates the lack of self-similarity property.

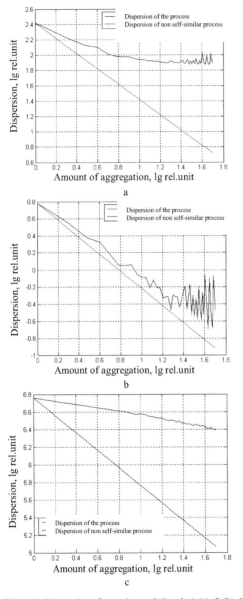

Figure 3. Dispersion of experimental signals 1 (a), 2 (b) & 3 (c)

The parent process of signal 3 has the properties peculiar to self-similar processes: hyperbolic decrease of AF with increase of the time delay (Fig. 1.c), power law of the graph of its spectral density (Fig. 2.c) and slowly decreasing dispersion (Fig. 3.c). Signal 3 has the 'tail burden' distribution, the histogram of experimental distribution according to Kolmogorov and Pearson's goodness-of-fit tests

corresponds to Pareto distribution, and the Herst exponent is (0,888-0,980).

The detrend fluctuations analysis (DFA) made on multi-fractal properties of the signal allowed to receive the following results. Signal 1 behavior (fig. 4, a) on the interval from 10 to 86 time steps is non-correlative as far as the scaling index is $\alpha 1 = 0,5$.

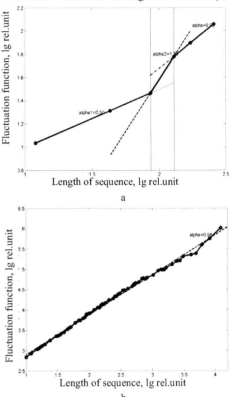

Figure 4. Fluctuation function of signals 1 (a) & 2 (b) for DFA

The $\alpha 2 = 1,79 > 1$ corresponds to the time interval from 87 to 129 time steps that indicates the correlations, however they don't have the power dependence.

On the interval more than 129 time steps, the index possesses the value $0,5 < \alpha = 0,92 < 1$ that is peculiar to the processes, having a trend. For signal 3 the value $\alpha = 0.98$ is close to 1 (Fig. 4, b) that is peculiar to the noise, the spectrum energy of which is inversely proportional to frequency.

It should be noted that for experimental signals the received values of scaling index α coincide with the Herst exponent value.

The results of multi-fractal DFA are shown on fig. 5. When limiting the values of order $-5 \le q \le 5$

on the scales less than $2^5 = 32$, signal 1 may be considered as a mono-fractal process and signal 3 – as a multi-fractal one (its fluctuation function is non-constant).

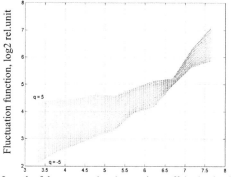

a

Length of the non-overlapping cuttings-off, log2 rel.unit

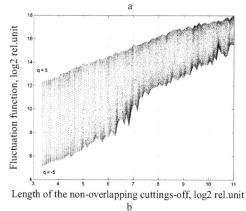

Length of the non-overlapping cuttings-off, log2 rel.unit

b

Figure 5. Fluctuation function of signals 1 (a) & 2 (b) for multifractal DFA

5 CONCLUSIONS

According to a complex method of estimation and identification of complex nonlinear processes, the study of procedure efficiency for traffics descriptions estimation in the ITN is fulfilled.

Based on the estimation analysis of the traffic signals descriptions of the switching systems and Internet network, it is established that the traffic of multiservice switching systems is a self-similar fractal process, and the traffic is multi-fractal.

REFERENCES

Crovella M. E. & Bestravos A. 1997. *Self-Similarity in World Wide Web Traffic: Evidence and Possible Causes*. IEEE Transactions on Networking, Issue 6.

Riedi R.H. & V'ehel J.L. 1997. *Multifractal properties of tcp traffic: A numerical study*. Technical Report 3129, INRIA.

Sheluhin O.I. 2011. *Multifractals. Infocommunicational application* (in Russian). Moscow: Hot line – Telecom: 576.

Kornienko V.I., Gulina I.G. & Budkova L.V. 2013. *Complex estimation, identification and prediction of difficult nonlinear processes* (in Ukrainian). Scientific Bulletin of National Mining University, Issue 6.

Power Engineering and Information Technologies In Technical Objects Control – Pivnyak, Beshta
& Alekseyev (eds)
© 2016 Taylor & Francis Group, London, ISBN 978-1-138-71479-3

The methods and algorithms for solving multi-stage location-allocation problem

S. Us, O. Stanina
*State Higher Educational Institution "National Min*ing University", *Dnipro, Ukraine*

ABSTRACT: The location problem of multi-stage production was considered in the paper. The combined model with continuous set placement points for one of the stages was proposed for its solving. Approaches to solving, based on discrete and continuous optimization methods, namely optimal set partition method and genetic algorithms, were proposed. Algorithms were formulated, the results of numerical experiments were shown.

1 INTRODUCTION

One of the topical problems, actively studied in operations research, is the problem of optimal facility location in a given area. A great number of papers are devoted to their researches where different types and classes of location problems have been formulated (Brimberg et al 2008; Drezner & Hamacher 2001; Farahani & Hekmatfar 2009).

All location problems can be divided into two classes: the location problem of linked facilities and the location-allocation problems (the problems of production location). The problems of the first type are characterized by previously defined communications between the objects. The second class problems do not have communications between the objects-production, which must be located and clients' allocation is carried out between them. They are p-median and p-center problems, simple location problem.

It is common to distinguish the discrete and continuous location problems. The discrete problems peculiarity is the finite number of potential location places. Continuous problems are characterized by non-discrete set of places for objects location and the metric for evaluating the distance between the objects. Moreover, the set of existing objects can also be discrete or continuous. For example, the location problem with customers, who are continuously filling in a given area is discussed in: (Kiselova & Koryashkina 2013).

Multi-stage location problems form another class of the facility location problem (Bischoff et al 2009; Paksoy et al 2010).

These problems assume existence of several groups of objects. Each group has its potential set of location places (sometimes their sets can be equal)

and the communication between the facilities is defined, for example some hierarchy between the objects is given.

The solution method of the location problem depends of the studied problem class.

In addition to the classical methods of discrete and continuous optimization, which are not always applicable and effective for this class of problems, algorithms of local optimization and approaches based on analogies with nature have been actively developed in recent years (genetic algorithms, simulated annealing, ant colony algorithms, etc.) (Farahani & Hekmatfar 2009; Ren 2011; Zhu et al 2008).

The present paper studied a multi-stage problem of enterprises location, where a set of potential placements of one stage is continuous, and a set of potential placements of another stage is discrete and finite. Solving methods and algorithms for this problem are developed.

2 FORMULATING THE PROBLEM

Multi-stage production and transport problems are understood as the problems, which reflect such sequential processes: manufacturing of one product type, its delivery to the point of its recycling into other products, manufacturing of this product and its delivery to final consumers. In the simplest cases, such problem includes two types of product – "raw materials" and "finished product". It is possible to use more names for the product: "raw", "semi-finished" and "finished product."

A great number of papers are devoted to solving multi-stage location problem, however, as a rule, problems of the discrete type are considered and communication circuit between different types of objects is given. Moreover, it is presumed that each point of demand for final product and each point of production are receiving the product from one supplier only, and the enterprises of level r get the product from enterprises of level $(r+1)$ (see Fig. 1).

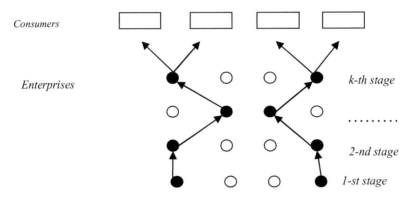

Figure 1. The location problem with a given communication circuit

The discrete multi-stage location problem was considered in (Gimadi 1995; Trubin & Sharifov 1992; Ageev et al 2009, You & Grossmann 2008; Paksoy et al 2010). But, continuous multi-stage problem is not studied, although a lot of practice areas generate such problems. Examples of such problems are the problems of pumping and distribution stations placement, which depends on the water level and the customers location, the problem of placing mines and concentrators according to mineral deposits and users location and others.

In this paper, the location problem is considered under conditions that the first stage enterprises can be located in any point of the area and they get the raw material from suppliers distributed continually in the given area. Finite number of potential places of second-stage enterprises location are predefined and the placement of finite number of customers is given.

The difference between this problem and a multi-stage location problem is that the proposed model allows to determine not only the place of enterprises of each stage, but also the quantity of product which must be transported from the enterprise of one production stage to the enterprise of another production stage, and to consumers, while the relationships between the enterprises are not rigidly defined (there are no pre-set chains of production).

3 MATERIALS UNDER ANALYSIS

3.1 Two-stage location-allocation problem with continuous set of potential placements of the first-stage enterprises

In this section, we present an appropriate model for the optimal location problem of multi-stage production in a given region. The following assumptions about the elements of the problem are made to describe the model.

– N first-stage enterprises can be located in any point of area Ω and they get raw material from suppliers continuously distributed in the given area;

– a finite number of potential placement points are given for the enterprises of the second stage $J = \left\{ \tau_1^{II}, \tau_2^{II}, \ldots \tau_{M_1}^{II} \right\}$;

– the cost of placing for enterprises of both stage is constant and is not related to the placement point;

– the location of a finite number K of customers of final product and their demands are known.

It is necessary to place production, which includes two types of enterprises (N first-stage enterprises and M second-stage enterprises) in the region Ω, to define regions of their service and the amount of product transported from first-stage enterprises to second-stage enterprises so that the cost of raw materials and products delivering is minimal.

The organizational scheme of production is shown in Fig. 2.

164

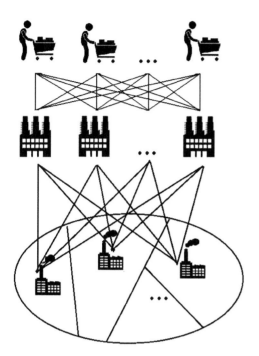

Figure 2. The scheme of two-stage continuous location problem

The above problem can be described by the following model:

Problem 1

Minimize

$$\sum_{i=1}^{N}\int_{\Omega_i} c_i^I(x,\tau_i^I)\rho(x)dx + \sum_{i=1}^{N}\sum_{j=1}^{M_1} c_{ij}v_{ij} +$$

$$+\sum_{k=1}^{K}\sum_{j=1}^{M_1} c_{jk}^{II}v_{jk}, \tag{1}$$

under constraints

$$\int_{\Omega_i}\rho(x)dx = b_i^I , \quad i=1,2,\dots N , \tag{2}$$

$$\sum_{j=1}^{M_1} v_{ij}^I = b_i^I , \quad i=1,2,\dots N , \tag{3}$$

$$\sum_{i=1}^{N} v_{ij}^I = b_j^{II}\lambda_j , \quad j=1,2\dots M_1 , \tag{4}$$

$$\sum_{j=1}^{M_1} v_{jk}^{II} = b_k , \quad k=1,2,\dots K , \tag{5}$$

$$\sum_{k=1}^{K} v_{jk}^{II} = b_j\lambda_j , \quad k=1,2,\dots K , \tag{6}$$

$$\bigcup_{i=1}^{N}\Omega_i = \Omega , \tag{7}$$

$$\Omega_i \bigcap \Omega_j = 0 , \quad i\neq j, \quad i,j=1,2,\dots N , \tag{8}$$

$$v_{jk}^{II}\geq 0, \quad v_{ij}^I\geq 0 , \quad k=1,2,\dots K , \quad j=1,2\dots M_1 ,$$

$$i=1,2,\dots N , \tag{9}$$

$$\sum_{j=1}^{M_1}\lambda_j = M , \quad \lambda_j\in\{0;1\} , \quad i=1,2,\dots N , \tag{10}$$

$$\tau^I = (\tau_1^I,\tau_2^I\dots\tau_N^I) , \quad \tau^I\in\Omega^N , \tag{11}$$

where b_i^r – the power of i-th enterprise of r-stage, $r=1,2$;

$\tau_i^r = (\tau_{i1}^r,\tau_{i2}^r)$ coordinates of i-th enterprise of r-stage; $r=1,2$;

$\tau_k = (\tau_1,\tau_2)$ – given coordinates of customer k,

b_k – demand of customer k;

$c_i^I(x) = c(x,\tau_i^I)$ – the cost of delivering a unit of raw material from supplier point $x\in\Omega$ to i-th enterprise of the first-stage;

$c_{ij} = c(\tau_i^I,\tau_j^{II})$ – the cost of transporting a unit of product from first-stage i-th enterprises to second-stage j-th enterprises;

$c_{jk}^{II} = c(\tau_j^{II},\tau_k)$ – the cost of delivering products from second-stage j-th enterprises to k-th consumer;

$\rho(x)$ – amount of raw material in the x point of Ω;

v_{ij}^I – the volume of products delivered from i-th of the first-stage enterprise to j-th second-stage enterprises;

v_{jk}^{II} – the volume of products delivered from j-th second-stage enterprises to k-th consumers;

$$\lambda_j = \begin{cases} 1, & \text{if the second-stage enterprise is placed} \\ & \text{in position } j, \\ 0, & \text{otherwise.} \end{cases}$$

The objective functional (1) describes the total cost of transportation of raw materials and final

165

products and the enterprise accommodation; the constraints (2) mean that the total reserves of the resource in the service area of the i-th enterprise of I stage are not less than the production capacity of the company; (3) – is the amount of product exported from the i-th enterprise of I stage, which is no more than the production capacity of this enterprise; (4) – is the amount of product delivered to the j-th enterprise of the II stage, which is no more than production capacity of this enterprise; (5) – is demand of consumers to be satisfied; the constraints (7), (8) express the requirement that each supplier shall be related to only one I stage enterprise.

Some of approaches based on optimal set partitioning method (OPS) (Kiselova & Koryashkina 2013) were proposed in works (Us S.A & Stanina 2014) to solving this problem.

3.2 The approaches to solving the problem and numerical algorithms

The existence of two components outlines the particularity of this problem. One of them is discrete and the other is continuous. That is why we propose to divide the process of solving into two steps, so two approaches are possible.

Approach 1

The solving of the problem is carried out in two steps.

During the first step some set N_1 of potential location places of first-stage enterprises is defined, taking into account the distribution of raw material, next the discrete location-allocation problem is solved.

The problem of the first step is the OPS problem under constrains (2), (7), (8), namely:

Problem 1.1

$$F\left(\Omega_1, \Omega_2, ..., \Omega_{N_1}, \tau_1, \tau_2, ..., \tau_{N_1}\right) =$$

$$= \sum_{i=1}^{N_1} \int_{\Omega_i} c_i^I\left(x, \tau_i\right) \rho(x) dx \to \min \tag{12}$$

$$\text{mes}\left(\Omega_i \cap \Omega_j\right) = 0 , \ i \neq j, \ i,j = 1,2,...N_1, \tag{13}$$

$$\bigcup_{i=1}^{N} \Omega_i = \Omega, \tag{14}$$

The second-step problem is formulated as follows:

Problem 1.2

$$\sum_{i=1}^{N_1} A_i^I x_i + \sum_{j=1}^{M_1} A_j^{II} \lambda_j + \sum_{i=1}^{N_1} \sum_{j=1}^{M_1} c_{ij}^{II} v_{ij} +$$

$$+ \sum_{j=1}^{M_1} \sum_{k=1}^{K} c_{jk} v_{kj} \to \min \tag{15}$$

under constraints

$$\sum_{j=1}^{M_1} v_{ij} = b_i^I x_i , \ i = 1,2,...N_1, \tag{16}$$

$$\sum_{i=1}^{N_1} v_{ij} = b_j^{II} \lambda_j , \ j = 1,2,...M_1, \tag{17}$$

$$\sum_{k=1}^{K} v_{jk} = b_j^{II} \lambda_j , \ j = 1,2,...M_1, \tag{18}$$

$$\sum_{j=1}^{M_1} v_{jk} = b_k , \ k = 1,2,...K , \tag{19}$$

$$v_{jk} \geq 0, \ v_{ij} \geq 0 , \ k = 1,2,...K , \ j = 1,2,...M_1,$$
$$i = 1,2,...N_1, \tag{20}$$

$$\sum_{j=1}^{M_1} \lambda_j = M , \ \lambda_j \in \{0;1\} , \ j = 1,2,...M_1, \tag{21}$$

$$\sum_{i=1}^{N_1} x_i = N , \ x_i \in \{0;1\} , \ i = 1,2,...N_1, \tag{22}$$

$$\lambda_j = \begin{cases} 1, & \text{if the second-stage enterprise is placed} \\ & \text{in position } j, \\ 0, & \text{otherwise,} \end{cases}$$

$$j = 1,2,...M ,$$

$$x_i = \begin{cases} 1, & \text{if the first-stage enterprise is placed} \\ & \text{in position } i, \\ 0, & \text{otherwise,} \end{cases}$$

$$i = 1,2,...N_1,$$

where set M_1 – is set of potential places for location of first-stage enterprises, which have been defined by step 1, A^r – costs of location for i-th enterprise of r-stage, $r = I, II$;

Let us assume, that conditions of existence of the problem solving are satisfied, namely

$$\sum_{k=1}^{K} b_k = \sum_{j=1}^{M_1} b_j^{II} \lambda_j = \sum_{i=1}^{N_1} b_i^I x_i .$$

It is noteworthy, that problem 1.1 is continual OSP problem with restrictions type equality.

166

The combined algorithm, based on using genetic algorithm and method of potentials, was proposed for solving a two-step problem.

The potential location places can be characterized by different properties. We proposed to take into consideration their features through quality coefficients β_i for objective function. Thus, we have the following problem:

Minimize

$$\sum_{i\in M_1}\sum_{j\in M_2}\beta_i c_{ij}^I v_{ij}^I + \sum_{j\in M_2}\sum_{k=1}^K c_{jk}^{II} v_{jk}^{II}, \qquad (23)$$

Under constraints $(16)-(22)$.

Let us formulate the algorithm for solving problem1.2.

Algorithm 1

1. Select the initial population consisting of arrays with possible initial placing of enterprises at I and II stages.

2. Calculate the distance from the I-st stage enterprises to II-nd stage enterprises and from the II-nd stage enterprises to the consumers.

3. Solve the transport problem by potential methods for known placement of enterprises at I and II stages.

4. Calculate of objective function by (23)

5. Save best of objective function value and corresponding enterprise's numbers into massive.

6. Select the "parents", crossover, and mutation of new options for placement.

7. Repeat points $2-4$ for a new solution.

8. Add a new solution to massive and delete the worst solution.

9. Check the criterion of the end: if it is satisfactory – go to the next step, otherwise return to point 6.

10. Display the best results and graphics.

End of the algorithm.

Approach 2

Its idea is also based on combined application of discrete and continuous optimization methods.

In that case we presume that placement of the second stage enterprises is primary and use discrete optimization methods or heuristic algorithms to solve their location problem. The inner problem is the location problem for the first-stage enterprises. It is solved by OPS method or its modification.

The inner problem in that case can be formulated as follows:

Problem 1.3

$$\sum_{i=1}^N \int_{\Omega_i} c_i\left(x,\tau_i\right)\rho(x)dx + \sum_{i=1}^N\sum_{j=1}^M c_{ij}^I\left(\tau_i^I,\tau_j^{II}\right)v_{ij} +$$

$$+\sum_{j=1}^M\sum_{k=1}^K c_{jk}^{II}v_{jk}^{II} \to \min \qquad (24)$$

under constraints

$$\int_{\Omega_i}\rho(x)dx = b_i^I, \qquad i=\overline{1,N} \qquad (25)$$

$$\mathrm{mes}\left(\Omega_i\cap\Omega_j\right)=0, \; i\neq j, \;\; i,j=1,2,\dots N, \qquad (26)$$

$$\bigcup_{i=1}^N\Omega_i=\Omega, \qquad (27)$$

$$\sum_{i=1}^N v_{ij}^I = b_j^{II}, \; j=1,2,\dots M, \qquad (28)$$

$$\sum_{j=1}^M v_{jk}^{II} = b_k, \; k=1,2,\dots K, \qquad (29)$$

$$v_{jk}^{II} \geq 0, \; v_{ij}^I \geq 0, \; k=1,2,\dots K, \; j=1,2,\dots M,$$
$$i=1,2,\dots N, \qquad (30)$$

$$\tau^I = \left(\tau_1^I,\tau_2^I\dots\tau_N^I\right)\tau^I\in\Omega^N, \qquad (31)$$

where $\tau^{II} = \left(\tau_1^{II},\tau_2^{II}\dots\tau_M^{II}\right)$ is the fixed placement for second-stage enterprises.

Last summand of functional (1.22) does not affect the first-stage enterprises's location, and can be excluded from the functional to solve location problem. As a result, we get the following problem.

Problem 1.4.

$$\sum_{i=1}^N\int_{\Omega_i} c_i\left(x,\tau_i^I\right)\rho(x)dx +$$

$$+\sum_{i=1}^N\sum_{j=1}^M c_{ij}^I\left(\tau_i^I,\tau_j^{II}\right)v_{ij} \to \min \qquad (32)$$

under constraints

$$\int_{\Omega_i}\rho(x)dx = b_i^I, \qquad i=\overline{1,N} \qquad (33)$$

$$\mathrm{mes}\left(\Omega_i\cap\Omega_j\right)=0, \; i\neq j, \;\; i,j=1,2,\dots N, \qquad (34)$$

$$\bigcup_{i=1}^N\Omega_i=\Omega, \qquad (35)$$

$$\sum_{i=1}^N v_{ij} = b_j^{II}, \; j=1,2,\dots M, \qquad (36)$$

167

$$v_{ij} \geq 0, \quad j = 1, 2....M, \quad i = 1, 2,...N, \quad (37)$$

$$\lambda_j \in \{0; 1\}, \quad \tau^I = \left(\tau_1^I, \tau_2^I...\tau_N^I\right), \quad \tau^I \in \Omega^N,$$

and next the problem:

$$\sum_{j=1}^{M} \sum_{k=1}^{K} c_{jk} v_{kj} \rightarrow \min, \quad (38)$$

$$\sum_{j=1}^{M} v_{jk} = b_k, \quad k = 1, 2,...K, \quad (39)$$

$$\sum_{k=1}^{K} v_{jk} = b_j^{II}, \quad j = 1, 2,...M, \quad (40)$$

must be solved to obtain the first-stage enterprise placement.

Problem 1.4 is a problem of continuous OPS with additional links. For its solution, a method based on the OPS method, and algorithm solutions, part of which is N.Z. Shore r-algorithm have been developed.

Now, we will describe the algorithm implementation of this approach in more detail.

Algorithm 2

Step 0

1. Select any arbitrary set of potential location places $\tau_0^{II} = \left(\tau_1^{II}, \tau_2^{II},...\tau_M^{II}\right)$ for second-stage enterprises.

2. Solve the location problem for the first-stage enterprises, presuming that second-stage enterprises have been already placed.

3. Calculate the value of the objective functional $F_1 = F(\tau^I, \tau^{II})$ of the original problem by (32), and suppose $F_{opt} = F_0$, $\tau^{opt} = \left(\tau_0^I, \tau_0^{II}\right)$.

Let step $(k-1)$ be carried out. Describe step k

Step k

1. Select any arbitrary set of potential location places $\tau_0^{II} = \left(\tau_1^{II}, \tau_2^{II},...\tau_M^{II}\right)$ for second-stage enterprises, which has not been examined;

2. Solve the location problem for the first-stage enterprises (32) – (37) by OPS methods modification, presuming that second-stage enterprises have been already placed;

3. Calculate the value of the objective functional $F_k = F(\tau^I, \tau^{II})$ of the original problem by (32);

4. If $F_{opt} \leq F_k$, then go to step 5, otherwise suppose $F_{opt} = F_k$, $\tau^{opt} = \left(\tau_k^I, \tau_k^{II}\right)$ and go to step 5;

5. If the condition of the end of the process* is achieved, then finish algorithm, otherwise – go to step $(k+1)$;

End of the algorithm.

*) Criterion for the end of the process is selected depending on the chosen method for solving the problem of second-stage enterprises location.

3.3 Computational results

The proposed algorithms were numerically implemented and tested on the following model problems.

Model problem 1. Supposing there is a set $\Omega = \{(x, y) | 0 \leq x \leq 30, 10 \leq y \leq 30\}$ of some product suppliers.

The coordinates of the final product consumers are known.

Production includes two stages. The set of potential places for I and II stage enterprises is given by experts. The quality of potential location places for first-stage enterprises has been evaluate by 10-point scale $a = \{5, 1, 10, 5, 8, 3, 5, 6, 5, 7, 10, 9, 10, 10, 6, 6, 7, 6, 10, 1\}$, where 1 – least preferred place, 10 – most preferred.

The enterprises capacity for each stage $b^I = \{50, 150, 150, 100, 50, 100\}$,

$b^{II} = \{100, 200, 250, 50\}$ and for consumer demand

$b = \{100, 100, 100, 150, 50, 100\}$ are known previously.

Six first-stage enterprises and four second-stage enterprises should be placed in this area Ω, the volume of products' delivery for each stage should be defined and partition $\Omega_1, \Omega_2, ... \Omega_N$ of the set Ω should be found, so that the cost of shipping raw materials and products is minimal under constraints (16) – (20).

Algorithm 1 was applied to solve this problem.

The time of program work was chosen as criterion to end the program: T = 30 sec.

Crossovering was carried out by the principle of «one-point», probability of mutations was taken to be 0.6, with 3% of the total number of genes mutated.

The results of algorithm application are shown in Table 1 and Fig. 3.

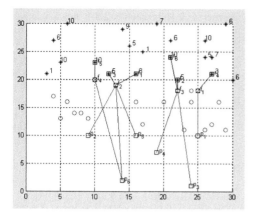

Figure 3. The location obtained by solving model problem 1

Asterisks indicate possible placement for I stage enterprises (r_i) and the assessments α_i, are related to them, circles – are the possible placements for II stage enterprises (f_j) and squares are the location of the consumers (p_k). The chosen places for enterprises location are indicated.

The obtained results show that application of algorithm gives acceptable solution for two-stage location problem and can be successfully used.

Now let us consider the inner problem of approach 2 (point 2 from algorithm 2). Remember, that it is the continuous OPS problem with additional links, the algorithm of its solving presented in (Us & Stanina 2015).

Table 1. The results of solving problem 1.2.

I-stage enterprises			II-stage enterprises			consumers	
coordinates of location	traffic volume	producers' capacities	coordinates of location	traffic volume	producers' capacities	coordinates of location	Demand
				100		(9, 10)	100
(16, 21)	50	50	(13, 19)	50	200	(16, 10)	50
(22, 20)	150	150	(22, 18)	150	250	(19, 7)	150
(12, 21)	150	150	(13, 19)	50	200	(14, 2)	100
(27, 21)	100	100	(25, 18)	100	100	(25, 10)	100
(10, 23)	50	50	(10, 20)	50	50	(14, 2)	100
(21, 24)	100	100	(22, 18)	100	250	(24, 1)	100
(16, 21)	50	50	(13, 19)	100	200	(9, 10)	100
Objective functional			9389,16				

Model problem 2. Suppliers of some product are located in area $\Omega = \{(x, y) | 0 \le x \le 1, 0 \le y \le 1\}$. The product is produced by three enterprises of first stage and processed by two enterprises of second stage.

The location places of second-stage enterprises are known: $\tau_1^{II} = (0,35;0,1)$, $\tau_1^{II} = (0,8;0,85)$. It is necessary to place the first-stage enterprises and to define the volume of transportation so that total delivery cost of raw material to first-stage enterprises and "semi-finished" product from first-stage to second-stage enterprises is minimal.

The maximum volumes of transportations have been determined in advance as $b_1^I = 1$, $b_2^I = 2$, and $b_3^I = 3$. Criterion for the end of algorithm is the time of program work, $T = 10$ sec.

The input data are shown in Fig. 4. The input location for first-stage enterprises are marked by

circles, the known places of second-stage enterprises are marked by crosses.

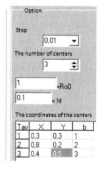

Figure 4. Input data for problem 1.4

As is shown in Figure 5 the optimal partitioning for service zone and placement for enterprises were obtained as a result. The traffic volumes are presented in Table 2.

Accordingly, optimal coordinates for first stage enterprises are $\tau_1^{I*} = (0,32;0,64)$,

$\tau_1^{II*} = (0,17;0,41)$, $\tau_1^{III*} = (0,78;0,67)$.

The optimal value of object functional is $F = 3,8461$.

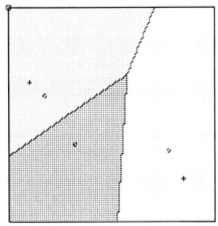

Figure 5. The result of solving problem 1.4

Table 2. The optimal traffic volume for model problem 2

	$\tau_1^{II} = (0,35;0,1)$	$\tau_2^{II} = (0,8;0,85)$
$\tau_1^{I*} = (0,32;0,64)$	1	0
$\tau_2^{I*} = (0,17;0,41)$	2	0
$\tau_3^{I*} = (0,78;0,67)$	0	3

4 CONCLUSION

In this paper, we considered the problem of multi-stage location. The appropriate mathematical model described in combined forms is proposed. The proposed model allows to consider cases when suppliers are located in any point of the given area and links between the enterprises are not predefined. The approaches and solution algorithms based on continuous set partitioning method are proposed for the combined model. Experiments show that the proposed algorithm is feasible, yields good results and can be used for solving cubting-edge problems.

REFERENCE

Brimberg J., Hansen P., Mladenović N. & Salhi. S. 2008. *A Survey of Solution Methods for the Continuous Location-Allocation Problem.* International Journal of Operations Research, Vol.5, Issue 1: 1–12.

Drezner Z. & Hamacher H. 2001. *Facility Location: Application and Theory.* Berlin: Springer.

Farahani R.Z. & Hekmatfar M. 2009. *Facility Location. Concept, Model, Algorithms and Case Studies.* Springer Dordrecht Heidelberg London New York.

Kiseleva E. & Koryashkina L.S. 2013. *Models and methods for solving continuous problems of optimal sets partitioning: linear, nonlinear, dynamic problems: monograph* (in Russian). Kiev: Naukova dumka.

Bischoff M., Fleischmann T. & Klamroth K. 2009. *The Multi-Facility Location-Allocation Problem with Polyhedral Barriers.* Computers and Operations Research, Vol. 36, Issue 5: 1376 – 1392.

Paksoy T., Özceylan E., Weber G.-W. 2010. *A Multi-Objective Mixed Integer Programming Model For Multi Echelon Supply Chain Network Design and Optimization.* System research and information technologies, Issue 4: 47–57.

Ren Y. 2011, *Metaheuristics for multiobjective capacitated location allocation on logistics Networks.* [See http://spectrum.library.concordia.ca/7269/1/Ren_MASc_S2011.pdf]

Zhu J., Huang J., Liu D., Han J. 2008. *Resources Allocation Problem for Local Reserve Depots in Disaster Management Based on Scenario Analysis.* The 7th International Symposium on Operations Research and Its Applications (ISORA'08): 395–407.

Gimadi, E.H. 1995. *Efficient algorithms for solving multistep problem in a string* (in Russian). Discrete Analysis and Operations Research, October–December, Vol. 2, Issue 4: 13–3.

Trubin V.A. & Sharifov F.A. 1992. *Simple multistage location problem on a treelike network* (in Russian). Cybernetics and Systems Analysis November–December, Vol. 28, Issue 6: 912–917.

Ageev A.A., Gimadi A.H. & Kurochkin A.A. 2009. *A polynomial algorithm for solving the problem of placing on the chain uniform capacities* (in Russian). Discrete Analysis and Operations Research, Vol. 16:5: 3–18.

You F. & Grossmann I.E. 2008. *Mixed-Integer Nonlinear Programming Models and Algorithms for Large-Scale Supply Chain Design with Stochastic Inventory Management.* Ind. Eng. Chem. Res. 2008, Vol. 47: 7802–7817.

Us S.A. & Stanina O.D. 2014. *On mathematical models of multi-stage problems of facility location* (in Russian). Issues of applied mathematics and mathematical modeling. Collected research works. Dnepropetrovsk: DNU: 258 – 267.

Us S.A. & Stanina O.D. 2015. *Algorithm to solving for problem of optimal set partition with additional relationships* (in Ukrainian). System analysis and information technology: 17-th International conference SAIT 2015, Kyiv, Ukraine, June 22 – 25, 2015. Proceedings. – ESC "IASA" NTUU "KPI": 116.

Power Engineering and Information Technologies In Technical Objects Control – Pivnyak, Beshta
& Alekseyev (eds)
© 2016 Taylor & Francis Group, London, ISBN 978-1-138-71479-3

Mathematical modelling of electric conductivity of dense and fluidized beds

M. Gubinskyi, S. Fiodorov, Ye. Kremniova
State Higher Educational Institution "National Metallurgical Academy of Ukraine", Dnipro, Ukraine

O. Gogotsi
Materials Research Centre, Kyiv, Ukraine

T. Vvedenska
State Higher Educational Institution "National Mining University", Dnipro, Ukraine

ABSTRACT: Carbon materials processed in electrical thermal furnaces with dense and fluidized beds are widely used in machine building and metallurgical industry (Gasik & Gasik 2007; Fedorov et al). Their operational principle is based on emitting joule heat in material volume by passing electric current through it. However, in spite of its commonality, the physical idea of the bed electric conductivity (Lakomskyi 2008; Borodulia 1973; Gupta, Sathiyamoorthy 1999), as the main mechanism of heating and processing carbon materials in high temperature electrothermal facilities, requires specification of a number of topical issues related to the impact produced by the dense phase, granulometric composition of raw material, the process temperature, the temperature of contact surfaces and geometry of the working space. To solve these tasks, it is necessary to apply a systematic approach comprising complementary elements of empirical and theoretical analyses. The aim of this work was to develop a generalized mathematical model of the bed electric conductivity.

1 GENERAL FORMULATION OF THE BED CONDUCTIVITY PROBLEM

Analysis of the known experimental results (Lakomskyi 2008; Borodulia 1973) shows that the processes of electrothermal heating in packed and fluidized beds may be presented by one model of conductivity considering specifics of either bed mode (Fig. 1). As in the case of packed bed the conductivity of emulsion phase of the fluidized bed can be defined as a result of a certain average actual number of contacts subjected to insufficient dynamic loads. The bubble phase can be considered as external insertions with relatively low conductivity.

Differential equation of instant electrical field in the bed with non-uniform properties for direct current can be generally presented in operator form (1)

under boundary conditions on the surfaces of phase boundary, electrodes and lining (2-4):

$$\text{div}\,(\sigma\,\text{grad}\,\varphi) = 0\,; \tag{1}$$

$$\left(\sigma\cdot\frac{\partial\varphi}{\partial n}\right)_{\text{I}} = \left(\sigma\cdot\frac{\partial\varphi}{\partial n}\right)_{\text{II}}\,; \tag{2}$$

$$\varphi = \varphi_e\,; \tag{3}$$

$$\frac{\partial\varphi}{\partial n} = 0, \tag{4}$$

where σ – medium conductivity related to space coordinates, properties and temperature of the substance, S/m; "e" – index related to the electrodes surfaces; n – normal to the surface.

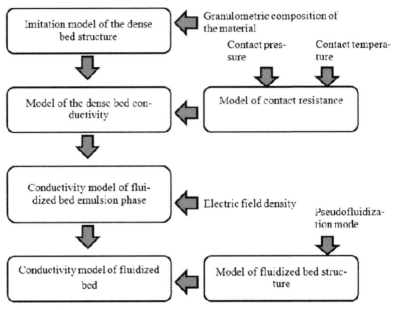

Figure 1. Structure diagram of the mathematical model of bed conductivity

In such universal form, equation (1) is solved numerically for arbitrary domains, including complex geometric shapes of bed facilities. Using the method of finite differences with partitioning of the computational domain into Cartesian grid, for the mean points we obtain the following:

$$\sigma_{i-1/2} \cdot \frac{\varphi_{i,j,k} - \varphi_{i-1,j,k}}{\Delta x} \cdot \Delta y \cdot \Delta z + \sigma_{i+1/2} \cdot \frac{\varphi_{i,j,k} - \varphi_{i+1,j,k}}{\Delta x} \cdot \Delta y \cdot \Delta z +$$

$$+ \sigma_{j-1/2} \cdot \frac{\varphi_{i,j,k} - \varphi_{i,j-1,k}}{\Delta y} \cdot \Delta x \cdot \Delta z + \sigma_{j+1/2} \cdot \frac{\varphi_{i,j,k} - \varphi_{i,j+1,k}}{\Delta y} \cdot \Delta x \cdot \Delta z +$$

$$+ \sigma_{k-1/2} \cdot \frac{\varphi_{i,j,k} - \varphi_{i,j,k-1}}{\Delta z} \cdot \Delta x \cdot \Delta y + \sigma_{k+1/2} \cdot \frac{\varphi_{i,j,k} - \varphi_{i,j,k+1}}{\Delta z} \cdot \Delta x \cdot \Delta y = 0,$$

wherefrom the potential of the current point is determined by:

$$\varphi_{i,j,k} = \left(\sigma_{i-1/2} \cdot \varphi_{i-1,j,k} \cdot \frac{\Delta y \cdot \Delta z}{\Delta x} + \sigma_{i+1/2} \cdot \varphi_{i+1,j,k} \cdot \frac{\Delta y \cdot \Delta z}{\Delta x} + \right.$$

$$+ \sigma_{j-1/2} \cdot \varphi_{i,j-1,k} \cdot \frac{\Delta x \cdot \Delta z}{\Delta y} + \sigma_{j+1/2} \cdot \varphi_{i,j+1,k} \cdot \frac{\Delta x \cdot \Delta z}{\Delta y} +$$

$$\left. + \sigma_{k-1/2} \cdot \varphi_{i,j,k-1} \cdot \frac{\Delta x \cdot \Delta y}{\Delta z} + \sigma_{k+1/2} \cdot \varphi_{i,j,k+1} \cdot \frac{\Delta x \cdot \Delta y}{\Delta z} \right) /$$

$$\left(\sigma_{i-1/2} \cdot \frac{\Delta y \cdot \Delta z}{\Delta x} + \sigma_{i+1/2} \cdot \frac{\Delta y \cdot \Delta z}{\Delta x} + \sigma_{j-1/2} \cdot \frac{\Delta x \cdot \Delta z}{\Delta y} + \sigma_{j+1/2} \cdot \frac{\Delta x \cdot \Delta z}{\Delta y} + \right.$$

$$\left. + \sigma_{k-1/2} \cdot \frac{\Delta x \cdot \Delta y}{\Delta z} + \sigma_{k+1/2} \cdot \frac{\Delta x \cdot \Delta y}{\Delta z} \right)$$

174

or for the constant subinterval ($\Delta x = \Delta y = \Delta z$):

$$\varphi_{i,j,k} = \left(\sigma_{i-1/2} \cdot \varphi_{i-1,j,k} + \sigma_{i+1/2} \cdot \varphi_{i+1,j,k} + \sigma_{j-1/2} \cdot \varphi_{i,j-1,k} \right.$$

$$+ \sigma_{j+1/2} \cdot \varphi_{i,j+1,k} + \sigma_{k-1/2} \cdot \varphi_{i,j,k-1} + \sigma_{k+1/2} \cdot \varphi_{i,j,k+1} \Big) /$$

$$\left(\sigma_{i-1/2} + \sigma_{i+1/2} + \sigma_{j-1/2} + \sigma_{j+1/2} + \sigma_{k-1/2} + \sigma_{k+1/2} \right).$$

Thus, stationary problem (1) is reduced to solving a system of IxJxK algebraic equations (7) under specified boundary conditions and known values of the medium conductivity in every elementary volume. The balance of currents on the electrodes' surfaces was taken as convergence criterion:

$$\sum \Box \cdot \frac{\partial \varphi}{\partial n}\bigg|_1 = \sum \Box \cdot \frac{\partial \varphi}{\partial n}\bigg|_2 , \qquad (8)$$

where indexes (1) and (2) correspond to the potential gradient on the cathode and anode.

Small fragments of the bed, which are of the same size range as particles forming the bed (Fig. 2), have a pronounced heterogeneous topological structure. For solution of (2.7), it is necessary to apply small grids of the research domain partitioning. It is noteworthy that computational problems arise when gas conductivity is by several orders of magnitude less than material conductivity: $\sigma_g \ll \sigma_m$.

As electric charge is mainly transferred in particles and their point contacts, the main gradient of the electric field potential is concentrated in the points of the particles' contacts. Such infinitely small contact surfaces act as a kind of "bottleneck" in solving the system of finite algebraic equations, which sufficiently reduces convergence of iterative computations. Furthermore, parameters of actual rough contacts of disperse particles in the bed (undulation, factual and nominal contact surfaces) are unknown, taken rather conventionally and act as adaptive coefficients of the model.

The above consideration explains the fact that for some fragments of the bed (Fig.2) it is not possible to apply formalized presentation of conductivity – which is common in modelling – in the form of a certain integral index related to the properties of material and gas, as this presentation turns very approximate. Such approach is not always applicable to different geometric and temperature conditions without considering bed properties. However, it works well for the known empirical values of isotropic medium conductivity, as well as for the case of grounded selection of the step for partitioning the bed into elementary volumes. Hence, a question follows what is meant under an elementary volume of the bed.

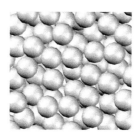

Figure 2. Fragment of monodisperse bed of spherical particles

Defining the limits of elementary volume existence in the bed is of great significance for a wide range of problems related to processes of heat and mass exchange. We assume that the elementary volume of the bed can be defined as a certain minimal fragment, containing a finite number of elements, on condition that physical and geometric properties of this fragment coincide with properties of the bed on the whole. From the practical perspective, further research should be focused on empirical determination of the bed elementary volume as a regularly shaped fragment: a parallelepiped or a cube.

2 CARCASS MODEL OF CONDUCTIVITY FOR A FRAGMENT OF THE PACKED (EMULSION PHASE OF THE FLUIDIZED) BED.

The adequate mathematical model of electric conductivity of the bed elementary volume (fragment) should take into account the bed carcass structure, electrocontact properties of interacting surfaces and conductivity of gas gaps. The main feature of this contact model is resistance of a single contact between grains, which leads to the formation of a certain electrically conductive carcass (Fig. 3). In such physical formulation, the sample spacing of the problem is an interval between geometric centers of the bed elements, potentials being located in the nodes of a polygonal grid, coinciding with these centers.

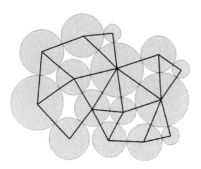

Figure 3. Diagram of electrically conductive carcass for the 2-dimensional bed of non-uniform particles

Conductivity equation for the carcass in respect to electric potential for such nodes can be presented in finite algebraic form:

$$\varphi_i = \frac{\sum_{j=1}^{N_i} \sigma_{i \cdot j} \cdot \varphi_j}{\sum_{j=1}^{N_i} \sigma_{i \cdot j}} \tag{9}$$

Where i – index number of the bed elements; N_i – number of contacts of i-th element; j – index number of the current contact for i-th bed element; $\square_{i \cdot j}$ – contact conductivity.

Solution of (9) is possible for a known bed structure and parameters of contact conductivity. The concept of the contact embraces all the diversity of the transfer considering the particles resistance, area of constriction, transient area and resistance of gas gaps. The contact resistance based on the developed diagram (Fig.4) is defined by:

$$R_\kappa = \frac{(R_1 + R_2 + R_3 + R_4 + R_5) \cdot R_6}{(R_1 + R_2 + R_3 + R_4 + R_5) + R_6} = \frac{R_M \cdot R_r}{R_M + R_r}, \tag{10}$$

where R_1, R_5 – resistances of the bed elements, Ohm; R_2, R_4 – resistances of constriction area; R_5 – transient contact resistance; $R_6 = R_g$ – resistance of a gas gap; $R_m = R_1 + R_2 + R_3 + R_4 + R_5 + R_6$ – total resistance between two bed elements.

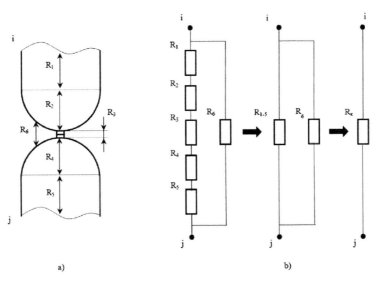

a)   b)

Figure 4. Equivalent replacement of resistances of a real contact: a) contact diagram; b) replacement diagram; i, j – indices of contacting elements coinciding with their geometric centers.

3 MODELLING MATERIAL CONDUCTIVITY

Temperature dependence of electric conductivity of graphitized carbon materials is well studied experimentally (Ostrovskyi et al 1986; Maltseva, Marmer 1963). Development and application of the known regression models of the type $\square(t)$ do not present any difficulty. However, for the research purposes it is necessary to consider the original crystal structure of the material, as well as it possible changes under the thermal effect. Hence, ungra-

phitized carbon material is characterized by mostly rising irreversible temperature dependence of conductivity because of graphitization. Graphite of different grades demonstrate inverse proportion $\square(t)$ starting from 500-1000°C. The major part of material in the working space of the electrical thermal furnace with fluidized bed (ETFFB) is comprised by already graphitized particles which is conditioned by the furnace volume capacity and operational time of material presence in the bed (about 20 minutes). Thus for the purpose of ETFFB modelling, it would be correct to use the processed material physical properties of the graphite type.

4 MODELLING GAS CONDUCTIVITY.

In modelling the electrical thermal fluidized bed, it is critical to understand the role of gas bubble phase in the general picture of bed conductivity. In view of this task, it is necessary to analyse electro-conductive properties of gas at relatively low temperatures (~3000K). According to numerous researches, nitrogen which is used as the fluidizing agent in ETFFB is ionized only insignificanly at temperatures below 3000K (Zdanov 2008; Krinberg 1965). Gas conductive properties in such conditions are determined mainly by small concentrations of mobile electrons. That is why a lot of plasma researchers treat nitrogen as a non-conductive medium in the given temperature range.

Modelling equilibrium thermodynamical state of gaseous nitrogen in «Terra» suite (Trusov 2012) for the conditions T = 2000-3500K, P = 0,1 MPa has yielded the following data on nitrogen composition (Table 2.1). As we see, N_2 molecules are rather stable, while the number of free electrons at 3500K does not exceed $0,72 \cdot 10^{-8}$ mole/kg.

Table 1. Nitrogen ionization at 2000-3500K, P = 0,1 MPa

Temperature, K	Mass composition, mole/kg					
	e^-	N^+	N_2^+	N	N_2	N_3
2000	$0,34 \cdot 10^{-18}$	0	$0,11 \cdot 10^{-17}$	$0,32 \cdot 10^{-7}$	35,70	$0,69 \cdot 10^{-13}$
2500	$0,15 \cdot 10^{-12}$	$0,55 \cdot 10^{-17}$	$0,15 \cdot 10^{-12}$	$0,10 \cdot 10^{-4}$	35,70	$0,14 \cdot 10^{-10}$
3000	$0,79 \cdot 10^{-10}$	$0,61 \cdot 10^{-13}$	$0,79 \cdot 10^{-10}$	$0,50 \cdot 10^{-3}$	35,70	$0,52 \cdot 10^{-9}$
3500	$0,72 \cdot 10^{-8}$	$0,48 \cdot 10^{-10}$	$0,72 \cdot 10^{-8}$	$0,79 \cdot 10^{-2}$	35,69	$0,68 \cdot 10^{-8}$

Both domestic and foreign scholars provide regrettably few experimental data of gas conductivity at 2000-3500K. Most research is done with computational methods based on thermodynamical modelling occasionally tested and confirmed experimentally. Gas conductivity model on the basis of approximate solution for Boltzmann kinetic equation [8,9] has become most widely held:

$$\sigma(T) = \frac{\sqrt{\pi} \cdot e^2}{\sqrt{8 \cdot m \cdot k \cdot T}} \cdot \frac{\alpha \cdot N_e}{N_i \cdot \langle Q_i \rangle + \sum N_l \cdot \langle Q_l \rangle} \, S/cm \quad (11)$$

where T – plasma temperature, K; m= $9,109 \cdot 10^{-31}$ kg and e = $-1,602 \cdot 10^{-19}$C – electron mass and charge; k=$1,381 \cdot 10^{-23}$ J/K – Boltzmann constant; N_e and N_i – concentrations of electrons and ions, cm^{-3}; N_l – concentration of neutral particles of the first kind, cm^{-3}; $\langle Q \rangle$ – so called electric conductivity sections for singly-charged ions and neutral par-

ticles и нейтральных частиц, cm^2; α – coefficient, accounting for electron-electron interaction.

Approximate solution (10) incorporates Maxwellian velocity distribution, absence of strong electric and magnetic fields, as well as of big temperature gradients, small concentrations of charged particles and presence of local thermal equilibrium. Results of computation using (10) for equilibrium compositions of nitrogen are given in Table 2. As the data obtained correlate with air conductivity values at 2000-3500K taken from literature, they can be used in further analysis of disperse systems' conductivity. Gas conductivity appears to be orders of magnitude less than that of fluidized bed of electrode graphite particles. Hence, in the system "suspended material-bubbles", gas phase can be treated as non-conductive medium.

Table 2. Nitrogen conductivity under weak ionization (P = 0,1 MPa)

T, K	2000	2500	3000	3500
σ, S/cm	$8,427 \cdot 10^{-17}$	$3,108 \cdot 10^{-11}$	$1,430 \cdot 10^{-8}$	$1,136 \cdot 10^{-6}$

However, the latter conclusion does not refer to point contacts between particles when high current density results in super-high local temperatures typical for plasma. This effect should be taken into account during modelling conductivity of a single contact. (Fig. 4).

5 IMPACT OF THE GAS BUBBLE PHASE ON FLUIDISED BED CONDUCTIVITY

To study two-phase fluidised bed, it is reasonable to divide task (1) and treat the accumulation of gas bubbles as uniformly distributed insertions in emulsion phase volume under the condition of constant physical gas properties within the limits of minimum one bubble. Set as such, the conductivity task for the system "emulsion phase-bubbles" goes in line with the classical theory of conductivity for multicomponent mixtures, elaborated by Maxwell, Rayleigh, Aiken, Burger etc (Davidson et al. 1985). Since the fluidized bed is a dynamically changing non-uniform system, it is well worth studying how relative bed conductivity is affected not only by the shape but also by the size and mutual orientation of gas bubbles. Given exsiting variety of analytical solutions for the conductivity problem of two-component mixtures, we will consider only a few transforming them to conditions of non-conductive insertions. The analysis below is based on solutions developed by Dulnev G.N. by way of partitioning the elementary volume by adiabatic (12) and isothermal planes (13), and Odelevskyi V.I. solution (14) (Davidson et al. 1985) which, as becomes obvious further, is most relevant to the results of numerous experiments:

$$\Lambda = \frac{\sigma}{\sigma_1} = 1 - \delta^{2/3}; \qquad (12)$$

$$\Lambda = \frac{\sigma}{\sigma_1} = \frac{1 - \delta^{2/3}}{1 - \delta^{2/3} \cdot \left(1 - \delta^{1/3}\right)}; \qquad (13)$$

$$\Lambda = \frac{\sigma}{\sigma_1} = \frac{2 \cdot (1 - \delta)}{2 + \delta}, \qquad (14)$$

Where $\square$ – electric conductivity of fluidized bed, S/cm; $\square_1$ – electric conductivity of emulsion phase, S/cm; δ – volume ratio of gas bubbles.

In solving the conductivity problem for the two-component mixture with distributed non-conductive insertions it is important to select an elementary unit cell with isotropic structure (Fig.5) possessing the properties of the system as a whole. Such selection for fluidized bed is defined by the frequency of gas bubbles repetition. The elementary cell dimensions may be commensurable with the appartus diameter which typical for big bubbles in the bed top portion, and for bed transition into the plug flow mode of pseudofluidization.

a) b) c)

Figure 5. Elementary cells of isotropic structure with volume insertions
a) Cube-like insertions with cube orientation
b) Sphere-like insertions with cube orientation
c) Sphere-like insertions with rhombohedral orientation

If the bubbles dimensions were sufficiently less than the apparatus diameter, a cubic cell with cube-like or sphere-like insertions was considered as an element of such structure. Numerical computations were made on the basis of spacial potential problem solution (7) for two types of isotropic functions: cubic and rhombohedral (Fig.6). In a certain computational series, we assumed that the bubble phase consists of uniformly distributed gas bubbles of d_1 and d_2 diameters of their ratio 2:1.

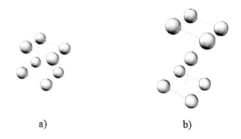

a) b)

Figure 6. Location of closed non-conductive insertions in isotropic medium
a) cubic structure
b) rhombohedral structure

The problem (8) was solved by the method of finite differences with the relative subinterval of the elementary cell $\Delta\bar{n} = 0,05$. The choice of research range in terms of bubbles volume ratio $\delta = 0$-30% was based on the joint analysis of expressions for the porosity of non-uniform fluidised bed (Gelperin et al. 1967) and two-phase system "gas bubbles-emulsion" (Davidson et al. 1985) considering physical properties of graphite ($\rho = 2050$ kg/m³, $d_q = 0,1$-2,0 mm), hydrodynamic modes and ETFFB geometry (Gasik & Gasik 2007):

$$\varepsilon = \varepsilon_{mf} \cdot \left(\frac{Re + 0,02 \cdot Re^2}{Re_{mf} + 0,02 \cdot Re_{mf}^2}\right)^{0,1} ; \qquad (15)$$

$$\varepsilon = \delta + (1 - \delta) \cdot \varepsilon_{mf}, \qquad (16)$$

where ε –porosity of non-uniform fluidised bed; $\varepsilon_{mf} = 0,40$-$0,45$ – porosity of dense bed at the time of passing the stability threshold; Re, Re_{mf} – Reynolds criteria for the fluidized bed and the passing of the stability threshold accordingly.

The research results are presented in Fig.7. Numerical computations testify that the relative conductivity of the fluidised bed lies in the intermediate region between the computational data of dependences (12-13) and correspond to Odelevskyi formula (14).

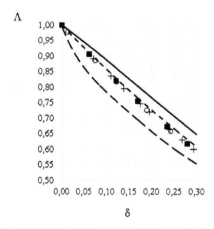

Figure 7. Relationship between relative conductivity of isotropic structure and the volume ratio of closed insertions

· — — - analytical dependence (1)
———— - analytical dependence (2)
- - - - - Odelevskyi formula(3)
■ - numerical solution (4-7) for cubic structure of sphere-like insertions of the same diameter
○ - numerical solution (4-7) for rhombohedral structure of sphere-like insertions of the same diameter
+ - numerical solution (4-7) for rhombohedral structure of sphere-like insertions with the diameters ratio 2:1.

Functional relationship $\Lambda(\delta)$ in the studied range approaches the linear kind. Increase in the volume ratio of bubbles is accompanied by decrease in conductivity by 41%, which is equivalent to growth of the bed resistance by 70% in respect to the "dense" phase. Nevertheless, if $\delta = 2\text{-}5\%$, which corresponds to the real structure of the bubble fluidised bed, the relative conductivity Λ is only 0,96-0,92.

Thus, gas bubbles contribution into conductivity reduction is quite insignificant, which is proved by experimental data. They testify that as pseudofluidization increases, the volume ratio of bubbles and specific electric resistance of the bubble bed stay practically unchanged (Borodulia 1973). Consequently, most of resistance growth (2-5times) during transition from dense bed to fluidised bed is conditioned by worsening of the contact interaction of emulsion phase particles both among themselves and with the electrodes. Changes in the mutual orientaton of the gas bubbles (transition from cubic structure to rhombohedral, as well as variations of non-uniform insertions diameters, do not change the picture in essence. Formally, this does not change the whole picture in principle. It allows to limit the computations of the specific electric conductivity of the fluidized bed to the volume ratio of bubbles, given their dimensions are constant and their distribution structure is isotropic. To be able to use these results, computational data (Fig. 7) were approx-

imated by the linear regression correlaton with the regression coefficient $R^2 = 0{,}99$:

$$\Lambda = 0{,}99 - 1{,}34 \cdot \delta. \tag{17}$$

The expression obtained could be used for assessment of the relative conductivity of fluidized bed under small sizes of gas bubbles within the range of

$$\delta = 0\text{-}0{,}3.$$

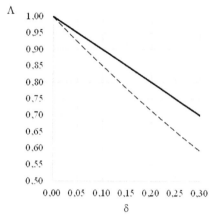

Figure 8. Dependence of relative conductivity on gas bubble distribution

· — — - uniform bubble group
———— - single bubble

Growth and coalescence of bubbles along the bed height is accompanied by the increase in their diameter, decrease in their total number and frequency. That is why, as was stated above, the bed top section can become expressly non-uniform when the horizon is crossed by separate bubbles whose size is commensurable with the apparatus dimensions. Figure 8 demonstrates how relative bed conductivity under uniform (isotropic) bubble distribution (17) is related to relative bed conductivity for a single bubble passing, which corresponds to the maximum non-uniformity of the bed. If $\delta = 2\text{-}5\%$, relative bed conductivity for a single bubble passing is 0,98-0,95.

Thus, for the bubble mode of carbon materials fluidization $\delta = 2\text{-}5\%$, decrease in the bed conductivity in respect to the "dense" phase is within 0,92-0,98, which corresponds to increase in the specific bed conductivity by 2-9%.

In conclusion, it is important to focus on certain characteristics of fluidizaiton which set limits to the so-called *active electrically conductive area*. The bubble forming zone located near gas distribution grid is characterized by a relatively high porosity and uniform disperse phase which creates resistance by several orders of magnitude higher. The same refers to the horizon of pistons formation for big ratios of fluidization as well as to the above-bed space where there is no consistent emulsion phase. These features should be necessarily considered in the research into ETFFB.

6 CONCLUSIONS

The research resulted in the development of the generalized mathematical model of electric conductivity of the disperse material bed which allows to take into account such bed characteristics as dense phase structure, presence of gas bubble phase, granulometric composition of the material, process temperature, contact surfaces properties and geometry of the working space.

Two-phase model describing the fluidized bed of carbon material particles helped to establish that if gas bubbles are distributed uniformly and their volume ratio δ = 2-5% , the bed relative electric conductivity Λ= 0,96-0,92. For the case of a single bubble, which corresponds to the maximum non-uniformity of fluidized bed, relative conductivity for the given range is 0,98-0,95. Thus , contribution of gas bubbles to the reduction of conductivity turns negligible. The resistance increase (2-5 times) during transition from dense phase to fluidized one occurs mostly due to poorer contact interaction of emulsion phase particles with each other and wth the electrodes.

Changes in the bubble size and in their relative position do not produce noteworthy impact on the electric conductvity of the bed.

Numerical computation applied to the potential problem of isotropic structures conductivity resulted in regressional dependence of bed conductivity from the bubbles volume ratio $\Lambda = 0,99 - 1,34 \cdot \delta$ with approximation validity $R^2 > 0,99$, which can be used for modelling specific electric conductivity of the bubble fluidized bed with uniformly distributed bubbles.

REFERENCES

Gasik M.M. & Gasik M.I. 2007. *Complex model for anthracite calcination in electric calcinator* (in Russian). Electrometallurgia, Issue 2.

Fedorov S.S., Gubynskyi M.V., Barsukov I.V. et al. *Modeling the Operation Regimes in Ultra-high Temperature Continuous Reactors.* Brookhaven National Laboratory, U.S. Department of Energy's Office of Science, USA. [See https://www.bnl.gov /isd/documents/86110.pdf]

Lakomskyi V.I. 2008. *Electric and electrocontact properties of electrode thermoanthracite* (in Russian). Kyiv: Akademperiodika: 106.

Borodulia V.A. 1973. *High-temperature processes in electrothermal fluidized bed* (in Russian). Minsk: Nauka i Technika: 173.

Gupta C.K., Sathiyamoorthy D. 1999. *Fluid bed technology in materials processing.* Boca Raton, Fla.: CRC Press: 528.

Ostrovskyi V. et al. 1986. *Artificial graphite* (in Russian). Moscow: Metallurgia: 272.

Maltseva L.F., Marmer E.N. 1963. *Determining electrical properties of graphite at high temperatures* (in Russian). Foreign technology division: 10.

Zdanov V. 2008. *Phenomena of transfer in gases and plasma* (in Russian). Moscow: MIPR: 240.

Krinberg I. 1965. *Electric conductivity of air in the presence of impurity* (in Russian). Applied mechanics and technical physics, Issue 1.

Trusov B. 2012. *Programme system of modelling phase and chemical equilibrium at high temperatures* (in Russian). Vestnyk of MSTU named after N. Bauman: Device Building Series: Special Issue 2: Programme Engineering: 240-249.

Chudnovskyi A. 1962. *Thermophysical properties of disperse materials* (in Russian). Moscow: Publishing House of physical and mathematical literature: 456.

Dulnev G., Novikov V. 1991. *Transfer processes in non-uniform media* (in Russian). Leningrad: Energoatomizdat: 248.

Gelperin N., Einshtein V., Kvasha V. 1967. *Fundamentals of fluidization technology* (in Russian). Moscow: Chemistry: 664.

Davidson J.F., Clift R., Harrison. D 1985. *Fluidization, 2nd ed.* New York: Academic Press: 733.

Power Engineering and Information Technologies In Technical Objects Control – Pivnyak, Beshta
& Alekseyev (eds)
© 2016 Taylor & Francis Group, London, ISBN 978-1-138-71479-3

Application of information technology for decrease of fine grinding power consumption

N. Pryadko
State Higher Educational Institution "National Mining University", Dnipro, Ukraine

ABSTRACT. Jet mill is widely provided in mining, chemical, ceramics and chemistry industries, where the pure fine products are used. Fine grinding is the most power-intensive preparation enrichment process in technological closed cycle. Therefore it is important to create grinding control systems for a mill productivity improvement and power consumption reduction. In this study the critical level of fine grinding power consumption is determined, that make Rittinger's hypothesis more exact for fine grinding. The numerical experiments on developed simulation dynamic model of the closed cycle demonstrate that use the grinding effectiveness ratio K, connected with the product size composition, can improve the mill productivity on 5,5% . The paper shows the possibilities to use information technologies for the grinding mode control and grinding power intensity on the basis of acoustic monitoring results. The sell model of jet grinding process differs from well known process models by description of mill operating condition according to the acoustic signal characteristics. The review of some developed grinding optimization methods is defined. They are based on the established connections of acoustic signal characteristics of grinding zone and technological process parameters. The results of theoretical study are proved on industrial mining plant of Ukraine.

1 INTRODUCTION

Efficiency of mineral enrichment is determined first of all by quality, the ready product volume and necessary technological and industrial resources. Fine crushing is the most power-intensive preparation enrichment process in technological closed cycle. Therefore it is the extremely important to create and improve grinding control systems for a mill productivity increase, that is for process power consumption decrease at observance of product required quality. Headway on a grinding optimization is impossible without wide attraction of the modern information technologies characterized by a wide spectrum of functions.

Complexity and multifactor processes of mineral processing at real manufacture imposes essential restrictions on an opportunity of adequate control model creation. The power feature research of the mineral fine grinding closed cycles has allowed to specify performance conditions of Rittinger's hypothesis and to establish dependence of proportionality factor change of new formed surface increase and spent energy for its formation on ready product fineness (Pyvnyak 2014). Earlier received acoustic monitoring results adopted for various methods and technologies of computer modelling of jet grinding processes, can be used for information system creation of decision-making support on management of continuous crushing productions (Pryadko 2013).

2 FORMULATION THE PROBLEM

For decrease of fine grinding power consumption the acoustic monitoring results can be used, but it's necessary to develop information technology basis.

3 MATERIALS FOR RESEARCH

Experimental researches of crushing laws in ball mills have allowed finding the main dependence which connects new formed specific surface at crushing with power inputs on this process (see Fig. 1), on which 3 sectors are possible to mark.

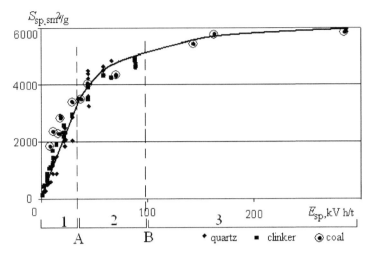

S_{sp}, sm²/g

6000

4000

2000

0

0   1   2   100   3   200 E_{sp}, kV h/t

A   B   ◆ quartz   ■ clinker   ◉ coal

0 - A – linear dependence area; B – a critical line.
A - B – storage site of fine particles;

Figure 1. Dependence of a specific surface on specific power consumption

The first area (0-A) linearly connects new formed specific surface with power inputs. It corresponds to a rather large particle grinding. The more fine particles are appearing the more energy is spent on destruction of these particles and consequently the mentioned characteristic sensitivity (an inclination angle) is reduced (site 2 on Fig. 1).

The third site corresponds to conditions when there are many fine particles, the sizes of them become nearer to micron, therefore it is necessary to destroy the particles in the sizes closed to molecules. Naturally it takes a huge energy for such destruction, therefore the characteristic $S_{sp}(E_{sp})$ becomes the least sensitive (the inclination angle approximates to zero).

Thus, the border between second and third sites for function $S_{sp}(E_{sp})$ is opportunity scope for mechanical disclosing of valuable components and, in general, applications of mechanical enrichment. We name this border (a straight B line on Fig. 1) as a critical level. Thus, the primary goal of fine grinding optimization is the control of power consumption transition over the established level, both aside increases, and the reduction side because that leads to infringement of an grinding optimality.

For dry jet grinding it is also observed three characteristic zones of dependence of a specific surface on process power consumption, and the critical level is allocated. But jet grinding allows getting finer product with higher dispersion parameter (S_{sp} is up to 8000 sm²/g), since grinding process in jets is more dynamical. It proves by the dependence equations of a specific surface on power consumption for fine grinding in jet and ball mills on Fig. 2, where the equation $S_{sp}^{1} = 182{,}4E_{sp} + 1057{,}5$ (a straight line 1) for jet grinding and $S_{sp}^{2} = 95{,}5E_{sp} + 28{,}2$ (a straight line 2) for ball mill grinding. Factors of straight line inclination show dynamism of grinding process ($k_1 = 182{,}4$; $k_2 = 95{,}5$).

The fine grinding product analysis has shown increase of mill power consumption at achievement of the certain product particle size. For ball grinding this size has estimated 40-60 microns. For superfine grinding, (on the example of jet grinding) the size test product analysis on device "Malvern" has shown, that prevailing (more than 90 %) the size of jet grinding product particles of various materials is 2-1 microns, and their average size - up to 10 microns. Thus, a critical level for superfine grinding is achievement of particle size about 10-25 microns.

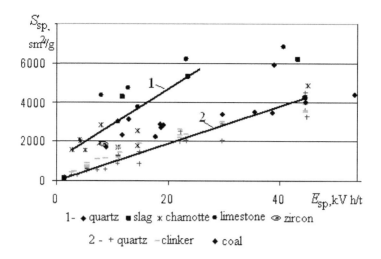

Figure 2. Dependences of a specific surface on power consumption at jet (1) and ball (2) grinding.

The critical level behind which the further grinding leads to substantial increase of power consumption is revealed. Excess of a critical level shows that the significant amount rather fine particles has collected and further their size decrease is not demanded. Reduction of a critical level indicates that grinding process has some reserve, and the mill productivity can be increased.

The developed simulation dynamic model of the closed cycle grinding kinetics has confirmed the received theoretical conclusions about kinetic features of fine grinding and connection with mill productivity, that is the process power consumption (Pryadko 2015), with particle-size distribution of a grinding product. The grinding effectiveness ratio K is deduced on the basis of that connection. It increased the mill productivity on 5,5% (Fig. 3b). Fig 3 shows main presentation window of working program written on basis of the AnyLogic software package.

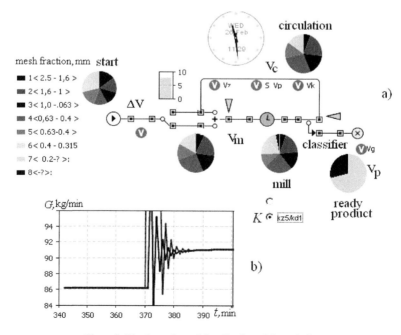

Figure 3. The dynamic model realization of fine grinding.

183

The acoustic information technology is developed for power consumption decrease of the grinding closed cycle and the size control regulation of ready product. The signals which are written down by the gauge in a grinding zone at continuous acoustic monitoring are used for the description of a material size and a grinding mode. Connections of the basic technological parameters and acoustic parameters of the grinding and classification zones are established (Pryadko 2015). On the acoustic monitoring result basis of mineral jet grinding the database including the basic information process characteristics is created.

Some fine grinding optimization approaches (Pryadko 2012), which have allowed creating the monitoring system of mill productivity and crushed product dispersion is developed.

On the basis of the established connections of acoustic signal characteristics of grinding zones, mode and process technological parameters some grinding optimization directions are developed. They have helped to increased mill productivity and decrease grinding power intensity.

1. The continuous control of the signal maximal amplitude during grinding (Pilov 2014);

2. The grinding process control on basis of the established technology – acoustic criteria and factors (Pryadko 2014);

3. Quality control of the crushed product (Pryadko 2013);

4. The continuous power analysis of acoustic signals (Pryadko 2015);

5. Wavelet-analysis system of the acoustic signals in the grinding zone (Mikhalyov 2014);

6. The Neural network analysis of grinding modes (Pryadko 2012);

7. The surface analysis of density function of signal amplitude probability distribution in characteristic zones (Pryadko 2015);

8. Modelling grinding process on the basis of Markov circuits (Pryadko 2013).

The developed information technologies used for jet grinding optimization is shown on Fig. 4.

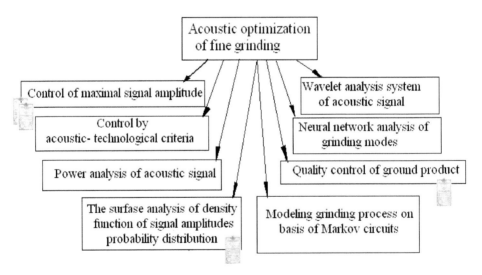

Figure 4. Information technologies used for jet grinding optimization

At grinding process modelling the particle distribution density is replaced with probabilities of a state. All possible system characteristics form space of conditions, and their probability makes a vector of a condition. It gives the basis to apply of Markov's circuit for jet grinding process modelling. The mathematical model is based on the amplitude analysis of the acoustic signals which have been written down during each discrete moment of time of process of grinding (Pryadko 2013). The grinding matrix formulas with the help of selective and distributive function of grinding are determined. Created the sell model of jet grinding process gives the information on the particle characteristic sizes in mill, the jet material loading degrees according to the acoustic radiation characteristics: signal amplitudes, frequencies and their distribution on size.

On the basis of the continuous power and amplitude - frequency analysis of the grinding zone signals control ways of jet mill grinding modes are developed. Performance of mill optimum operation conditions allows to realize timeliness of material loading and to support stably high level productivity. These ways are covered by patents of Ukraine (Pilov 2014).

The wavelet (Mikhalyov 2014) and neuronal analysis of the signals which have been written down by the gauge during monitoring of grinding installation characteristic zones has allowed to identify optimum modes of various loose material grinding. The created control and process optimization program enables to increase the grinding efficiency at product quality compliance on the basis of the signal characteristics analysis.

The research results approbation in industrial mill (Volnogorsk mining plant, Ukraine) conditions proved efficiency of the created optimization system of fine grinding process. By the example of an industrial jet mill operation with production of crushed zircon concentrate sized 63 microns it's established the opportunity on the one hand of power consumption decrease in 1,5 - 2 times (Fig.5) by optimization of jets loading by material and on the other hand on maintenance of a maximum mill productivity.

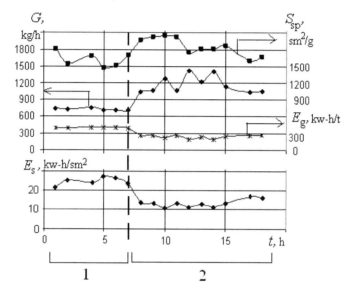

Figure 5. The results of jet grinding optimization

On Fig.5 the industrial mill productivity (G, kg/h) improve and power consumption reduction due to optimization on acoustic monitoring results are shown, where E_g is power consumption (for one ton of ready product) and E_s - power intensity of zircon dispersion for one sm². The first time period (Fig.5) the standard grinding mode is on operation, the second period the optimization according to acoustic monitoring results is applied.

4 CONCLUSIONS

Process modelling and application of the information technologies based on acoustic monitoring results, has allowed solving a main fine grinding task – the control of a grinding mode and keeping parameters over a power consumption critical level. It results in decrease of grinding power consumption at the product dispersion and mill high productivity compliance.

REFERENCES

Mikhalyov A., Pryadko N., Suhomlin R., Kotyra A. 2014. *Application of wavelet transform in analysis of jet grinding process.* Lublin: Elektronika, Issue 8: 20-22.
Pilov P.I., Gorobets L.J., Pryadko N.S. 2014. *The monitoring way of jet grinding and gas jet mill* (in Ukrainian). UA Patent № 104427.
Pivnyak G.G., Pilov P.I., Pryadko N.S. 2014. *Decrease of Power Consumption in Fine Grinding of Minerals.* Springer International Publishing Switserland: Mine Planning and Equipment Selection, C. Drebenstedt and R. Singhal (eds): 1069 -1079.
Pryadko N.S. 2012. *Acoustic-emission monitoring of jet grinding process grinding* (in Russian).

Cambridge International Science Publishing: Technical Diagnostics and Non-Destructive Testing, Issue 4: 46 – 52.

Pryadko N.S. 2013. *Acoustic researches of jet grinding* (in Russian). Saarbrucken Germany: LAP LAMBERT Academic Publishing: 172.

Pryadko N., Strelnikov G. 2014. *Information technology of fine grinding managment* (in Russian). Dnipropetrovsk: Technical Mechanics, Issue 4: 118 – 125.

Pryadko N.S. 2015. *Optimization of fine grinding on the acoustic monitoring basis.* Taylor & Francis Group: CRC Press: Power Engineering, Control & Information Technologies in Geotechnical Systems: 99 – 108.

Power Engineering and Information Technologies In Technical Objects Control – Pivnyak, Beshta
& Alekseyev (eds)
© 2016 Taylor & Francis Group, London, ISBN 978-1-138-71479-3

The algorithm of automatic lump material's granulometric composition monitoring

I. Udovik, S. Matsyuk & V. Kornienko
State Higher Educational Institution "National Mining University", Dnipro, Ukraine

ABSTRACT: The modified algorithm of images processing to automatic granulometric composition monitoring of the lump material in a stream using the morphological filtration of binary images has been developed. Efficiency of its use at monitoring of iron ore granulometric composition has been evaluated by modeling.

1 INTRODUCTION

A granulometric composition of lump material is a key indicator of the material crushing and grinding technological processes in construction and mining industries.

The operational control of technological processes, for example, iron ore crushing and grinding at automated monitoring of its granulometric composition at different processing stages improves the performance of these processes for the final product to 5-7% (Herbst & Blust 2000).

2 FORMULATION OF THE PROBLEM

The noncontact methods and control devices of lump material fineness in a stream, especially, optical monitoring devices are prospective for problems of automatic control and operating (Kozin 2005).

The theoretical basis for optical monitoring methods realized by the systems of technical sight (STS) is Ackerman's theorem (Nazarov 2006) about equality of relations of areas and volumes of lump material. It proves the use of photoplanimetry method in such STS (Herbst & Blust 2000; Nazarov 2006).

Algorithms of information processing in STS contain, as a rule, the following stages (Pratt 2001; Kornienko 2009):

- formation of the image;
- preliminary image processing;
- coding - transfer - receipt - decoding of the image;
- restoration, segmentation and classification of the image.

Formation of the image includes adjustment of optoelectronic system, at which the issues of choice of the vision field, focusing, illumination conditions, exposition, protection of vision field from dust, etc. are solved. Thus, the choice of vision field has to exclude the influence of segregation in the stream of lump material at the controlled image (that can be reached, for example, by installation of optical camera in the material overload zone), and to provide staying of demonstrative sample in the vision field (Kozin 2005).

The preliminary processing of images is directed at improvement (restoration, smoothing) of the image corrupted by noise and other factors.

The sources of noise is non-ideality of camera and algorithms of image formation, poor conditions of shooting (insufficient illumination and dust), and noise in communication channels.

The operations of coding – transfer – receipt – decoding of the image are used in the distributed STS, in which the blocks of image formation and its processing are separated spatially. During their fulfillment the standard decisions (Haffman channel coding), compression of display frames JPEG, MPEG, etc. are used to solve them. (Gonzalez, Woods & Eddins 2006).

The segmentation is to break down the image into homogeneous areas, which are classified further according to purpose of STS (Pratt 2001; Kornienko 2009; Pilov, Alekseev & Udovik 2013).

In the work (Kornienko 2009), the algorithm of image processing for lump materials automatic control is considered. However, it has insufficiently high precision and efficiency owing to use of Wiener frequency filtration and logical filtration of binary images.

The object of the article is development and research of efficiency of the modified algorithm for images processing and technical solutions to realize automatic granulometric composition monitoring of lump material in a stream.

3 THE PROCEDURES OF CONTROL ALGORITHM

The structure of the offered algorithm of image processing for automatic control of lump material granulometric composition includes the following procedures:
1) formation of the image;
2) preliminary image processing;
3) segmentation of the image;
4) binary image filtration;
5) description of the image elements and calculation of granulometric composition.

1. Formation of the image is carried out in the form of two-dimensional matrix of values of brightness $\hat{P}(x, y)$ in coordinates (x, y) - a halftone image (we consider that the levels of brightness of the background is less than the brightness of material pieces).

For technological reasons the points of control are the material overload places (unloading of dump trucks and dump cars), and its transportation through the conveyor.

In this regard, while forming the image there are distortions due to its blurriness, when material moves during exposition.

The images of original and coarsely crushed ore on the conveyor received in the conditions of Inguletsky and Lebedinsky ore mining and processing works (MPW) are shown on Fig. 1.

Figure 1. The images of ore on the dump truck (a), dump car (b) and conveyor (c).

It is possible to limit and compensate the blurriness and defocusing, for example, by limiting the exposition time and by using the restoring filtration.

2. The preliminary image processing. The distorted images may be presented in the spatial domain in the following form:

$$\hat{P}(x, y) = h(x, y) * P(x, y) + n(x, y), \qquad (1)$$

where $*$ – a sign of convolution operation; $h(x,y)$ – the point spread function (PSF) of distorting system; $\widehat{P}(x,y)$, $P(x,y)$ – the observed and original images; $n(x,y)$ – spatial noise.

The Wiener optimum filter takes into account the existence of the noise in the initial signal (image). At low frequencies, the Wiener filter coincides with inverse filter, but unlike it, the Wiener filter is steady at high frequencies as well.

The realized researches have shown that the Wiener filter using correlation functions, which carries out the deconvolution operation, is preferable (in terms of accuracy).

The solutions of the problem of image restoration using the deconvolution operation according to distortions model (1) with Gausse model of PSF are shown on the Fig. 2.

Figure 2. The restored images of ore on the dump truck (a), dump car (b) and conveyor (c).

The size of minimum controlled fineness class (diameter of minimum piece, which can be distinguished on the restored image) is the efficiency (quality) index of performance of formation stages and image restoration.

3. Segmentation of the image. The traditional way of image segmentation is to distinguish the borders of its elements. The gradient methods realized by algorithms of Sobel, Canny, laplacian gaussian, and watershed (Kornienko 2009; Gonzalez Woods & Eddins 2006) are widely used for this purpose. There the segmentation problem is formulated as the problem of the search of the area borders, which are corresponded to the maxima of gradient of brightness function.

The analysis of use of gradient methods showed (Kornienko 2009) that in our case they led to image re-segmentation, and it significantly complicated the subsequent processing (classification of image elements into pieces and background).

The other way of segmentation is a threshold binarization assuming that the background and pieces on the image have in average the different brightness. The histogram methods based on the following idea (Pratt 2001; Gonzalez, Woods & Eddins 2006) are widely used for choosing the value of a threshold. As far as the distribution of

189

probabilities for each class of the image is unimodal, and points of borders are not numerous, the histogram is polymodal (with the number of modes by quantity of objects classes), and failures correspond to the borders.

It is expedient to fulfill the previous distinguishing of borders on the image for realization of such method. It reduces the size of borders in space and respectively reduces their level on a histogram that simplifies the choice of optimum threshold.

The increase of image sharpness by underlining the borders is carried out according to expression (Kornienko 2009):

$$P_1(x,y) = \hat{P}(x,y) + c \cdot \nabla^2 \hat{P}(x,y), \tag{2}$$

where $\nabla^2 \hat{P}(x,y) = \partial^2 \hat{P}(x,y)/\partial x^2 + \partial^2 \hat{P}(x,y)/\partial y^2$ – the second derivative (the Laplace scalar operator – laplacian); $c = sign[\nabla^2 \hat{P}(x,y)]$ – the coefficient caused by a sign of the central element of laplacian mask.

Laplace operator increases the image sharpness (the steepness of brightness transitions from background to pieces and vice versa), but transfers areas with constant brightness to zero. Therefore, the addition of initial image to laplacian in expression (2) restores the brightness levels of these areas (fig. 3).

a
b
c

Figure 3. The images of ore on the dump truck (a), dump car (b) and conveyor (c) after distinguishing the borders.

After distinguishing the image limits its binariszation is carried out, i.e. it means transformation of a halftone image into binary one. The purpose of operation is to divide the image elements into two levels of brightness (background and pieces).

In the general case, the binary image is formed according to expression (Pratt 2001; Kornienko 2009):

$$P_2(x,y) = \begin{cases} 1 & for\ P_{fr}(x,y) \geq \Pi[P_1(x,y)]; \\ 0 & if\ else. \end{cases} \tag{3}$$

where Π - limit function; $P_{fr}(x,y)$ – brightness level of division of image elements on background and pieces.

Otsu`s method (Gonzalez, Woods & Eddins 2006) realizes the choice of threshold value according to image histogram. Thus, the limit function Π is chosen as the parameter, the value of which maximizes the interclass dispersion:

$$\Pi[P_{fr}(x,y)] = \Pi^{O} = \underset{\Pi^{O*}}{\arg\ \max}\ \sigma^{2}(\Pi^{O*}) \qquad (4)$$

where σ^{2} – dispersion between brightness of background and pieces defined by probability of belonging to classes of background and pieces, and mathematical expectations of brightness of background, pieces and the whole image (Kornienko 2009; Gonzalez, Woods & Eddins 2006).

The results of Otsu`s method use are shown on Fig. 4.

Figure 4. The image of ore on the dump truck (a), dump car (b) and conveyor (c) after using Otsu`s method.

4. Filtration of binary images. Binarization of an image does not allow to suppress its high-frequency noise (minor defects on the Fig. 4) through the integrated character of limit function (4). Therefore, the filtration (smoothing) of a binary image is necessary.

The use of algorithm of binary images logical filtration showed (Kornienko 2009) that its use demanded the large volumes of calculations (time expenses). Therefore, the morphological filtration of binary images is used in the offered algorithm. The morphological operations of (Gonzalez, Woods & Eddins 2006): dilatation, erosion, disconnection and closing over processed image and structural element are used in such filtration.

The results of morphological filtration when using the structural element 'disk' with the size of 3x3 are given on Fig. 5.

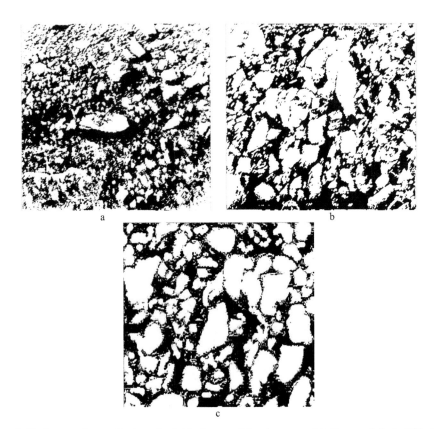

Figure 5. The images of ore on the dump truck (a), dump car (b) and conveyor (c) after morphological filtration.

5. The result description of the image elements are the values of their areas. The differential characteristic of granulometric composition (distribution of pieces by fineness) of the monitoring test (image shot) is determined similar to the photoplanimetry method as the ratio of the total area of pieces of a certain size (equivalent diameter) S_n to the total area of pieces in a shot:

$$\gamma_i = \sum_{n=d_{i-1}}^{d_i} \cdot S_n \Big/ \sum_{z=1}^{Z} S_z , \tag{5}$$

where d_i – equivalent diameter of material pieces of the i fineness class (range); Z – total number of pieces on the image.

The characteristics of lump material granulometric composition for some time interval are defined by averaging the characteristics of tests (images of lump material) passed through the monitoring zone during that time.

For example, the differential characteristics of ore granulometric composition averaged by 30 image frames received in the conditions of Inguletsky and Lebedinsky MPW are given on Fig. 6.

4 ESTIMATION OF ALGORITHM EFFICIENCY

The automatic monitoring of granulometric composition may be realized both within telecommunication system of video monitoring of a company, and in the form of autonomous STS.

The system of video monitoring when using, for example, the equipment of Advantech company may be constructed using the video recorders (VR) VBOX-3200 and industrial video cameras (VC) VBOX-3900. The system elements are connected by the Ethernet network, which provides the access to video information on the basis of the different technologies: modem, ADSL, ISDN, and Internet.

The system includes up to 10 VR, to each of which it may be connected up to 16 VC. The record speed is from 25 to 480 frames per second of NTSC/PAL video standards. While compressing MPEG-4, the size of shot is 2-4 Kb at 320x240 resolution.

192

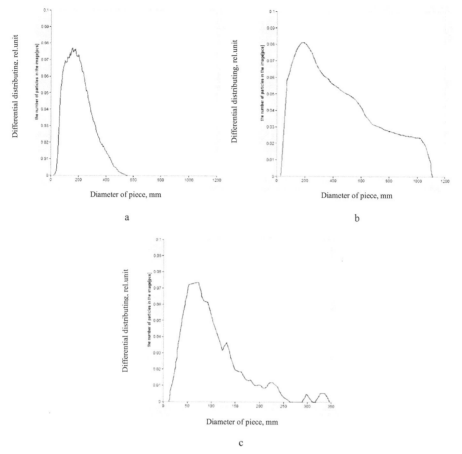

Figure 6. Differential characteristics of ore granulometric composition on the dump truck (a), dump car (b) and conveyor (c).

The autonomous STS may be realized on the digital signal processor TMS320C5xx of Texas Instruments company with a halftone video camera with the resolution of 1200x800.

The simulation result has revealed that the offered algorithm of image processing provides the relative error of determination of the image area of pieces (instrumental error) less than 1% at realization in the video monitoring system and less than 0,1% - for autonomous STS. At that, the error of granulometric composition definition in comparison with the direct screen method is 3-6% relational that meets the technical requirements (Kozin 2005).

The processing time of the image frame in autonomous STS is no more than 0,7 s, and in the video monitoring system - no more than 10 s.

4 CONCLUSIONS

The algorithm for image processing, that provides the problem solving as for automatic granulometric composition monitoring of lump material in a stream with sufficient precision automatic control systems and efficiency has been developed.

The realization of monitoring for granulometric composition in video monitoring system demands less expenses than in autonomous STS, but as well, it has the lower accuracy and efficiency.

The subsequent researches will have to be directed at the development of the control system for processes of ores crushing and grinding using the offered algorithm of automatic lump material`s granulometric composition monitoring.

193

REFERENCES

Herbst J.A. & Blust S.L. 2000. *Video Sampling for Mine to Mill Performance Evaluation, Model Calibration and Simulation.* Control 2000 Annual SME Meeting and Exhibit. – Salt Lake City, Utah, February 28-March 1: 1-17.

Kozin V.Z. 2005. *Monitoring of concentrate technological processes* (in Russian). Ekaterinburg: Higher school: 303.

Nazarov A.S. 2006. *Photogrammetry* (in Russian). Moscow: Tetra of Systems: 368.

Pratt W.K. 2001. *Digital Image Processing.* New York; Chichester; Weinheim; Brisbane: John Wiley and Sons Inc.: 723.

Kornienko V.I. 2009. *Automatic monitoring of particles-sizes of crushed materials in a stream* (in Ukrainian). Dnipropetrovsk: Science Bulletin of The National Mining University, Vol. 6: 69-73.

Gonzalez R., Woods R. & Eddins S. 2006. *Digital image processing in MATLAB* (in Russian). Moscow: Technosphera: 616.

Pilov P.I., Alekseev M.A. & Udovik I.M. 2013. *Binarization algorithm of rock photo images on inhomogeneous background.* Taylor & Francis Group: CRC Press: Energy Efficiency Improving of Geotechnical Systems: 101-108.

Power Engineering and Information Technologies In Technical Objects Control – Pivnyak, Beshta & Alekseyev (eds)
© 2016 Taylor & Francis Group, London, ISBN 978-1-138-71479-3

The distribution of energy consumption in wireless networks to anycast protocol

Vik. Hnatushenko
National Metallurgical Academy of Ukraine, Dnipro, Ukraine

ABSTRACT: Analysis for the probabilistic network lifetime in WSNs, the probabilistic energy consumption analysis is represented, when using discrete channel model for the anycast protocol. The developed model to analyze the distributions of energy consumption in WSNs is presented, and the distributions of node lifetime and network lifetime are derived using the energy consumption distribution. A case study for the anycast protocol is provided, and the relationships between network parameters and the lifetime distribution are investigated. The analysis is validated by extensive simulations.

1 INTRODUCTION

Wireless sensor networks (WSNs) have been a topic of much interest to researchers due to their wide-ranging applications: military applications, environmental applications, health applications, and home applications (Shen 2001; Shi 2004). However, a fundamental problem in WSNs is the limited lifetime of sensor nodes.

To this end, a significant amount of work has been carried out across the protocol stack to prolong the lifetime of WSNs. Examples of which include energy-efficient medium access control (MAC) protocols (Dam 2003), duty-cycling strategies (Gu 2007), energy-efficient routing (Heinzelman 2000), and topology control mechanisms (Kang 2012; Koutsonikolas 2010; Dimokas 2010).

WSN elements interact with each other and depend on each other so that failures in the individual points in the network can reduce the informational value of data transmitted during this operation of the system is maintained. In real situations, there are a variety of natural and man-made interference effects which lead to failures at certain points of the WSN and to the distortion of the transmitted information.

The main requirements for WSN nodes are low power consumption (providing long battery life), high performance, small size and low cost. An important requirement is the ability to WSN their self - nodes must be able to self-associate to the network and to relay data packets to each other, subject to the exchange of information packets only between nodes that are in the field of radio visibility of each other, defines the probability of delivery of information packets between nodes. Shipping routes data from each of the nodes to the PAN-coordinator should be determined dynamically based on the possible failure of the relays or communication disorders.

These requirements regulate the wireless communication standard IEEE 802.15.4 (IEEE Standards 802.15.4. 2006; IEEE 802.15.4 2010). This standard, as well as IEEE 802.11 technology for Wi-Fi, defines two lower levels the Open Systems Interconnection (ISO-OSI) - Physical (PHY) and medium access control (MAC) - sublayer lower OSI data link layer. The features of the standard are low power, short time to connect to the network, to support a large number of clients, the ability to implement the requirements of the standard in low-cost devices.

Methods of preserving energy					
Based on the cycles work		Based on the the quantity of information		Based on mobility	
Control topology	Management power	Data reduction	Optimizing data	Mobile stock	Mobile retransmission

Figure 1. Methods of energy conservation in wireless networks

There are many ways to save energy units, their classification shown in Fig. 1. All methods of preservation can be divided into three groups - is conserving energy by using a series of works based on the amount of information transmitted and mobility. For a series of works include topology control and power management. Control topology aims to use or reduction of redundant links in a network in order to save resources. Managing consumption can apply a variety of energy-efficient MAC protocols and modes of operation. A second class of methods is based on conservation of energy resource amount information to be transmitted, as well as to receive this information an economical way. The energy spent on processing of the information is not comparable less - required for transmission, so the use of intra-network data processing, data compression or prediction. Just use your mobile drains or repeaters to save energy units of sensor networks.

The majority of existing work on energy and lifetime analysis in WSNs is focused on the average measures. Average energy efficiency is evaluated for specific protocols (Buettner 2006; Polastre 2004; Ye 2004), and average energy consumption models are proposed in analytical studies (Dargie 2010; Muhammad 2011). In (Dargie 2010), an analytical energy model is provided to estimate the energy consumption, assuming SMAC (Ye 2004) and Directed Diffusion (Intanagonwiwat 2003) as the MAC and routing protocols. Stochastic characteristics of energy consumption, however, are not captured by these models.

The effects of routing strategies on energy consumption have also been investigated recently. These models, while providing a detailed estimation of average energy consumption for certain protocols, do not offer higher-order statistics of energy consumption for generic MAC or routing protocols (Wang 2006; Haapola 2005).

For single node and network lifetime analysis, most of the existing work only investigates average measures. An analytical model is provided in (Duarte-Melo 2002) to study the energy consumption and lifetime of two-tier cluster WSNs. Lifetime is measured in terms of data collection "rounds" in which an average amount of energy is consumed for each node (Mihalyov 2015). In (Kumar 2005), the lifetime is analyzed for an always-on network, where the energy management problem.

The node lifetime distribution is modeled for a cluster-based network topology, using a TDMA MAC protocol (Noori 2010). However, the applications of this analysis to other network topologies, such as mesh topologies and ad hoc networks, and other protocols, such as CSMA-based protocols, have not been shown.

Anycast technology is utilized widely to solve the data transmission problems for multiple sinks deployed in WSNs. The paradigm of anycast communications (Hu 2005; Thepvilojanapong 2005), also termed one-to-any communications, become very important to a network with multiple sinks: when a sensor node produces data, it has to send it to any neighbor node or sink available. A neighbor node or a sink selection strategy is to choose for each source a neighbor or a sink arbitrarily. An alternative strategy is to route data to the closest neighbor node or sink. Assuming that the sources and the sinks are uniformly distributed throughout the network, this simple strategy is assumed to improve data transmission efficiency and reduce latency.

2 FORMULATION THE PROBLEM

Presented by probabilistic method the distribution of energy consumption in wireless networks using the anycast protocol and calculation based on his lifetime of the network.

3 MATERIALS FOR RESEARCH

The randomness in energy consumption and associated network lifetime depends on two main components. First, the communication protocol is a coincidence of the wireless channel errors and queueing operation. Second, the differences in the network topology leading to different amounts of energy consuming in the network nodes. Consider the distribution of energy consumption for the random deployment, and the special case of deterministic deployment of an example grid topology in which the randomness due to topology can be ignored.

3.1. Problem definition and model systems

In WSNs, energy is consumed by each node for various activities including sensing, data processing, and communication. We assume that each node is equipped with K sensors, and each sensor $k \in \{1, ..., K\}$ is used to sense the physical environment every $T_{s,\kappa}$ seconds with to the moment when energy consumption $e_{s,k}$ to sensing. On the basis of the technical requirements for its intended purpose, the packet is generated locally, if it considers the information meets the specific event. For each received and locally generated package, the node processes the data with an energy consumption of e_p. In addition, the energy consumption for the connection e_c, is a variable depending on the network parameters and protocols running on each node.

Each node x is characterized by speed input traffic $\lambda(x)$, a long line of $M(x)$ and the battery capacity $C(x)$. The wireless channel between the nodes is modeled according to the lognormal fading channel models.

Lifetime distribution node depends on the distribution of energy consumption during any given period T and the total capacity of its battery lifetime C. Distribution the lifetime network depends on the distribution of each node and, how lifetime network is defined. For different applications and network topologies, the lifetime network may be defined differently. The network lifetime is defined as the duration before the battery depletion of the first node.

3.2. Consumption energy to the random topology

We consider the distribution of energy consumption of a single unit with a discrete time Markov model, which is the basis for the analytical model.

Experiments show that in WSNs with random deployment, the randomness of the topology introduces changes to the energy consumption. This change is small, if T is short and increases quadratically with T. It is simulated by a zero mean normal distribution r.v. $E_{tc}(T)$.

Dispersion $E_{tc}(T)$:

$$\Box^2_{tc}(T) = cT^2 , \tag{1}$$

where c - a scaling coefficient, which is determined by the network parameters: the nodes density ρ, the speed of the locally generated traffic λ_{lc} and protocols. For given network parameters and protocols, c is constant.

For the calculations c semi-empirical approach is used for a given set of parameters network and protocols. Simulation is carried out for a finding c as a function of network parameters ρ and λ_{lc} . The obtained c then used to calculate the energy consumption distribution of nodes using anycast protocol.

3.3. Time to live for used protocol Anycast

For the analysis of energy and lifetime with anycast protocol, we assume that nodes are deployed in a circular plane of radius R, have a homogeneous battery capacity C and generate a homogeneous amount of local traffic to a sink located at the center of the plane. Because of symmetry, node dependent variables are the same for each of the narrow ring of radius R.

The techniques for the anycast protocol can be used to calculating the energy consumption and lifetime distribution for other protocols. For this Markov chain for $\{X_n\}$ should be constructed according to the specific behavior of the protocol. Then, the distribution of energy consumption of the single node can be obtained by (2).

$$f_{Ecp(T)}(e) = \pi h^{(T)}(e)\mathbf{1},$$

$$F_{Ecp(T)}(e) = \int_0^e f_{Ecp(T)}(e)de \tag{2}$$

where $\mathbf{1}$ is the appropriately dimensioned column vector containing all $\mathbf{1}$'s.

The distribution of energy consumption at time T depends on the initial state of the system during this period, which is usually randomly chosen. The initial probabilities vector of the system represented by the equilibrium state probability vector π.

Distribution for the single node and network lifetime are using (3) and (4), (5), respectively. Detailed solutions for other protocols is beyond the scope of this work.

The *cdf* of single-node lifetime is approximated as:

$$F_{L(C)}(t) \approx Q\left(\frac{\mu(t) - C}{\sqrt{\Box^2(t)}}\right) \tag{3}$$

Since each node needs to be alive during the lifetime network, the network lifetime distribution (NL) is obtained for a WSN with random deployment as:

$$F_{NL}(t) \approx 1 -$$
$$- \prod_{x \in A} (1 - p_{ex}(x)\Pr\{L(x,C(x)) \le t\}) \tag{4}$$

where $L(x, C(x))$ is the lifetime node located at the point x, if any, with capacity of the battery $C(x)$. Using approximation (3) for the distribution of the lifetime single-node, the network lifetime distribution is approximated by:

$$F_{NL}(t) \approx 1 -$$
$$- \prod_{x \in A} (1 - p_{ex}(x)\Pr\{L(x,C(x)) \le t\}) \tag{5}$$

where $\mu(x, t)$, $\sigma^2(x, t)$ are given by the node located at x. A is the network area. To calculate the

197

area A is discretized small areas of Δx, $p_{ex}(x)$ is the probability that a node exists in a small area around x, the network density ρ, $P_{ex}(x) = \rho\Delta x$.

Presented by semi-empirical approach to obtain compensation component the topology for anycast protocol. Randomly deployed networks with PPP node locations are considered. Modeling are conducted to identify the relationship between the scaling coefficient c and network parameters, such as node density ρ, the locally generated traffic speed λ_l and the duty cycle ξ.

Measurement of energy consumption and the lifetime distribution represented by topology randomness represented zero-mean r.v. $E_{tc}(T)$. The scaling coefficient c is empirically obtained for the anycast protocol for the node located in the ring r.

The energy consumption for each node is measured and recorded dispersion. Regression least squares method was used to determine the scale coefficient c. For these values of node density ρ is locally generated traffic λ_{lc} and performance ξ are the empirical expression for the scale coefficient c is obtained as:

$$c(\lambda_{lc}, \rho) = \alpha(\lambda_{lc})^2 \rho^{-2} \tag{6}$$

where α is a constant irrelevant to λ_{lc}, ρ and ξ. Hundred realizations of the random network topology are generated with a density ρ^* and locally generated traffic speed λ_{lc}^*. The energy consumption and energy dispersion are fixed $(\sigma^*)^2(T)$, during the period T.

3.4. Simulation model

Simulation and modeling are important approaches in the development and evaluation of the systems in terms of time and costs. The simulation shows the expected behavior of the system based on its simulation model under different conditions (Fg. 2). Hence, the purpose of this simulation model is to determine the exact model and predict the behavior of the real system. For the purpose of simulation, we will use OPNET (Marghescu 2011). This simulation tool provides a comprehensive development environment to support modeling of communication networks and distributed systems. Simulation of ZigBee based networks by providing several components of a ZigBee network (ZigBee coordinator, ZigBee router, ZigBee end device, these components can be fixed or mobile, figure 1). The objects are defined according to the standard. The OPNET ZigBee model uses process models (Hammoodi 2009):

• ZigBee MAC model which implements a model

of the IEEE 802.15.4 MAC protocol. The model implements channel scanning, joining and failure/recovery process of the protocol in the un slotted operation mode.

• ZigBee Application model. The process model initiates network joins and formations, generates and receives traffic and generates different simulation reports.

• ZigBee (CSMA/CA) model which implements the media access protocol of the MAC layer. ZigBee parameters used in simulation are:

• Mesh routing Disabled
• Transmit power 0.05
• Transmit band 2.4GHz
• ACK mechanism Enable
• Destination: random
• Packet size: 1024 bytes
• Packet inter-arrival time: constant (1.0)
• Start Time: uniform (20, 21)

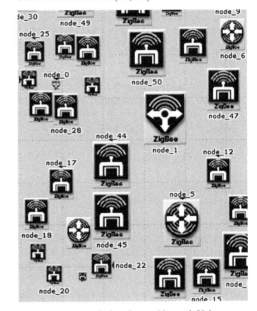

Figure 2.Fragment Wireless Sensor Network Using OPNET

The simulation are conducted with with the installed duty cycle and the speed of traffic, to assess the relationship between the c and the network density ρ. In the network with radius R in total deployed uniformly (equivalent to Poisson point process) N=50,100,...,250 nodes. The duty cycle is ξ=0.2 and locally generated traffic rate at the node is λ_{lc}= 0.05 pkt/min. The value of c in each simulation is shown in Fig. 3(a), the value cN^2 shown in Fig. 3(b). It is revealed that the value cN^2 is constant and thus c proportionately to N^{-2} or ρ^{-2}, when ξ and λ_{lc} are fixed. Note that the peaks in Fig. 3(b) are due to the relatively small total number of random nodes (100).

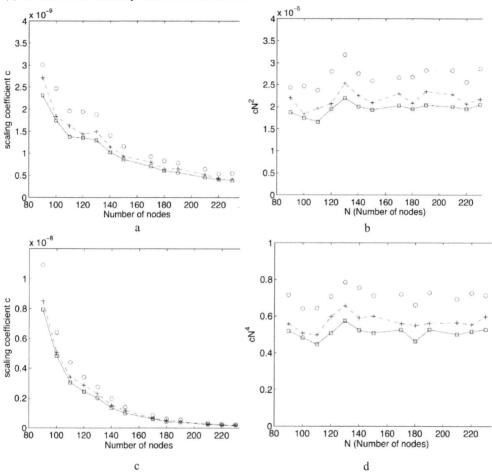

Figure 3. The scaling coefficient c (o – ξ=0.2, + – ξ=0.1, – ξ=0.05)

A higher number of point enables smoothen the peaks, but leads to a significantly higher value calculations in the simulation. Then, for each of the different total number nodes, on the node traffic rate varies, so that the total locally generated traffic rate constant. The values of c and cN^4 shown in Figure 3(c) and 3(d), respectively. It can be seen that the value of cN^4 is constant, assuming that c is proportional to N^4 or ρ^{-4}, when fixed ξ and λ_{lc}.

Simulations are also conducted for various values of the speed of traffic at the node λ_{lc} with fixed duty cycle $\xi = 0.2$ and the total number of nodes N=160. After analyzing these data we can conclude that c is a constant, and thus c is proportional to $(\lambda_{lc})^2$.

The effect of different values of performance ξ is investigated and can be seen, that the variance of the energy consumption does not change too much for different duty cycle values. Consequently, in an empirical approach can be ignored the duty cycle when calculating the energy consumption variance.

Then, taking into account all the above, the empirical expression for c obtained in (5). Using the simulation result for ρ^* =0.057 nodes/m², λ^*_{lc}=0.05 pkt/min and ξ^*=0.2, the scaling coefficient is expressed as

$$c = 3.1 \times 10^3 \rho^2 \lambda_{lc}^2$$

Developed structure can be used to help network design, to investigate the relationship between the probability of achieving a given node and the network lifetime and various network parameters, using the lifetime distribution is obtained in (2). In each of the following tests is considered network architecture. The network density ρ, performance ξ for all nodes, the speed of traffic λ_{lc} for all nodes are different. The default values are for these parameters 0.052, 0.2 and 0.1 pkt/min. The radius of a network is 20 m. The capacity of batteries for all the nodes C=2000mA·H. Other parameters remain unchanged from the previous experiments.

The probability that the lifetime of the node at distance r=12 m is more than 500 hours. To maximize the probability of reaching this lifetime, the density must not be less than 0.053. It can be seen that the probability increases dramatically from around 0 to 1 with the density changes within 0.002 node/m² from 0.095 node/m² to 0,097 node/m². It is a steep change due to the fact that the change in the lifetime of the network is small when the topology is deterministic.

Moreover, the reduction of the duty cycle directly reduces the energy consumption. Increasing or decreasing the speed of traffic on the packet 0.05/min reduces the likelihood of achievement of lifetime. The relationship between probability and network parameters are due to the relationship between energy consumption and the parameters that are represented.

4 CONCLUSIONS

The paper presents a new approach to solving the problems of energy saving in sensor networks, allowing to increase the lifetime of the network using the anycast protocol. The approach is based on a probabilistic analysis of energy consumption throughout the network and received by the distribution of energy consumption for each node. It is shown that, when the length is large, the energy converges to a normal distribution, also provided the mean and variance of the distribution. On the basis of the distribution of energy derived lifetime distribution for each node and the network. The developed model is confirmed by simulation. The results show that the model accurately reflects the distribution of energy consumption an multi-hop WSN.

REFERENCES

Shen C.-C., Srisathapornphat C. and Jaikaeo C. 2001. *Sensor information networking architecture and applications.* IEEE Personal Communications, Vol. 8, Issue 4: 52–59.

Shi E. and Perrig A. 2004. *Designing secure sensor networks.* IEEE Wireless Communications, Vol. 11, Issue 6: 38–43.

Van Dam T. and Langendoen K. 2003. *An adaptive energy-efficient MAC protocol for wireless sensor networks.* Proceedings of the 1st International Conference on Embedded Networked Sensor Systems (SenSys '03), Vol. 11: 171–180.

Gu Y., He T. 2007. *Data forwarding in extremely low dutycycle sensor networks with unreliable communication links.* Proceedings of the 5th ACM International Conference on Embedded Networked Sensor Systems (SenSys '07): 321–334.

Heinzelman W.R., Chandrakasan A. & Balakrishnan H. 2000. *Energy-efficient communication protocol for wireless microsensor networks.* Proceedings of the 33rd Annual Hawaii International Conference on System Siences (HICSS '00): 223.

Kang S., Lee S., Ahn S. and An S. 2012. *Energy efficient topology control based on sociological cluster in wireless sensor networks.* KSII Transactions on Internet and Information Systems, Vol. 6, Issue 1: 361–380.

Koutsonikolas D., Das S.M., Hu Y.C. and Stojmenovic I. 2010. *Hierarchical geographic multicast routing for wireless sensor networks.* Wireless Networks, Vol. 16, Issue 2: 449–466.

Dimokas N., Katsaros D. and Manolopoulos Y. 2010. *Energy efficient distributed clustering in wireless sensor networks.* Journal of Parallel and Distributed Computing, Vol. 70, Issue 4: 371–383.

IEEE Standards 802.15.4. Wireless Medium Access Control (MAC) and Physical Layer (PHY) Specifications for Low-Rate Wireless Personal Area Networks (LR-WPANs). — IEEE Computer Society, 2006.

IEEE 802.15.4 WPAN-LR Task Group. IEEE 802.15 WPAN™ Task Group 4 (TG4), 2010. [See http://www.ieee802.org/15]

ZigBee Alliance [See http://www.zigbee.org]

Buettner M, Yee G.V., Anderson E. and Han R. 2006. *X-MAC: a short preamble MAC protocol for duty-cycled wireless sensor networks.* Proc. of ACM SenSys 2006: 247-261

Dargie W., Chao X. and Denko M.K. 2010. *Modelling the energy cost of a fully operational wireless sensor network.* Telecommunication Systems, Vol. 44(1-2): 3-15.

Duarte-Melo E.J. and Liu M. 2002. *Analysis of energy consumption and lifetime of heterogeneous wireless sensor networks.* Proc. of IEEE GLOBECOM 2002, Taipei, Taiwan: 413-430.

Haapola J., Shelby Z., Pomalaza-Raez C. and Mahonen P. 2005. *Cross-layer energy analysis of multihop wireless sensor networks.* Proc. of IEEE EWSN 2005: 345-367.

Intanagonwiwat C., Govindan R., Estrin D., Heidemann J. & Silva F. 2003. *Directed diffusion for wireless sensor networking.* IEEE/ACM Trans. on Networking, Vol. 11(1): 2-16.

Mihalyov A., Hnatushenko Vic., Hnatushenko V., Vladimirska N. 2015. *Optimization Model Lifetime Wireless Sensor Network.* Proceedings of the 2015 IEEE 8th International Conference on Intelligent Data Acquisition and Advanced, Warsaw, Poland, Vol. 2: 867-871.

Kumar S., Arora A. and Lai T.H. 2005. *On the lifetime analysis of always-on wireless sensor network applications.* Proc. of IEEE MASS 2005, Washington, DC, 274-286, Nov.

Muhammad A., Olivier B., M. Daniel, Thomas A. and Olivier S. 2011. *A Hybrid Model for Accurate Energy Analysis of WSN Nodes.* EURASIP Journal on Embedded Systems, 527-543, Jan.

Noori M. and Ardakani M. 2010. *Lifetime Analysis of Random Event-Driven Clustered Wireless Sensor Networks.* IEEE Trans. on Mobile Computing, Vol. 10(10): 1448-1458.

Polastre J., Hill J. and Culler D. 2004. *Versatile low power media access for wireless sensor networks.* Proc. of ACM SenSys 2004, Baltimore, MA: 289-295.

Wang Q., Hempstead M. and Yang W. 2006. *A realistic power consumption model for wireless sensor network devices.* Proc. of IEEE SECON 2006, Reston, VA: 286-295.

Ye W., Heidemann J. and Estrin D. 2004. *Medium access control with coordinated adaptive sleeping for wireless sensor networks.* IEEE/ACM Trans. Networks, Vol. 12(3): 493-506.

Hu W., Bulusu N. and Jha S. 2005. *A communication paradigm for hybrid sensor/actuator networks.* International Journal of Wireless Information Networks, Vol. 12, Issue 1: 47–59.

Thepvilojanapong N., Tobe Y. and Sezaki K. 2005. *HAR: hierarchybased anycast routing protocol for wireless sensor networks.* Proceedings of the 5th Symposium on Applications and the Internet (SAINT '2005): 204–212.

Hammoodi I.S., Stewart B.G., Kocian1 A., McMeekin S.G. 2009. *A Comprehensive Performance Study of OPNET Modeler For ZigBee Wireless Sensor Networks.* Third International Conference on Next Generation Mobile Applications, Services and Technologies: 357 – 362.

Marghescu P., Pantazica M., Brodeala A. and Svasta P. 2011. *Simulation of a Wireless Sensor Network Using OPNET.* IEEE 17th 1nternational Symposium for Design and Technology in Electronic Packaging (SIITME), Timisoara, Romania: 249–252.

Power Engineering and Information Technologies In Technical Objects Control – Pivnyak, Beshta & Alekseyev (eds)
© 2016 Taylor & Francis Group, London, ISBN 978-1-138-71479-3

The mathematic model of the 3D generalized Neumann boundary problem for the rotating empty cylinder

M. Berdnyk
State Higher Educational Institution "National Mining University", Dnipro, Ukraine

ABSTRACT: In the paper, a generalized 3D equation for energy balance in a cylinder, which is rotating with constant angular velocity ω, is presented with taking into account finiteness of the heat-conduction velocity value. The 3D temperature field in the rotating empty cylinder has been defined with taking into account finiteness of the heat-conduction velocity value.

1 INTRODUCTION

The phenomenological theory of thermal conductivity assumes that heat conduction velocity is infinitely large (Podstrigac 1976). However, when highly intensive non-stationary processes occur, for example, at explosions or high rotation velocities or in supersonic flows, impact of finiteness of the heat conduction velocity value on the heat exchange becomes dramatic (Podstrigac 1976). The (Podstrigac 1976) states that in case of Fourier thermal conductivity law, equation of the energy transfer is true for the one-dimensional, homogeneous and stationary space.

A problem of possibility to extend a generalized energy-transfer equation for the 3D cases is considered in (Podstrigac 1976).

Objective of this work was to develop new generalized 3D mathematic models of temperature distributions in the moving media in the form of boundary problems of mathematical physics which would be further used in a thermal conductivity equation, and to solve the obtained boundary problems for the thermal conductivity equation in order to employ these solutions for controlling the temperature fields.

The overviewed literature has shown that today understanding of the heat-exchange process occurring in rotating cylinders and liquids is not comprehensive enough. The (Kuwashimo 1978) states that numerical methods used for studying non-stationary nonaxisymmetric problems of the heat exchange in rotating cylinders and liquids are not economically justified for calculations where rotation velocity is high.

According to the (Kuwashimo 1978), conditions for the calculation accuracy are specified by similar characteristics for the finite element method and finite difference method which both are applied for computing non-stationary non-axisymmetric temperature fields in rotating cylinders and liquids. These conditions can be expressed as:

$$1 - 2\frac{\Delta F_O}{\Delta\varphi^2} \geq 0 \quad \text{and} \quad \frac{1}{\Delta\varphi} - \frac{Pd}{2} \geq 0,$$

where Fo is Fourier criterion, and Pd is Predvoditelev criterion.

If $Pd = 10^5$ corresponds to angular velocity of the steel cylinder rotation $\omega = 1{,}671\,\text{s}^{-1}$, with radius 100 mm, then variables $\Delta\varphi$ and ΔF_O should be subject to the following conditions:

$$\Delta\varphi \leq 2\,10^{-5} \quad \text{and} \quad \Delta F_O \leq 2\,10^{-10}.$$

In the case of uniformly cooled cylinder, and if Bi = 5 (Bi is Bio criterion), a time period needed in order the temperature can reach 90% of stationary state should be $Fo \approx 0.025$. It means that, within this period of time, at least $1.3\cdot10^8$ operations should be fulfilled in order to obtain stationary temperature distribution.

Moreover, it should be mentioned that it would be necessary to make $3.14\ 10^5$ calculations within one cycle of computation as the inside state of the ring should be characterized by $3.14\ 10^5$ points. It is obvious that this number of calculations needed for getting a numerical result is unrealistic.

Therefore, we will employ integral transformations for solving boundary problems which occur during mathematic modeling of the 3D non-stationary heat-exchange processes in rotating cylinders and liquids.

2 BODY TEXT

Let's consider a calculation of nonstationary non-axisymmetric temperature field in an empty cylinder with finite length L, which rotates with constant angular velocity ω round the axis OZ, with taking into account finite velocity of the heat conduction and with an assumption that the cylinder's heat-transfer properties do not depend on temperature and no internal sources of heat are available. At zero time, temperature of the cylinder is constant G_0, and heat flows on the external and internal surfaces of the cylinder are known and do not depend on time $G(\varphi,z)$ and $G_1(\varphi,z)$, correspondingly.

The (Berdnyk 2005) presents a generalized energy-transfer equation for a moving element in the continuum medium with taking into consideration finiteness of the heat-conductivity velocity value. According to the (Berdnyk 2005), for the case of a solid body, which rotates with constant angular velocity ω round the axis OZ, and whose heat-transfer properties do not depend on temperature, and when no internal sources of heat are available, a generalized energy-balance equation in cylindrical system of coordinates (ρ,φ,z) can be written in the following way:

$$
\gamma c\left\{\frac{\partial T}{\partial t}+\omega\frac{\partial T}{\partial \varphi}+\tau_r\left[\frac{\partial^2 T}{\partial t^2}+\omega\frac{\partial^2 T}{\partial \varphi \partial t}\right]\right\}=
$$
$$
\lambda\left[\frac{\partial^2 T}{\partial r^2}+\frac{1}{r}\frac{\partial T}{\partial r}+\frac{1}{r^2}\frac{\partial^2 T}{\partial \varphi^2}+\frac{\partial^2 T}{\partial z^2}\right] \quad (1)
$$

where γ – is continuum medium; c is specific heat capacity; $T(\rho,\varphi,z)$ is temperature of the medium; t is time; τ_r - is relaxation time; and λ is heat conductivity coefficient.

Mathematically, a problem of defining a temperature field in a cylinder consists of integration of the differential thermal-conductivity equation (1) into the domain

$$
D=\left\{(\rho,\varphi,z,t)\,|\,\rho\in(\rho_0,1),\,\varphi\in(0,2\pi),\right.
$$
$$
\left. z\in(0,1),t\in(0,\infty)\right\},
$$

which, with taking into consideration the accepted assumptions, can be written down in the following way:

$$
\frac{\partial \theta}{\partial t}+\omega\frac{\partial \theta}{\partial \varphi}+\tau_r\frac{\partial^2 \theta}{\partial t^2}+\tau_r\omega\frac{\partial^2 \theta}{\partial \varphi \partial t}=
$$
$$
\frac{a}{R^2}\left[\frac{\partial^2 \theta}{\partial \rho^2}+\frac{1}{\rho}\frac{\partial \theta}{\partial \rho}+\frac{1}{\rho^2}\frac{\partial^2 \theta}{\partial \varphi^2}+\chi\frac{\partial^2 \theta}{\partial z^2}\right] \quad (2)
$$

with initial conditions:

$$
\theta(\rho,\varphi,z,0)=0,\quad \frac{\partial \theta(\rho,\varphi,z,0)}{\partial t}=0 \quad (3)
$$

and boundary conditions:

$$
\theta(\rho,\varphi,0,t)=0\,,\,\theta(\rho,\varphi,1,t)=0\,, \quad (4)
$$

$$
\int_0^t\frac{\partial \theta}{\partial \rho}\bigg|_{\rho=1}e^{\frac{\zeta-t}{\tau_r}}\,d\zeta=V(\varphi,z),
$$
$$ \quad (5) $$
$$
\int_0^t\frac{\partial \theta}{\partial \rho}\bigg|_{\rho=\rho_0}e^{\frac{\zeta-t}{\tau_r}}\,d\zeta=W(\varphi,z),
$$

where $\theta=\dfrac{T(\rho,\varphi,z,t)-G_0}{T_{max}-G_0}$ is relative temperature in the cylinder; $\rho=\dfrac{r}{R}$; R is external radius of the cylinder; $a=\dfrac{\lambda}{c\gamma}$ is thermal diffusivity; $\chi=(R/L)^2$; $V(\varphi)=\dfrac{G(\varphi)\tau_r}{\lambda}$; $W(\varphi)=\dfrac{G_1(\varphi)\tau_r}{\lambda}$; $G(\varphi,z),G_1(\varphi,z)\in C(0,2\pi)$.

Then, solution of the boundary problem (2)-(5) $\theta(\rho,\varphi,z,t)$ is double continuum differential by ρ,z and φ, once differential by t in domain D and continuum in domain $\overline{D}$, i.e. $\theta(\rho,\varphi,z,t)\in C^{2,1}(D)\cap C(\overline{D})$, and functions $\theta(\rho,\varphi,z,t)$ and $V(\varphi,z),W(\varphi,z)$ can be decomposed into the Fourier complex series :

$$
\begin{Bmatrix}\theta(\rho,\varphi,z,t)\\ V(\varphi,z)\\ W(\varphi,z)\end{Bmatrix}=\sum_{n=-\infty}^{+\infty}\begin{Bmatrix}\theta_n(\rho,z,t)\\ V_n(z)\\ W_n(z)\end{Bmatrix}\cdot\exp(in\varphi),\quad (6)
$$

where

$$\begin{Bmatrix} \theta_n(\rho,z,t) \\ V_n(z) \\ W_n(z) \end{Bmatrix} = \frac{1}{2\pi} \int_0^{2\pi} \begin{Bmatrix} \theta(\rho,\varphi,z,t) \\ V(\varphi,z) \\ W(\varphi,z) \end{Bmatrix} \exp(-in\varphi)d\varphi,$$

where

$$\theta_n(\rho,z,t) = \theta_n^{(1)}(\rho,z,t) + i\theta_n^{(2)}(\rho,z,t),$$

$$V_n(z) = V_n^{(1)}(z) + i\,V_n^{(2)}(z),$$

$$W_n(z) = W_n^{(1)}(z) + i\,W_n^{(2)}(z),$$

and i is imaginary unit.

In view of the fact that $\theta(\rho,\varphi,z,t)$ is a real function, let's confine ourselves by considering only $\theta_n(\rho,z,t)$ for n=0,1,2,..., because $\theta_n(\rho,z,t)$ and $\theta_{-n}(\rho,z,t)$ are complexly conjugate . By putting values of functions from (6) into (2) – (5) we can compose the following system of differential equations:

$$\frac{\partial \theta_n^{(i)}}{\partial t} + \vartheta_n^{(i)}\theta_n^{(m_i)} + \tau_r\frac{\partial^2\theta_n^{(i)}}{\partial t^2} + \tau_r\vartheta_n^{(i)}\frac{\partial\theta_n^{(m_i)}}{\partial t} =$$

$$\frac{a}{R^2}\left[\frac{\partial^2\theta_n^{(i)}}{\partial\rho^2} + \frac{1}{\rho}\frac{\partial\theta_n^{(i)}}{\partial\rho} - \frac{n^2}{\rho^2}\theta_n^{(i)} + \chi\frac{\partial^2\theta_n^{(i)}}{\partial z^2} \right] \quad (7)$$

with initial conditions:

$$\theta_n^{(i)}(\rho,z,0) = 0, \qquad \frac{\partial\theta_n^{(i)}(\rho,z,0)}{\partial t} = 0 \quad (8)$$

with boundary conditions:

$$\theta_n^{(i)}(\rho,0,t) = 0, \qquad \theta_n^{(i)}(\rho,1,t) = 0, \quad (9)$$

$$\int_0^t \left.\frac{\partial\theta_n^{(i)}}{\partial\rho}\right|_{\rho=1} e^{\frac{\zeta-t}{\tau_r}}\,d\zeta = V_n^{(i)}(z)$$

$$\int_0^t \left.\frac{\partial\theta_n^{(i)}}{\partial\rho}\right|_{\rho=\rho_0} e^{\frac{\zeta-t}{\tau_r}}\,d\zeta = W_n^{(i)}(z) \quad (10)$$

where $\vartheta_n^{(1)} = -\omega n$; $\vartheta_n^2 = \omega n$; $m_1 = 2$; $m_2 = 1$; i=1,2.

Let's employ the Laplace integral transformation for the system of differential equations (7):

$$\tilde{f}(s) = \int_0^\infty f(\tau)e^{-s\tau}\,d\tau.$$

As a result, we receive the following system of differential equations:

$$s\tilde{\theta}_n^{(i)} + \vartheta_n^{(i)}\left(\tilde{\theta}_n^{(m_i)} + \tau_r s\tilde{\theta}_n^{(m_i)}\right) + \tau_r s^2\tilde{\theta}_n^{(i)} =$$

$$\frac{a}{R^2}\left[\frac{\partial^2\tilde{\theta}_n^{(i)}}{\partial\rho^2} + \frac{1}{\rho}\frac{\partial\tilde{\theta}_n^{(i)}}{\partial\rho} - \frac{n^2}{\rho^2}\tilde{\theta}_n^{(i)} + \chi\frac{\partial^2\tilde{\theta}_n^{(i)}}{\partial z^2} \right] \quad (11)$$

with boundary conditions

$$\tilde{\theta}_n^{(i)}(\rho,0,t) = 0, \qquad \tilde{\theta}_n^{(i)}(\rho,1,t) = 0, \quad (12)$$

$$\left.\frac{d\tilde{\theta}_n^{(i)}}{d\rho}\right|_{\rho=1} \tilde{V}_n^{(i)}(z), \qquad \left.\frac{d\tilde{\theta}_n^{(i)}}{d\rho}\right|_{\rho=\rho_0} \tilde{W}_n^{(i)}(z) \quad (13)$$

where

$$\tilde{V}_n^{(i)}(z) = V_n^{(i)}(z)\left(1 + \frac{1}{s\tau_r}\right),$$

$$\tilde{W}_n^{(i)} = W_n^{(i)}(z)\left(1 + \frac{1}{s\tau_r}\right), \quad (i=1,2)$$

Let's employ the Fourier integral transformation (Galitsin 1979) for the system of the differential equations (11):

$$\hat{f}(\lambda_m) = \int_0^1 f(x)\sin(\lambda_m x)\,dx,$$

where $\lambda_m = \pi\cdot m$; m=1,2,..., and formula of the inverse transformation can be expressed in the following way:

$$f(x) = 2\sum_{m=1}^\infty \sin(\pi\,m\,x)\hat{f}(\lambda_m).$$

As a result, we receive the following system of differential equations:

$$s\hat{\tilde{\theta}}_n^{(i)} + \vartheta_n^{(i)}\left(\hat{\tilde{\theta}}_n^{(m_i)} + \tau_r s\hat{\tilde{\theta}}_n^{(m_i)}\right) + \tau_r s^2\hat{\tilde{\theta}}_n^{(i)} =$$

$$\frac{a}{R^2}\left[\frac{d^2\hat{\tilde{\theta}}_n^{(i)}}{d\rho^2} + \frac{1}{\rho}\frac{d\hat{\tilde{\theta}}_n^{(i)}}{d\rho} - \frac{n^2}{\rho^2}\hat{\tilde{\theta}}_n^{(i)} - \lambda_m\hat{\tilde{\theta}}_n^{(i)} \right] \quad (14)$$

with boundary conditions

$$\left.\frac{d\widehat{\bar{\theta}}_n^{(i)}}{d\rho}\right|_{\rho=1} = \widehat{\bar{V}}_n^{(i)}, \qquad \left.\frac{d\widehat{\bar{\theta}}_n^{(i)}}{d\rho}\right|_{\rho=\rho_0} = \widehat{\bar{W}}_n^{(i)}$$

Let's employ the Hankel integral transformation (Galitsin 1979) for the system of the differential equations (14):

$$\bar{f}(\mu_{n,k}) = \int\limits_{\rho_0} \rho\, f(\rho)\Psi_{n,k}(\xi_{n,k}\rho)d\rho,$$

where
$$\Psi_{n,k}(\xi_{n,k}\rho) =$$
$$Y_n'(\xi_{n,k}\rho_0)J_n(\xi_{n,k}\rho) - J_n'(\xi_{n,k}\rho_0)Y_n(\xi_{n,k}\rho);$$

$J_n(x), Y_n(x)$ – are Bessel functions of the 1st and 2nd order, correspondingly; $\xi_{n,k}$ are roots of the transcendental equation:

$$Y_n'(\xi_{n,k}\rho_0)J_n'(\xi_{n,k}) - J_n'(\xi_{n,k}\rho_0)Y_n'(\xi_{n,k}) = 0,$$

where $\xi_{0,0} = 0$; $\Psi_{00} = 1$.

The inverse transformation formula can be expressed by:

$$f(\rho) = \sum_{k=0}^{\infty} H_{n,k}\,\bar{f}(\xi_{n,k}).$$

where $H_{n,k} = \dfrac{\Psi_{n,k}(\xi_{n,k}\rho)}{\|\Psi_{n,k}\|^2}$

$$\|\Psi_{n,k}\|^2 = \frac{1}{2\xi_{n,k}^2}\left[\left(\xi_{n,k}^2 - n^2\right)\cdot\Psi_{n,k}^2(\xi_{n,k}) - \right.$$
$$\left.\left(\xi_{n,k}^2\rho_0 - n^2\right)\cdot\Psi_{n,k}^2(\xi_{n,k}\rho_0)\right];$$

$$\|\Psi_{0,k}\|^2 = \frac{2\left[J_1^2(\xi_{0,k}\rho_0) - J_1^2(\xi_{0,k})\right]}{\pi^2\xi_{0k}^2 J_1^2(\xi_{0,k})}, \text{ if } k>0;$$

$$\|\Psi_{0,0}\|^2 = \frac{1-\rho_0^2}{2}.$$

As a result, we receive the following system of equations relatively to $\widehat{\bar{\theta}}_n^{(i)}$:

$$s\widehat{\bar{\theta}}_n^{(i)} + \vartheta_n^{(i)}\left(\widehat{\bar{\theta}}_n^{(m_i)} + \tau_r s\widehat{\bar{\theta}}_n^{(m_i)}\right) + \tau_r s^2\widehat{\bar{\theta}}_n^{(i)} =$$
$$q_{n,k}\left(\frac{\widetilde{\Omega}_{n,k}^{(i)}}{\mu_{n,k}^2} - \widehat{\bar{\theta}}_n^{(i)}\right), \qquad (15)$$

where

$$i=1,2; \quad q_{n,k} = \frac{a}{R^2}\cdot\mu_{n,k}^2; \quad \mu_{n,k}^2 = \xi_{n,k}^2 + \lambda_m^2,$$

$$\widetilde{\Omega}_{n,k}^{(i)} = \Psi_{n,k}(\mu_{n,k})\cdot\widehat{\bar{V}}_n^{(i)} - \rho_0\Psi_{n,k}(\mu_{n,k}\rho_0)\widehat{\bar{W}}_n^{(i)}$$

Having solved the system of equations (15), we receive:

$$\widehat{\bar{\theta}}_n^{(i)} = \alpha_{n,k}\frac{\widetilde{\Omega}_n^{(i)}\Theta_{n,k} + (-1)^{i+1}\omega n\widetilde{\Omega}_n^{(m_i)}P}{\Theta^2_{n,k} + \omega^2 n^2 P^2}, \qquad (16)$$

where $\Theta_{n,k} = \tau_r s^2 + s + q_{n,k}; P = 1 + s\tau_r$;

$$\alpha_{n,k} = \frac{q_{n,k}}{\mu_{n,k}} \; ; i=1,2.$$

By employing the Laplace inverse transformation formula for expression of the functions (16) we can receive originals of the functions:

$$\bar{\theta}_n^{(1)}(t) = \sum_{j=1}^{2}\zeta_{n,k}(s_j)\left\{\Omega_n^{(1)}[P_1 + \delta i] + \Omega_n^{(2)}[\delta - P_1 i]\right\}\Lambda$$
$$+ \sum_{j=3}^{4}\zeta_{n,k}(s_j)\left\{\Omega_n^{(1)}[P_1 - \delta i] + \Omega_n^{(2)}[\delta + P_1 i]\right\}\Lambda, (17)$$

$$\bar{\theta}_n^{(2)}(t) = \sum_{j=1}^{2}\zeta_{n,k}(s_j)\left\{\Omega_n^{(2)}[P_1 + \delta i] - \Omega_n^{(1)}[\delta - P_1 i]\right\}\Lambda$$
$$+ \sum_{j=3}^{4}\zeta_{n,k}(s_j)\left\{\Omega_n^{(2)}[P_1 - \delta i] - \Omega_n^{(1)}[\delta + P_1 i]\right\}\Lambda, (18)$$

where $\zeta_{n,k}(s_j) = \dfrac{0.5 s_j^{-1}}{P_1^2 + (\tau_r\omega n)^2}$;

$P_1 = 2\tau_r s_j + 1; \delta = \tau_r\omega n; \Lambda = e^{s_j t}$, and values of s_j for the $j=1,2,3,4$ are expressed by the following formulas:

$$s_1, s_2 = \frac{(\tau_r \omega n i - 1) \pm \sqrt{(1 + \tau_r \omega n i)^2 - 4\tau_r q_{n,k}}}{2\tau_r},$$

$$s_3, s_4 = \frac{(\tau_r \omega n i + 1) \pm \sqrt{(1 - \tau_r \omega n i)^2 - 4\tau_r q_{n,k}}}{2\tau_r}.$$

Thereby, by taking into account the inverse transformation formulas, we receive a temperature field in the empty cylinder with finite length, which rotates with constant angular velocity ω round the axis OZ, with taking into account finite velocity of the heat conduction:

$$\theta(\rho, \varphi, z, t) =$$

$$\sum_{n=-\infty}^{+\infty} \left\{ \sum_{k=0}^{\infty} \left\langle \sum_{m=1}^{\infty} \Xi(t)_{n,k,m} \sin(\pi m z) \right\rangle H_{n,k} \right\} \Lambda_1,$$

where

$$\Xi(t)_{n,k,m} = \bar{\theta}_n^{(1)}(t) + i \bar{\theta}_n^{(2)}(t); \Lambda_1 = \exp(in\varphi),$$

values of the $\bar{\theta}_n^{(1)}(t)$ and $\bar{\theta}_n^{(2)}(t)$ are defined by the formulas (17) and (18).

CONCLUSIONS

The generalized energy transfer equation has been formulated for the moving element in the continuum medium (1). The temperature field in the empty cylinder with finite length L, which rotates with constant angular velocity ω round the axis OZ, has been defined with taking into account finite velocity of the heat conduction and has been shown in the form of convergent orthogonal series by Bessel and Fourier functions . The proposed analytical solution of the generalized boundary problem of the heat exchange in the rotating empty cylinder with taking into account finite value of the heat conductivity velocity can be applied for modeling temperature fields occurred in numerous technical systems (satellites, rolling mills, turbines, etc.).

REFERENCES

Berdnyk M.G. 2005. *Mathematical simulation of temperature field in cylinder, taking into account the inite velocity of propagation of heat* (in Russian). Dnipropetrovsk: DNU: Questions of applied mathematics and mathematical modeling: 37- 44.
Kuwashimo K. & Yamada T. 1978. *Temperature distribution within a rotatinq cylindrieal body.* Bull. JSME, Vol. 21, Issue 152: 266 – 272.
Galitsin A.S. & Zhukovsky A.I. 1979. *Integral transforms and special functions in problems of thermal conductivity* (in Russian). Kiev: Naukova dumka: 561.
Podstrigac Y.S. & Kolano Y.M. 1976. *Synthesis of thermal mechanics* (in Russian). Kiev: Naukova dumka: 310.

Power Engineering and Information Technologies In Technical Objects Control – Pivnyak, Beshta
& Alekseyev (eds)
© 2016 Taylor & Francis Group, London, ISBN 978-1-138-71479-3

Small amount of random fluctuations of the underwater hydraulic transport pipeline

V. Kozlov, I. Gulina, I. Shedlovsky & V. Gubkina
State Higher Educational Institution "National Mining University", Dnipro, Ukraine

ABSTRACT: The numerical method for solution of a small amount of fluctuations of the underwater hydraulic transport pipeline is proposed. The vibrations of the three-dimensional pipeline, which is suspended at both ends in the water flow, are caused by random changes in the density of the transported pulp. The problem is solved in terms of concentrated forces and masses according to the float's presence.

1. INTRODUCTION

The lengthy flexible hydraulic transport pipelines (HP) are widely used in various areas of engineering. In particular, they are applied when developing underwater fields of solid minerals and in other industries. Usually such service conditions cause the random changes of the pumped-over pulp density. It leads to random pipeline fluctuations and, therefore, to emergence of additional dynamic efforts which can be very considerable. The specified occurrence circumstance promotes fatigue damage accumulation and decrease in hydro transport reliability (HT) (Svetlitsky 2001; Svetlitsky 2005).

2. MAIN PART

We have developed the mathematical model of the spatial flexible HT that is fixed for both ends and makes the compelled parametrical fluctuations relative to a balance position caused by casual changes of the pumped-over pulp density. The problem of linear statistical dynamics is solved by Monte-Carlo method. The model considers the action of external water flow as well as the concentrated forces and masses stipulated by the existence of floats intended for contributing positive floatability to HT.

The vector nonlinear equations of the HT movement are received in a motionless system of coordinates (Fig. 1). For this purpose separately allocated pipeline element consisting of a pulp element and floats and having the length of ds was considered. The element of the pipeline considered to be as an absolutely flexible rod (hose) in this work is affected by the following forces:
$-m_1 g \bar{i}_z ds$ — weight;

$-m_{B1} g \bar{i}_z ds$ is buoyancy; $-m_1 \dfrac{d^2 \bar{r}}{dt^2} ds$ is inertia;

$$d\bar{J}_{np} = -m_{np} \left[\frac{d^2 \bar{r}}{dt^2} - \left(\frac{d^2 \bar{r}}{dt^2} \cdot \bar{e}_1 \right) \bar{e}_1 \right] ds \quad \text{is inertia}$$

stipulated by the attached water mass presence; $\bar{q}_a ds$ is a hydrodynamic force from an external water flow; $\bar{f} ds$ is interactions between a pulp and a pipeline; $Q_1^{(1)} \bar{e}_1$ is an axial force. Here m_1 is a hose bulk weight; m_{np} is the attached water mass. The expression for $d\bar{J}_{np}$ is received by assuming that the inertia force of the attached water mass is defined by pipe acceleration in the direction being orthogonal to the axial line. In this case $m_{np} = \rho_B \pi D^2 / 4$ is for a core of a round cylindrical form. Here ρ_B is water density; D is the outer pipeline diameter.

Forces operating on a pulp element are: $-m_2 g \bar{i}_z ds$ is weight; $m_{B2} g \bar{i}_z ds$ is buoyancy; $P \bar{e}_1$ is the force connected with the action of pulp excessive pressure relative to surrounding water. Here $m_2(s,t) = m_{20} + m_{21}(s,t)$ is pulp bulk weight including stationary and non-stationary components. The speed of pulp movement relative to pipeline w is accepted to be constant.

Forces operating on float k are the following: $\bar{F}^{(k)}$ is hydrodynamic; $\bar{R}^{(k)}$ is the interactions between the hose and the float k ($k = 1, 2, \ldots, n$);

$$-M^{(k)} \frac{d^2 \bar{r}}{dt^2} \bigg|_{s = s_k} \quad \text{is inertia;} \quad -M_{np}^{(k)} \frac{d^2 \bar{r}}{dt^2} \bigg|_{s = s_k}$$

is the inertia of attached water mass; $\bar{p}^{(k)}$ is floatability. Here $M^{(K)}$, s_k are weight and fastening coordinate of the float k ; $M_{np}^{(k)}$ is attached water mass identified for it. For a separately taken float of a spherical shape $M_{np}^{(k)} = \rho_B \pi (D_n^{(k)})^3 /12$, where $D_n^{(k)}$ is its diameter.

For transition to a dimensionless record form the following ratios were used:

$$\tau = t p_0 , p_0 = (g/L)^{1/2}, \varepsilon = s/L ,$$

$$\widetilde{\widetilde{r}} = \bar{r}/L, n_0 = (m_1 + m_{20})g/q ,$$

$$n_{11} = m_{21}g/q, n_{np} = m_{np}g/q, \widetilde{M}^{(k)} = M^{(k)}g/(qL), \widetilde{M}_{np}^{(k)} = M_{np}^{(k)}g/(qL) ,$$

$$\widetilde{Q}_1^{(1)} = Q_1^{(1)}/(qL), \widetilde{P} = P/(qL) ,$$

$$\widetilde{\bar{q}}_a = \bar{q}_a/q, \widetilde{w} = w/(p_0 L) ,$$

where L is the length of HT; q is the weight per unit length of the hose with pulp of average density with account of water buoyancy. A symbol "~" that designates dimensionless values is omitted further.

The equation of the HT movement transporting a pulp of variable density in a dimensionless form of record is written as:

$$(n_0 + n_{11} + n_{np})\frac{\partial^2 \bar{r}}{\partial \tau^2} - n_{np}(\frac{\partial^2 \bar{r}}{\partial \tau^2}\cdot\bar{e}_1)\bar{e}_1 + \sum_{k=1}^{n}(M^{(k)} + M_{np}^{(k)})\delta(\varepsilon - \varepsilon_k)\frac{\partial^2 \bar{r}}{\partial \tau^2} +$$

$$+2w(n_1 + n_{11})\frac{\partial^2 \bar{r}}{\partial \tau \partial \varepsilon} + (n_1 + n_{11})w^2 \frac{\partial^2 \bar{r}}{\partial \varepsilon^2} = \frac{\partial[(Q_1^{(1)} - P)\bar{e}_1]}{\partial \varepsilon} + \bar{q}_a + \sum_{k=1}^{n}(\bar{p}^{(k)} + F^{(k)})\delta(\varepsilon - \varepsilon_k) - (1 + n_{11})\bar{i}_2 \qquad (1),$$

where δ is Dirac delta function

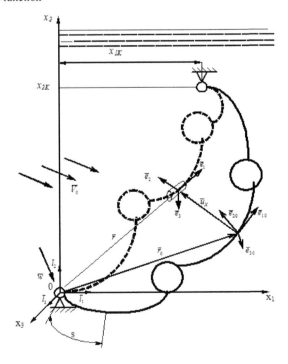

Figure 1. Pipeline design scheme

The equation of the pulp movement, which is presented with projections to the tangent direction, is shown as follows:

$$(n_1 + n_{11})(\frac{\partial^2 \overline{r}}{\partial \tau^2} \cdot \overline{e}_1) = -\frac{\partial P}{\partial \varepsilon} - (n_1 + n_{11} - n_{B2})x_2' - f_{10}, \tag{2}$$

where $n_{B2} = m_{B2}g/q$; f_{10} is a distributed pulp friction force on an internal pipeline surface.

Geometrical ratio encloses the system of the equations (1), (2)

$$\sum_{i=1}^{3}(x_i')^2 = 1. \tag{3}$$

The equations of small number fluctuations of HT relative to an equilibrium state are resulted by assuming that the number of the values included into the nonlinear equations of the movement (1) — (3) can be presented in the form:

$$\overline{r} = \overline{r}_0(\varepsilon) + \overline{u}_x(\varepsilon, \tau),$$

$$\overline{r}' = \overline{r}_0'(\varepsilon) + \overline{u}_x'(\varepsilon, \tau),$$

$$\Delta \overline{q}_a = (A^{(1)} + A^{(3)})\overline{u}_x' + (A^{(2)} + A^{(4)} + A^{(5)})\dot{\overline{u}}_x. \tag{5}$$

Linear dependence for the hydrodynamic force operating on a float at small quantity fluctuations in a stream in a dimensionless form of record is as follows [1]:

$$\Delta \overline{F}^{(k)} = B^{(k)}\frac{\partial \overline{u}_x}{\partial \tau}\bigg|_{\varepsilon = \varepsilon_k}, \tag{6}$$

where

$$B^{(k)} = -q_0 \begin{bmatrix} (1 + \cos^2 \alpha) & 0 & \sin \alpha \cos \alpha \\ 0 & 1 & 0 \\ \sin \alpha \cos \alpha & 0 & (1 + \sin^2 \alpha) \end{bmatrix};$$

$$Q_1^{(1)} = Q_{10}^{(1)}(\varepsilon) + \Delta Q_1(\varepsilon, \tau),$$

$$P = P_0(\varepsilon) + P_1(\varepsilon, \tau), \tag{4}$$

$$\overline{q}_a = \overline{q}_{a0}(\varepsilon) + \Delta \overline{q}_a(\varepsilon, \tau),$$

$$\overline{F}^{(k)} = \overline{F}_0^{(k)}(\varepsilon_k) + \Delta \overline{F}^{(k)}(\varepsilon_k, \tau),$$

where $\overline{u}_x, \overline{u}_x', \Delta Q_1, P_1, \Delta \overline{q}_a, \Delta \overline{F}^{(k)}$ are dynamic components; summands with an index zero are predefined static components.

A dynamic component vector of the hydrodynamic force operating on the pipeline is defined as follows (Svetlitsky 2001):

$$q_0 = \frac{c_x \rho_e \pi p_0 V_0 (D_\cdot^{(k)})^2}{8q};$$

c_x is a coefficient of hydrodynamic resistance of the float of spherical shape without considering the hose influence; α is an angle between the vector $\overline{V}_0$ and the plane $x_1 0 x_2$.

The equations of small quantity fluctuations of HT relative to an equilibrium position are as the follows:

$$\overline{L}_1(\overline{u}_x, \Delta \overline{Q}_x^{(1)}) = (n_0 + n_{11} + n_{np})\frac{\partial^2 \overline{u}_x}{\partial \tau^2} - n_{np}(\frac{\partial^2 \overline{u}_x}{\partial \tau^2} \cdot \overline{e}_{10})\overline{e}_{10} +$$

$$+ \sum_{k=1}^{n}(M^{(k)} + M_{np}^{(k)})\delta(\varepsilon - \varepsilon_k)\frac{\partial^2 \overline{u}_x}{\partial \tau^2} + 2w(n_1 + n_{11})\frac{\partial^2 \overline{u}_x}{\partial \tau \partial \varepsilon} -$$

$$- (A^{(2)} + A^{(4)} + A^{(5)})\frac{\partial \overline{u}_x}{\partial \tau} - (n_{11}x_{20}'E + A^{(1)} + A^{(3)})\frac{\partial \overline{u}_x}{\partial \varepsilon} - \tag{7}$$

$$- \sum_{k=1}^{n}B^{(k)}\delta(\varepsilon - \varepsilon_k)\frac{\partial \overline{u}_x}{\partial \tau} + (n_{11}w^2 + P_1)\frac{\partial^2 \overline{u}_x}{\partial \varepsilon^2} - \frac{\partial \Delta \overline{Q}_x^{(1)}}{\partial \varepsilon} - n_{11}x_{20}'\frac{\partial \overline{r}_0}{\partial \varepsilon} +$$

$$+ (n_{11}w^2 + P_1)\frac{\partial^2 \overline{r}_0}{\partial \varepsilon^2} + n_{11}\overline{i}_2 = 0,$$

$$\overline{L}_2(\overline{u}_x, \Delta\overline{Q}_x^{(1)}) = \frac{\partial \overline{u}_x}{\partial \varepsilon} + C_0^{(1)}\Delta\overline{Q}_x^{(1)} = 0, \tag{8}$$

where $\quad \Delta\overline{Q}_x^{(1)} = Q_1\dfrac{\partial \overline{u}_x}{\partial \varepsilon} + \Delta Q_1\dfrac{\partial \overline{r}_0}{\partial \varepsilon}; \qquad Q_1 = Q_1^{(1)} - P_0 - n_1 w^2;$

$$C_0^{(1)} = \begin{bmatrix} -\dfrac{1-x_{10}'^2}{Q_1} & \dfrac{x_{10}'x_{20}'}{Q_1} & \dfrac{x_{10}'x_{30}'}{Q_1} \\ \dfrac{x_{10}'x_{20}'}{Q_1} & -\dfrac{1-x_{20}'^2}{Q_1} & \dfrac{x_{20}'x_{30}'}{Q_1} \\ \dfrac{x_{10}'x_{30}'}{Q_1} & \dfrac{x_{20}'x_{30}'}{Q_1} & -\dfrac{1-x_{30}'^2}{Q_1} \end{bmatrix}; \qquad P_1(\varepsilon,\tau) = -\int_0^\varepsilon n_{11}x_2'd\eta \cdot$$

The system of the vector differential equations (7), (8) is closed. The axial thrust $Q_1(\varepsilon,\tau)$ entering into the expressions for matrix elements $C_0^{(1)}$ can be defined by iteration. However, it is sufficient to use the thrust value obtained at the previous stage of the numerical solution to make practical calculations.

As a result of the system solution (7), (8) the dynamic component of the pipeline axial thrust is defined by a ratio:

$$\Delta Q_1 = (\Delta\overline{Q}_x^{(1)}\,\overline{e}_{10}). \tag{9}$$

The generalized principle of possible movements was applied for the numerical solution of the linear system of the equations (7), (8) (Svetlitsky 2001; Svetlitsky 2005). At the same time the solution for equations was tried in the following forms:

$$\overline{u}_x(\varepsilon,\tau) = \sum_{i=1}^m \overline{\varphi}^{(i)}(\varepsilon)f_i^{(1)}(\tau), \tag{10}$$

$$\Delta\overline{Q}_x^{(1)}(\varepsilon,\tau) = \sum_{i=1}^m \overline{\psi}^{(i)}(\varepsilon)f_i^{(2)}(\tau), \tag{11}$$

where $\overline{\varphi}^{(i)}(\varepsilon), \overline{\psi}^{(i)}(\varepsilon)$ are characteristic vectors; $f_i^{(1)}, f_i^{(2)}$ are unknown time-varying functions.

In the supplement to the studied pipeline the generalized principle of possible movements is written as:

$$\int_0^1 (\overline{L}_1 \cdot \overline{\varphi}^{(j)})d\varepsilon = 0, \quad j = 1, 2, \dots m, \tag{12}$$

$$\int_0^1 (\overline{L}_2 \cdot \overline{\psi}^{(j)})d\varepsilon = 0, \quad j = 1, 2, \dots m. \tag{13}$$

The systems (12) and (13) can be presented in the form of the vector ordinary differential equation:

$$\dot{\overline{z}} + A\overline{z} = \overline{b}, \tag{14}$$

where $\overline{z} = \begin{vmatrix} \dot{\overline{f}}^{(1)} \\ \overline{f}^{(1)} \end{vmatrix}; \quad A = \begin{bmatrix} M^{-1}H & R_1 \\ -E & 0 \end{bmatrix}; \quad \overline{b} = \begin{vmatrix} M^{-1}\overline{s} \\ 0 \end{vmatrix};$

$R_1 = M^{-1}(R - GD^{-1}C);$

E is a single matrix having the size of $m \times m$.

Elements of matrixes M, H, R, G, D, C and components of a vector $\overline{s}$ are written as:

$$m_{ji} = \int_0^1 [(n_0 + n_{11} + n_{np})(\overline{\varphi}^{(i)}\cdot\overline{\varphi}^{(j)}) - n_{np}(\frac{\partial \overline{r}_0}{\partial \varepsilon}\cdot\overline{\varphi}^{(i)})(\frac{\partial \overline{r}_0}{\partial \varepsilon}\cdot\overline{\varphi}^{(j)})]d\varepsilon +$$

$$+ \sum_{k=1}^n (M^{(k)} + M_{np}^{(k)})(\overline{\varphi}^{(i)}(\varepsilon_k)(\overline{\varphi}^{(j)}(\varepsilon_k)),$$

212

$$h_{ji} = \int_0^1 [2w(n_1 + n_{11})(\overline{\varphi}'^{(i)} \cdot \overline{\varphi}^{(j)}) - (A^{(2)} + A^{(4)} + A^{(5)})\overline{\varphi}^{(i)} \cdot \overline{\varphi}^{(j)}]d\varepsilon - \sum_{k=1}^{n} B^{(k)} \overline{\varphi}^{(i)}(\varepsilon_k) \cdot \overline{\varphi}^{(j)}(\varepsilon_k),$$

$$r_{ji} = \int_0^1 [(n_{11}w^2 + P_1)(\overline{\varphi}''^{(i)} \cdot \overline{\varphi}^{(j)}) - (n_{11}x_{20}'E + A^{(1)} +$$

$$+ A^{(3)})\overline{\varphi}'^{(i)} \cdot \overline{\varphi}^{(j)}]d\varepsilon + \sum_{k=1}^{n} [n_{11}(\varepsilon_k)w^2 + P_1(\varepsilon_k)](\Delta\overline{\varphi}_k'^{(i)} \cdot \overline{\varphi}^{(j)}(\varepsilon_k)),$$

$$g_{ji} = -\int_0^1 (\overline{\psi}'^{(i)} \cdot \overline{\varphi}^{(j)})d\varepsilon - \sum_{k=1}^{n} (\Delta\overline{\psi}_k^{(i)} \cdot \overline{\varphi}(\varepsilon_k)),$$

$$d_{ji} = \int_0^1 (C_0^{(1)} \overline{\psi}^{(i)} \cdot \overline{\psi}^{(j)})d\varepsilon,$$

$$c_{ji} = \int_0^1 (\overline{\varphi}'^{(i)} \cdot \overline{\psi}^{(j)})d\varepsilon,$$

$$s_i = \int_0^1 [n_{11}x_{20}'(\frac{\partial \overline{r}_0}{\partial \varepsilon} \cdot \overline{\varphi}^{(j)}) - (n_{11}w^2 + P_1)(\frac{\partial^2 \overline{r}_0}{\partial \varepsilon^2} \cdot \overline{\varphi}^{(j)}) -$$

$$- n_{11}\varphi_2^{(j)}]d\varepsilon - \sum_{k=1}^{n} [n_{11}(\varepsilon_k)w^2 + P_1(\varepsilon_k)](\Delta\overline{r}_0'^{(k)} \cdot \overline{\varphi}^{(j)}(\varepsilon_k)),$$

where $\Delta\overline{\varphi}_k'^{(i)} = \overline{\varphi}_+'^{(i)}(\varepsilon_k) - \overline{\varphi}_-'^{(i)}(\varepsilon_k)$; $\Delta\overline{\psi}_k^{(i)} = \overline{\psi}_+^{(i)}(\varepsilon_k) - \overline{\psi}_-^{(i)}(\varepsilon_k)$;

$\Delta\overline{r}_0'^{(k)} = \overline{r}_{0+}'(\varepsilon_k) - \overline{r}_{0-}'(\varepsilon_k)$ (the values of corresponding functions are marked by indexes "+" и "-" to the right and to the left of ε_k).

The numerical solution of the equation (14) solved by Runge-Kutt's method results in

calculating a dynamic component of the movement u_x in accordance with the expression (10), and the dynamic component of the axial thrust of HT is defined as follows:

$$\Delta Q_1(\varepsilon, \tau_j) = \sum_{i=1}^{3} [\Delta\hat{Q}_{xi}^{(1)}(\varepsilon, \tau_j) + c_i(\tau_j)]x_{i0}' \quad (15)$$

where $\Delta\hat{Q}_{xi}^{(1)}(\varepsilon, \tau_j) = \int_0^\varepsilon \frac{\partial\Delta Q_{xi}^{(1)}(\eta, \tau_j)}{\partial\eta}d\eta$; τ_j $(j = 1, 2, ...)$

are the grid nodes of the numerical solution of the equation (14). Derivatives $\partial\Delta Q_{xi}^{(1)}(\eta, \tau_j)/\partial\eta$ are defined from the equation (7). Arbitrary constants $c_i(\tau_j)$ are calculated while solving the system of the linear algebraic equations.

$$a_{i1}c_1 + a_{i2}c_2 + a_{i3}c_3 + b_i = 0, i = 1, 2, 3, \quad (16)$$

where $a_{ik} = \int_0^1 \frac{(\delta_{ik} - x_{i0}'x_{k0}')}{Q_1(\varepsilon, \tau_j)}d\varepsilon$;

$$b_i = \int_0^1 \sum_{k=1}^{3} \frac{(\delta_{ik} - x_{i0}'x_{k0}')\Delta\hat{Q}_{xk}^{(1)}}{Q_1(\varepsilon, \tau_j)}d\varepsilon.$$

As an example, we will consider HT located in a water flow where the pulp has being pumped for the long period of time. Key design and operational parameters of the pipeline are the following. Pipeline length is 160 m, an outer diameter is 0,435 m, a hose bulk weight without pulp is 126,1 kg/m. The speed of external water flow is 0,5 m/s, and a pulp internal is 3 m/s. At an arbitrary point of time pulp density on a pipeline entrance is a random variable distributed under the normal law with the average value of 1160 kg/m3 and a mean square deviation of 36,7 kg/m3. Poisson's flow with the parameter 0,01 c^{-1} is formed due to the sequence of time points of discontinuous pulp density change.

The results of solving the problem of the constrained casual fluctuations of the studied pipeline are received in the form of histograms for its various cross-sections. Histogram appearance allows putting forward a hypothesis of a normality of laws of casual dynamic component distribution. Checking by criterion χ^2 has shown that this hypothesis can be accepted with the probability not lower than 0,95. Fig. 2a demonstrates the histogram and corresponding density of the normal law of distribution $f(\Delta Q_1)$ of a casual dynamic component of axial thrust$\Delta Q1$ (in a dimensionless form) for the cross-section located at the pipeline exit. Distribution law of the module of a vector $\overline{u}$ of a casual dynamic component of the hose movement relative to the balance position for the cross-section located in the pipeline centre is given in Fig. 2b. The maximum values of casual dynamic components for the chosen hose cross-sections can be determined by the rule of "three sigma".

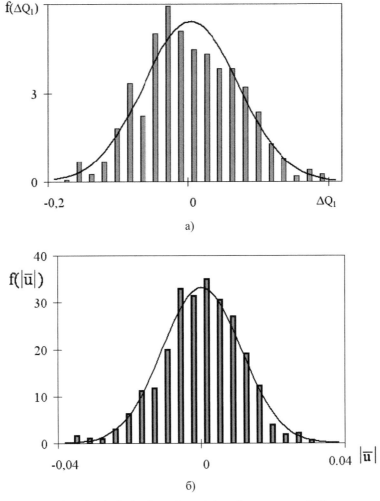

a)

б)

Figure 2. Distribution laws of casual dynamic components of HT

214

The analysis of simulation results has shown that the value of a casual dynamic component of an axial thrust can reach nearly 30% in relation to a maximum (along a pipeline length) static value of an axial thrust, and the vector module of a casual dynamic movement component relative to an equilibrium state is about 4% of a hose length.

3 CONCLUSIONS

The simulation results have shown that the values of casual dynamic components can be very considerable in comparison with the values of the corresponding static components. Therefore, while computing theendurance strength of hydraulic transport pipelines these values should be taken into consideration.

REFERENCES

Svetlitsky V.A. 2001. *Mechanics of absolutely flexible rods* (in Russian). Moscow: MAI Press: 432.

Svetlitsky V. 2005. *Dynamics of rods.* Berlin: Springer: 448.

Power Engineering and Information Technologies In Technical Objects Control – Pivnyak, Beshta
& Alekseyev (eds)
© 2016 Taylor & Francis Group, London, ISBN 978-1-138-71479-3

Optimal control of underground conveyor transport system in coal mines

R. Kiriya
M.S. Polyakov Institute of Geotechnical Mechanics under the National Academy of Sciences of Ukraine, Dnipro, Ukraine

ABSTRACT: The control system of conveyor transport is represented as a two level hierarchical system on first level of which an operator sets certain parameters for the lower level of control. A global criterion of efficiency of this system is a complex value, which characterizes energy efficiency of coal mass transportation, but the local criteria are minimum values of average volumes of cargo in accumulating hoppers of underground conveyor transport system. We developed the algorithm of adaptive control of accumulating hoppers and underground conveyor transport system.

1 INTRODUCTION

The main objective of control of conveyor transport in coal mines is to provide its maximum current capacity with minimum power consumption for transportation of coal mass.

The coal mines conveyor transport system has complex branched structure consisting of conveyors and hoppers which are connected together with the help of batchers, loaders and unloaders.

Failures of conveyors often lead to downtime of lava and as result to poor productivity of conveyor transport system.

To increase current capacity of conveyor transport system in coal mines the accumulating hoppers (temporal redundancy) (Cherkesov 1974; Ponomarenko, Kreymer & Dunaev 1975) have received wide application.

2 FORMULATION OF THE PROBLEM

The accumulating hoppers permit to increase the current capacity of an underground conveyor transport system by means of the accumulation of a certain amount of cargo in a hopper during idle conveyors.

However, the functional effectiveness of the underground conveyor transport system in coal mines with hoppers is low. This is connected with frequent downtime of conveyor lines due to the overflow of hoppers, as well as with the loss of electricity because of underload of conveyors.

One method of improving the efficiency of underground conveyor transport in coal mines is controlling of accumulating hoppers by means of controllers and controlling of the conveyor lines speed using frequency-controlled motors.

At the same time in accumulating hoppers with the help of speed control of batchers we support a set amount of cargo in hopper, in particular, the switching-in of the batchers on reaching the given maximum volumes of cargo in hoppers V_{2i} and switching-off the batchers on reaching the minimum volume of cargo in hoppers V_{1i} (Fig. 1). In this case, the cargo flows coming into the hoppers are not switched off, and their volumes should be smaller than the batchers' capacity. This mode of the hoppers operation allows not to disable above-hopper conveyor lines in the case of hoppers' overflow, and significantly reduces conveyors' downtime because of their underload (Kiriya, Maksyutenko & Braginets 2012; Kiriya, Mischenko & Babenko 2012).

On Fig. 1 V_1 и V_2 – volumes of minimum and maximum rate of cargo in accumulating hopper; V_{max} – maximum volume of hopper; Q_3 – productivity of cargo flow coming into hopper; Q_n – productivity of batcher; h – height of slit of outlet hopper.

However, the local control of each accumulating hopper is still insufficient for significantly improvement of the efficient operation of the underground conveyor transport system in a coal mines. To do this, it is necessary to adjust each hopper's control with operation of the entire system of conveyor transport. At this the maximum efficiency of the conveyor transport system should be provided. I.e. this control system should provide maximum current capacity and minimum charges for transportation of the coal mass.

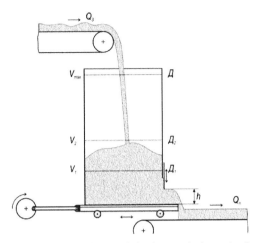

Figure 1. Control of accumulating hoppers in the mode of maintaining set cargo amount

Therefore, the creation of the algorithm of control of the underground conveyor transport system in coal mines is an urgent task.

The operation of underground conveyor transport system in coal mines can be represented as a two-level hierarchical system. On the upper level of the system an operator, on the assumption of the current information about the coal cargo flows m_{Qi}, coming from the faces, about the state of conveyors and technological equipment of underground conveyor transport system, determines the conveyors' speeds v_{ni} and productivities of batchers Q_{ni}. Then he transmits this information, i.e. values v_{ni} and Q_{ni}, to the lower level of control system of hoppers - to the controllers, which determine the maximum volumes of cargo in hoppers V_{2i} (Fig. 2).

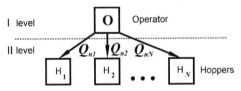

Figure 2. Structural scheme of underground conveyor transport system control

The input variables of the control system of underground conveyor transport are the average values of cargo flows m_{Qi} coming from lava.

And the output variables of the control system are the average value of the capacity m_c and the average value of energy consumption of transportation w_c of the underground conveyor transport system in coal mines (Kiriya, Mischenko & Babenko 2014).

Global objective function, i.e., criterion of efficiency of this two level control system is a complex quantity (Moiseev 1981)

$$K = c_1 m_c - 60 c_2 w_c \; (hrn/min), \qquad (1)$$

where c_1 – cost price of one mass unit of coal, hrn/t; c_2 – cost price of one unit of electricity, hrn/ kW.

The law of underground conveyor transport system control is dependence of m_c and w_c on the input phase and controlled parameters, as well as on the parameters of the conveyors, on the accumulating hoppers' volumes and velocities of the batchers (Kiriya, Mischenko & Babenko 2014):

$$m_c = f_m(m_{Qi}, \lambda_i, \mu_i, Q_{ni}, V_i, V_{1i}, V_{2i}); \qquad (2)$$

$$w_c = f_w(m_{Qi}, \lambda_i, \mu_i, Q_{ni}, V_i, V_{1i}, V_{2i}, N_i), \qquad (3)$$

where λ_i, μ_i – parameters of failure flows and recoveries of conveyors, 1/min; V_i – accumulating hoppers' volumes, m³; N_i – powers used by drivers of conveyors, kW.

The form of these functions depends on the structure of the underground conveyor transport system, work mode of accumulating hoppers, in particular, the mode of maintaining a set cargo amount in a hopper, on the parameters of cargo flows coming from lava m_{Qi}, reliability of conveyors – parameters of failure flows and recoveries of conveyor transport system λ_i, μ_i, productivity of batchers Q_{ni}, volumes of hoppers V_i, and also minimum and maximum volume rate of cargo in accumulating hoppers V_{1i} and V_{2i} correspondingly. For their determination it is necessary to develop an algorithm for calculating the average current capacity m_c and the average energy consumption of transporting w_c of conveyor transport system in the case of accumulating hoppers control in the mode of maintaining a set cargo amount in them.

Structural analysis of underground conveyor transport system in coal mines has shown that they have self-similar dendritic structure (Kiriya, Mischenko, Babenko 2014).

In this work (Kiriya, Mischenko & Babenko, 2014), it was obtained the algorithm for calculating m_c and w_c for underground conveyor transport systems with self-similar dendritic structure (Fig. 3) in the mode of maintaining a set cargo amount in hopper.

On Fig. 3 $\gamma_i^{(c)}, \gamma_{ij}^{(3)}$ – coefficients of failures of shaft and face paths of conveyor lines of conveyor transport system; $V_i^{(c)}, V_{ij}^{(3)}$ – volumes of accumulating hoppers of shaft and face paths, m³; $Q_{ni}^{(c)}, Q_{nij}^{(3)}$ – productivities of batchers of shaft and face paths hoppers, t/min.

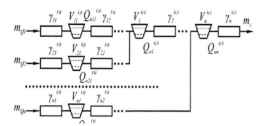

Figure 3. Calculation scheme of dendritic self-similar structure of hoppers connection

This algorithm is a recursive relation.

The average current capacity of conveyor transport system shown on Fig. 3 is determined by formula

$$m_c = m_{c_n},$$ (4)

where

$$m_{c_i} = \left(1 - \frac{\gamma_{i+1}^{(c)}}{1+\gamma_{i+1}^{(c)}} e^{-\frac{\rho\Delta V_i^{(c)}}{m_i^{(s)}}\mu_i}\right) m_i^{(s)},$$ (5)

$$\gamma_{9_i}^{(c)} = \frac{\sum_{k=1}^{i} m_{Q_k}}{m_i^{(s)}} - 1; \quad m_i^{(s)} = m_{c_{i-1}} + \frac{m_{Q_i}}{1+\gamma_{9_i}^{(3)}}; \quad m_{c0} = 0;$$

$$\Delta V_i^{(c)} = V_i^{(c)} - V_{2i}^{(c)}; \; (\mu_i = \mu_c; \; i = 1,n; \; m_i^{(s)} < Q_{n_i}).$$

Here $\gamma_i^{(c)}$ – coefficients of conveyors' accidents of shaft path are determined by formula: $\gamma_i^{(c)} = \lambda_i^{(c)}/\mu_i^{(c)}$, where $\lambda_i^{(c)}, \mu_i^{(c)}$ – parameters of failure flows and recoveries of conveyor lines of shaft path, 1/min; $\gamma_{9_i}^{(c)}$ – equivalent coefficients of accidents of shaft paths with hoppers; $\gamma_{9_i}^{(3)}$ – equivalent coefficients of accidents of face paths with hoppers; $m_i^{(s)}$ – average values of cargo flows coming into accumulating hoppers of shaft path, t/min; ρ – volume mass of transported cargo, t/m³.

The efficient coefficients of accidents of face paths with hoppers are determined by formulas:

$$\gamma_{9_i}^{(3)} = \frac{m_{Q_i}}{m_{c_i}^{(3)}} - 1, \; (\gamma_{9_1}^{(c)} = \gamma_{9_1}^{(3)}, \; i = 1,n),$$ (6)

where $m_{c_i}^{(3)}$ – average current capacity of i-th face path of conveyor transport system with serial

connection of face path hoppers (Kiriya, Mischenko & Babenko 2014).

The average energy cost of transporting in this case is determined by formulas

$$w_c = w_n^{(s)},$$ (7)

where

$$w_i^{(s)} = \frac{w_{i-1}^{(s)}}{1+\gamma_{9_{i-1}}^{(c)}} + w_{i+1}^{(c)} + w_{i+1}^{(3)}, \; (i=1,n),$$ (8)

$$\gamma_{9_{i-1}}^c = \frac{m_i^{(s)}}{m_{c_i}} - 1; \; m_i^{(s)} = m_{c_{i-1}} + \frac{m_Q}{1+\gamma_{9_i}^{(3)}}; \; \gamma_{9_i}^{(3)} = \frac{m_Q}{m_{c_i}} - 1$$

$$(w_0^{(s)} = w_1^{(3)}; \; w_i^{(c)} = N_i^{(c)}).$$

Here $w_i^{(s)}$ – electricity power that is used by conveyor transport system on transporting of coal mass on the area till $i+1$ of shaft line hopper; $w_i^{(c)}$ – electricity power used by i-th conveyor of shaft line on transporting of coal mass; $w_i^{(3)}$ – electricity power that is used by i-th face line on transporting of coal mass; $\gamma_{9_i}^{(3)}$ – equivalent coefficient of accidents of i-th face conveyor line; $N_i^{(c)}$ – power used by driver of i-th conveyor of shaft line on transporting of coal mass, kW; m_{Q_i} – average productivity of i-th face, t/min.

Besides, $w_i^{(3)}$ for each face conveyor line can be determined according recursive relations, which were obtained for serial connection of conveyors and hopper (Kiriya, Mischenko & Babenko 2014).

In case of serial connection of hoppers and conveyors in formulas (4)–(8) we should use

$$m_{Q1} = m_Q; \; \gamma_{9_1}^{(3)} = \gamma_1^{(c)}; \; w_1^{(3)} = w_1^{(c)}; \; m_{Qi} = 0;$$

$$m_i^{(s)} = m_{c_{i-1}}; \; \gamma_{9_i}^{(3)} = 0; \; w_i^{(3)} = 0 \; (i = 2,...,n).$$

The controlled parameters of conveyor transport system in coal mines are productivities of batchers Q_{ni}, which are set by operator on the upper level of control and maximum cargo volumes in hoppers V_{2i}, which are calculated by controllers in subsystems of accumulative hoppers control in case of their control in the mode of maintaining a set cargo amount in them.

That's why for control of accumulating hoppers of conveyor transport system in coal mines it is necessary to know dependencies V_{2i} on productivities of batchers Q_{ni}.

In this work (Kiriya, Mischenko & Babenko 2012), it was solved the problem of determining the optimal average value of cargo volume in the accumulating hopper which works in a mode of maintaining a set cargo amount in hoppers.

In this case the optimal average values of cargo volumes in accumulating hoppers V_{imin} are equal to the semi sum of maximum and minimum set cargo amount values in hopper, i.e.

$$V_{i\min} = \frac{V_{1i} + V_{2i}}{2} + \frac{\lambda_i}{(\lambda_i + \mu_i)\mu_i} \cdot \frac{m_i^{(s)}}{2\rho}, \qquad (9)$$

where V_{1i}, V_{2i} – maximum and minimum values of cargo volumes in accumulating hoppers.

Therefore, the local functions of subsystem objective of hoppers control are the average cargo value in accumulating hopper.

Therefore, the average cargo values in accumulating hoppers V_{ci} should reach to the average minimum values of cargo volumes in hoppers V_{imin}, i.e.

$$V_{ci} \to V_{imin}. \qquad (10)$$

On the other hand, according to work (Kiriya, Mischenko & Babenko 2012), the average values of cargo volumes in accumulating hoppers, which work in the mode of maintaining a set cargo amount in hoppers, can be determined by the formula

$$V_{ci} = \frac{V_{1i}t_{3i} + V'_{2i}t_{pi}}{t_{3i} + t_{pi}} + \frac{m_i^{(s)}t_{3i}^2 - \left(Q_{ni} - m_i^{(s)}\right)t_{pi}^2}{2\rho\left(t_{3i} + t_{pi}\right)}, \quad (11)$$

where $V'_{2i} = V_{2i} + \dfrac{\lambda_{i+1}}{\lambda_{i+1} + \mu_{i+1}} \dfrac{Q_{ni}}{\gamma} t_{pi}$; λ_i, μ_i – parameters of failure flows and recoveries of above-hoppers conveyor lines, min^{-1}; λ_{i+1}, μ_{i+1} – – parameters of failure flows and recoveries of under-hoppers conveyor lines, min^{-1}.

Here t_{3i}, t_{pi} – average time periods of loading and unloading of hoppers, which are determined by

$$t_{3i} = \frac{\rho(V_{2i} - V_{1i})}{m_i^{(s)}} + \frac{\lambda_i}{(\lambda_i + \mu_i)\mu_i}, \qquad (12)$$

$$t_{pi} = \frac{\rho(V_{2i} - V_{1i})}{\left(\bar{Q}_{ni} - m_i^{(s)}\right)} + \frac{\lambda_i}{(\lambda_i + \mu_i)\mu_i} \\ \times \frac{Q_{ni}m_i^{(s)}}{\left(\bar{Q}_{ni} - m_i^{(s)}\right)\left(Q_{ni} - m_i^{(s)}\right)}, \qquad (13)$$

where $\bar{Q}_{ni} = \dfrac{\mu_{i+1}}{\lambda_{i+1} + \mu_{i+1}} Q_{ni}$.

Equating the average values of cargo volume in hoppers V_{ci} to minimum values of volume V_{cimin}, we will get equation for the relatively unknown maximum values of set cargo amounts in hoppers V_{2i}, at which the average volume of cargo in hoppers takes the minimum values. As a result, according to (11), we will get the equation according to V_{2i} :

$$\frac{V_{1i}t_{3i} + V'_{2i}t_{pi}}{t_{3i} + t_{pi}} + \frac{m_i^{(s)}t_{3i}^2 - \left(Q_{ni} - m_i^{(s)}\right)t_{pi}^2}{2\rho\left(t_{3i} + t_{pi}\right)} \\ = \frac{V_{1i} + V_{2i}}{2} + \frac{\lambda_i}{(\lambda_i + \mu_i)\mu_i} \cdot \frac{m_i^{(s)}}{2\rho} \qquad (14)$$

In obtained equation (14), according to (12) and (13), t_{3i} and t_{pi} are functions from V_{2i}. Besides, $m_i^{(s)}$ is determined by difficult recursive relations (see (5), Kiriya, Mischenko & Babenko 2014).

Thus determining of V_{2i} from equation (14) is a complex mathematical problem.

To simplify the problem of determining the set maximum cargo amount in accumulating hoppers, you can use the scale to determine the values of cargo flows coming into accumulating hoppers $m_i^{(s)}$. However, this requires high financial costs.

Therefore, for the settling of cargo in the hopper with the help of controller the adaptive control algorithm is applied (Kiriya, Maksyutenko & Braginets 2012).

In this case to determine $m_i^{(s)}$ at first with the help of time sensors, the time periods of loading T_{3i} and unloading T_{pi} are determined in accumulating hoppers of conveyor transport system in time of stopping and work of batcher accordingly.

Substituting into equation (12) instead of t_{3i} the value T_{3i}, from the obtained expression we can determine the average values of cargo flows coming into accumulating hoppers by the formula:

$$\bar{m}_i^{(s)} = \frac{\rho\left(V_{2i}^{(0)} - V_{1i}\right)}{T_{3i} - \dfrac{\lambda_i}{(\lambda_i + \mu_i)\mu_i}}. \qquad (15)$$

where $V_{2i}^{(0)}$ – values of initial maximum cargo volumes in accumulating hoppers when time of loading T_{3i} and unloading T_{pi} of hoppers are determined.

The value of equation standing in the denominator on the right side of equality (15), can be expressed by m_{Qi} and $\bar{m}_i^{(s)}$ by formula (Kiriya, Mischenko & Babenko 2014):

$$\frac{\lambda_i}{(\lambda_i + \mu_i)\mu_i} = \frac{1}{\mu_c}\left(1 - \frac{\bar{m}_i^{(s)}}{\sum\limits_{k=1}^{i} m_{Qk}}\right), \quad (16)$$

where $\mu_c = \mu_i$.

By substituting (16) into (15), we will obtain

$$m_i^{(s)} = \frac{\rho\left(V_{2i}^{(0)} - V_{1i}\right)}{T_{3i} - \left(1 - m_i^{(s)} \Big/ \sum\limits_{k=1}^{i} m_{Qk}\right)\dfrac{1}{\mu_c}} \quad (17)$$

Let's determine from the last equation $m_i^{(s)}$, as result we have:

$$m_i^{(s)} = \frac{\sqrt{(\mu_c T_{3i} - 1)^2 m_{\varepsilon i}^2 + 2\rho(V_{2i} - V_{1i})\mu_c m_{\varepsilon i}}}{2} - \frac{(\mu_c T_{3i} - 1)m_{\varepsilon i}}{2}, \quad (18)$$

where $m_{\varepsilon i} = \sum\limits_{k=1}^{i} m_{Qk}$ ($m_{\varepsilon i} = m_Q$ – when we have serial connection of hoppers).

To determine the average values of cargo flows $\bar{Q}_{ni}$, unloaded from accumulating hoppers, let's substitute into equation (13) taking into account (16) instead of value t_{pi} the value T_{pi}. As a result after transformation we will have

$$\bar{Q}_{ni} = \bar{m}_i^{(s)} + \frac{\rho\left(V_{2i}^{(0)} - V_{1i}\right)}{T_{pi}} + \frac{\dfrac{1}{\mu_c}\left(1 - \dfrac{\bar{m}_i^{(s)}}{m_{\varepsilon i}}\right)\cdot\dfrac{Q_{ni}\bar{m}_i^{(s)}}{\left(Q_{ni} - \bar{m}_i^{(s)}\right)}}{T_{pi}}, \quad (19)$$

Hence, knowing time of loading T_{3i} and unloading T_{pi} of accumulating hoppers, the average values of cargo flows m_{Qi} coming from lava, and also set values of minimum V_{1i} and maximum $V_{2i}^{(0)}$ of cargo volumes in accumulating hoppers of conveyor transport system in coal mines, it can be determined by formulas (18) and (19) the average values of cargo flows $\bar{m}_i^{(s)}$ coming into hoppers and the

average values of cargo flows $\bar{Q}_{ni}$ unloaded from hoppers.

To determine maximum volumes of cargo in accumulating hoppers V_{2i}, which work in mode of maintaining a set cargo amount in them, when the average cargo volumes in hoppers V_{ci} have minimum values, let's substitute into equation (14) instead of t_{3i} and t_{pi} values T_{3i} and T_{pi} correspondingly. As a result we will get equation of relatively unknown maximum volumes of cargo in hoppers V_{2i}:

$$\frac{V_{1i}t_{3i} + V_{2i}t_{pi}}{t_{3i} + t_{pi}} + \frac{\bar{m}_i^{(s)}t_{3i}^2 - \left(Q_{ni} - \bar{m}_i^{(s)}\right)t_{pi}^2}{2\rho\left(t_{3i} + t_{pi}\right)}$$

$$+ \frac{\left(Q_{ni} - \bar{Q}_{ni}\right)}{\rho}\cdot\frac{t_{3i}^2}{t_{3i} + t_{pi}} \quad (20)$$

$$= \frac{V_{1i} + V_{2i}}{2} + \left(1 - \frac{\bar{m}_i^{(s)}}{m_{\varepsilon i}}\right)\cdot\frac{1}{\mu_c}\cdot\frac{\bar{m}_i^{(s)}}{2\rho}$$

In obtained equation (20) values $\bar{m}_i^{(s)}$ and $\bar{Q}_{ni}$ are determined by formulas (18) and (19).

The simplest and most effective method of solving the equation (20) is the method of dichotomy (bisection) or sounding method. In this case, the unknown parameter V_{2i} changes in limits $V_{1i} \leq V_{2i} \leq V_i$.

Hence, when we have set values of cargo flows m_{Qi}, coming from lava, the productivities of batchers Q_{ni}, volumes of accumulating hoppers V_i and minimum values of cargo volumes V_{1i} in accumulating hoppers, and also current time values of loading T_{3i} and unloading T_{pi} of accumulating hoppers of conveyor transport system in coal mines, from equation (20) we can find values of maximum cargo volumes V_{2i}^*, when the average volumes of cargo in hoppers V_{ci} take minimum values equal to

$$V_{ci} \to V_{i\,min} = \frac{V_{1i} + V_{2i}^*}{2} + \left(1 - \frac{\bar{m}_i^{(s)}}{m_{\varepsilon i}}\right)\cdot\frac{1}{\mu_c}\cdot\frac{\bar{m}_i^{(s)}}{2\rho}. \quad (21)$$

Hence the algorithm of adaptive control of underground conveyor transport system in coal mines can be represented as follows.

1. The generating of alternatives.

For this on set average values of cargo flows m_{Qi} coming from lava and structure of underground conveyor transport system we determine several alternatives of conveyors' speeds v_{ai} and productivities of batchers Q_{ni}, where a – number of

221

alternative variants ($a = 1,...,N;$ N – quantity of alternative variants). At the same time conditions must be satisfied (restrictions)

$$\sum_{1}^{i} m_{Qi} < Q_{ni}^{(a)} \le Q_{mi}, \ (i = 1,...,n; \ a = 1,...,N), \quad (22)$$

where Q_{mi} – maximal productivities of under-hopper conveyor lines of conveyor transport system, t/min.

Moreover, for accumulating hoppers which work in the mode of maintaining a set cargo amount in them should be performed such limitations as

$$m_i^{(s)} < Q_{ni}^{(a)};$$

$$V_{1i} < V_{2i} < V_i, \ (i = 1,...,n; \ a = 1,...,N). \quad (23)$$

2. Estimation of the average values of cargo flows loaded into the hoppers and unloaded from them.

For this purpose, with the help of time sensors during the loading and unloading time periods of accumulating hoppers the values of loading time $T_{зi}$ and unloading time T_{pi} are determined. By this values $T_{зi}$ and T_{pi} according to formulas (18) and (19) we determine the valuations of average values of cargo flows $m_i^{(s)}$, coming into hoppers, and the average values of cargo flow $\bar{Q}_{ni}^{(a)}$, unloaded from hoppers.

3. Determination of the maximum set cargo amount values in accumulating hoppers.

To do this, by the obtained values of estimations $\bar{m}_i^{(s)}$ and $\bar{Q}_{ni}^{(a)}$ from equation (20) we determine maximum values of cargo volumes in accumulating hoppers V_{2i}^*, at which the average cargo volumes in hoppers V_{ci}, working in the mode of maintaining a set cargo amount in them, have minimum values.

4. Determination the criteria of efficiency and the objective function.

For this by calculated values V_{2i}^* and according to recursive formulas (4)–(8) we determine current capacity $m_c^{(a)}$ and energy consumption $w_c^{(a)}$ of the entire underground conveyor transport system.

By obtained values $m_c^{(a)}$ and $w_c^{(a)}$ according to formula (1) we can determine the objective function K_a.

5. Repeating this process for various alternatives, i.e. determining the objective function K_a for different productivities of batchers $\bar{Q}_{ni}^{(a)}$, we select from obtained objective functions K_a the minimum value K_{min}, i.e.

$$K_{min} = \min_{1 \le a \le N} \{K_a\},$$

Values of productivities of batchers Q_{ni} and maximum cargo volumes in accumulating hoppers V_{2i}, that correspond to this minimum criterion of efficiency $K = K_{min}$, are optimal.

If you change the structural diagram of underground conveyor transport associated with the promotion of lava or changing its amount, as well as with changes in values of the average cargo flows coming from lava m_{Qi}, optimal values Q_{ni} and V_{2i}, at which the criterion of efficiency K takes the minimum value are determined anew at the above given algorithm.

Calculations have shown that the energy efficiency of transportation of coal mass, i.e, the criterion of efficiency K of underground conveyor transport system in coal mines at the optimal values Q_{ni} and V_{2i}, obtained on the basis of made algorithm, increases up to 30 %.

3 CONCLUSIONS

We set and solved the problem of optimal control of underground conveyor transport system in coal mines, which is represented as two-level hierarchical system. We defined a global criterion of efficiency of the system which is a complex value consisting of the current capacity and energy consumption of conveyor transport system, as well as local criteria which are the minimum value in the average volume of cargo in accumulating hoppers. An algorithm for optimal control of conveyor transport system, including the algorithm of adaptive control of accumulating hoppers, was developed. The use of this algorithm of control of underground conveyor transport system in coal mines will increase the energy efficiency of transportation of mined rock by belt conveyors up to 30%.

REFERENCES

Cherkesov G.N. 1974. *Reliability of the Technical Systems with Temporal Redundancy* (in Russian). Moscow: Soviet Radio: 296.

Ponomarenko V.A., Kreymer E.L. and Dunaev G.A. 1975. *Underground Transport Systems on Coal Mines* (in Russian). Moscow: Nedra: 309.

Kiriya R.V., Maksyutenko V.J. & Braginets D.D. 2012. *Management of hoppers which work in conveyor transport systems of coal mines* (in Russian). Dnipropetrovsk: National Mining University: Compedium of scientific works of NMU, Vol. 37: 230–236.

Kiriya R.V., Mischenko T.F. & Babenko J.V. 2012. *Mathematical model of functioning of accumulative hoppers in mode of keeping in it set cargo amount* (in Russian). Dnipropetrovsk: NMetAU: Modern problems of metallurgy, Issue 15: 85–96.

Kiriya R.V., Mischenko T.F. & Babenko J.V. 2014. *Determining of efficiency criteria of underground conveyor transport systems of coal mines in mode of keeping set cargo amount in accumulative hoppers* (in Russian). Dnipropetrovsk: System technologies: Regional interuniversity compendium of scientific works, Vol 1: 135–141.

Moiseev N.N. 1981. *Mathematical problems of system analysis* (in Russian). Moscow: Science: 488.

Kiriya R.V. Mischenko T.F., Babenko J.V. 2014. *Mathematical models of conveyor transport systems functioning of coal mines* (in Russian). Dnipropetrovsk: System technologies: Regional interuniversity compendium of scientific works, Vol 1: 146–158.

Power Engineering and Information Technologies In Technical Objects Control – Pivnyak, Beshta
& Alekseyev (eds)
© 2016 Taylor & Francis Group, London, ISBN 978-1-138-71479-3

Fractal analysis for forecasting chemical composition of cast iron

O. Gusev, V. Kornienko, O. Gerasina & O. Aleksieiev
State Higher Educational Institution "National Mining University", Dnipro, Ukraine

ABSTRACT: By the methods of fractal analysis the properties of temporal rows, presenting the results of cast-iron chemical analysis on issue were investigational and the recommendations on creation of tools for its forecasting were developed.

1 INTRODUCTION

The most important properties of cast iron include the chemical composition, which largely determine the properties of the main product of the blast furnace (cast iron).

Fierce competition in the markets of the metallurgical production necessitates a given quality of cast iron. Quality of cast iron depends on the efficiency of the management process of its production, which is impossible without the operational forecasting of the main characteristics of the products for a period determined by the number of smelts. Operational forecast allows you to identify unwanted trends in the behavior of the blast furnace process and to develop corrective actions aimed at their elimination (Kornienko, Gulina & Budkova 2013).

2 FORMULATING THE PROBLEM

The characteristic features of blast furnace are:
- the random character of time changes in the physical and chemical properties of the blast furnace charge materials;
- a large number of factors (including uncontrolled) affecting the final result of the blast furnace.

These features determine the need for studies of the properties of time series that presents the results of chemical analysis of iron release. Such research is needed to develop recommendations on creation of methods of forecasting of the chemical composition of cast iron in the conditions of production are adequate to the nature of the predictable process.

The time series represents a sequence of values of investigated quantities measured in consecutive moments of time. Usually, time series are submitted to random changes in the values, which permit to represent the evolution of complex systems on the basis of the data obtained (Boffetta, Cencini, Falconi & Vulpiani 2002). Such an analysis reduces to the calculation of correlation functions of the vectors of the states – a time series of the variables characterizing the system.

The most common methods use correlation and spectral analysis, smoothing and filtering of data, models of autoregression and forecasting (Tyurin & Makarov 1998; Kornienko, Gerasina & Gusev 2013). In most cases statistical analysis is based on the assumption that the studied system is random, i.e., the causal process that generated the time series has many component parts or degrees of freedom. The interaction between these components is so complex that a deterministic explanation is not possible. The object of consideration is the class of models corresponding to the case of a Gaussian random process.

However, many real time series are characterized by relative invariance of scale transformation (the property of self-similarity), in connection with which standard Gaussian statistics is untenable and the problem of investigation of time series is reduced to the analysis of stochastic self-similar processes that can be described by fractal sets (Mandelbrot 2002; Feder 1991).

3 THE PURPOSE OF THE WORK

Study of properties of time series representing the results of chemical analysis of cast iron in the production and development of recommendations on creation of instrumental means of forecasting.

4 EXPERIMENTAL RESEARCH

The present study involves the justification of the hypothesis on the fractal nature of time series, which presents the results of chemical analysis of cast iron release.

The main quantitative characteristic of fractals is topological dimension D introduced by Hausdorff. For most natural time series analytical finding topological fractal dimension is impossible, therefore, D is determined numerically in the form of correlation estimates, or through the values related by simple ratios (for example, through the Hurst exponent (H)). To calibrate the time measurements Hurst introduced a dimensionless ratio by dividing the magnitude by the standard deviation of the observations of R/S. This method of analysis became known as the method of normalized range or R/S analysis (Feder 1991).

Thus, to test the hypothesis about fractal properties of the studied processes is first necessary to estimate the Hurst exponent H.

Time series analysis on self-similarity was performed on the basis of the silicon content in the cast iron (Fig. 1) obtained in different time periods of blast furnace No. 3 (BF-3), Mariupol metallurgical combine named Ilyich (MMC).

Analysis of fractal properties of time series, in particular the properties of self-similarity, were performed using the Fractan 4.4 (Sychev V.).

One of the main properties of fractal (self-similar) process is a slow decrease of the autocorrelation function (ACF) (Fig. 2). This property is of key importance in the theory of self-similar processes and, in fact, determines the most important from the point of view of predicting the characteristics of random process – duration of the process memory.

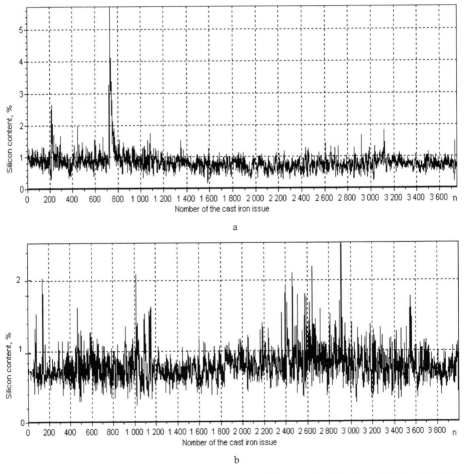

Figure 1. Dynamics of change of silicon content in cast iron on the issue: series 1 for the period from 01.01.2011 to 31.12.2011 (a) & series 2 for the period from 01.01.2012 to 10.12.2012 (b)

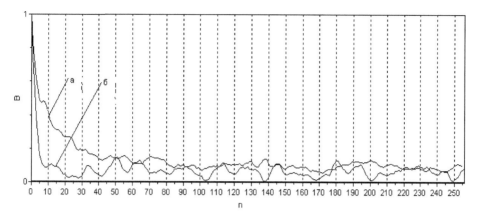

Figure 2. The ACF of the investigated time series 1 (a) & series 2 (b)

From Fig. 2 shows that the ACF are slowly decreasing, and in their "tails" missing trend stemline to zero. This behavior of the ACF is typical for self-similar processes.

The Hurst exponent H is interpreted as follows (Feder 1991; Kuznetsov 2002):
- H = 0,5 assumes the time series to white noise (independent, random process);
- $0 \le H \le 0,5$ means pink noise or antipersistence (the time series changes direction more often than a random series of independent variables);
- $0,5 \le H \le 1$ means black noise or persistence (the time series is characterized by the effect of long-term memory and tends to follow the trends). Trendologist behavior of the process increases when approaching H to 1.

Note that the Hurst exponent is linked with topological fractal dimension by relation $D = 2 - H$.

The calculated values of the Hurst exponent have made H1 = 0,9866 ± 0,1329 and H2 = 0,8963 ± 0,2916 for time series 1 and 2, respectively, which also confirms the self-similar nature of the processes.

The indicator H were carried out by R/S analysis. Graphics of R/S statistics of analyzed series is shown in Fig. 3.

It was thus established that the investigated time series are fractal in nature and possess the property of self-similarity. Therefore, further research should be carried out using not classical stochastic methods, and using fractal methods and methods of stochastic dynamics that are adequate to the nature of the studied processes. The idea of applying the methods of chaotic dynamics to time series analysis is that the structure of the chaotic system that contains all the information about the system, namely, its attractor can be reconstructed through the measurement of only one observable of this dynamical system, fixed as the time series (Kuznetsov 2002).

According to the method of Grassberger and Procaccia (Grassberger & Procaccia 1983) the procedure of phase space reconstruction and restoration of the chaotic attractor of the system while dynamic analysis of the time series, is reduced to the construction of the so-called lag or refurbished space using the method of delays.

Vectors $\overline{S}_k$ in the new space attachments formed from the values of a time series of scalar measurements with time-delay:

$$\overline{S}_k = (S_{k-(m-1)\tau}, S_{k-(m-2)\tau}, ..., S_k),$$

where k - the size of the time series; m - is embedding dimension; τ - is the delay (lag).

In the theorems of Takens (Takens 1981) and Sauer (Sauer, Yorke & Casdagli 1991) it is shown that if the sequence {Sk} consists of scalar measurements of the structure of a dynamical system, then under certain assumptions, this recovery phase portrait is an accurate picture of the real set {x}, (if m is suciently large).

In other words, the real attractor of a dynamical system and «attractor», restored in lag space on the base of time series according to the rule (pseudoattractor), adequate selection of dimensions of attachment m are topologically equivalent and have the same generalized fractal dimensions and other numerical characteristics.

In that case, if the analyzed time series is a realization of a random process, then restored pseudoattractor is a structureless cloud of points that are in sequential infinite increase in the dimension of the lag space attachment m, like a gas, fills the entire allotted volume (Kuznetsov 2002).

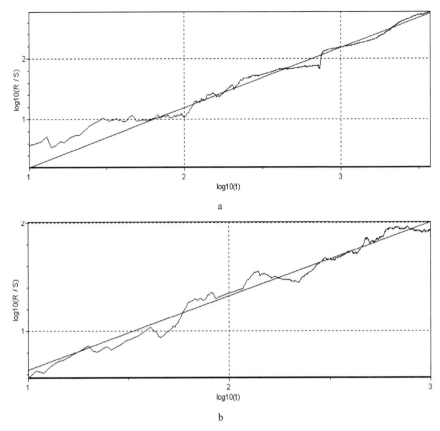

Figure 3. The Hurst exponent for series 1 (a) & series 2 (b)

View strange attractors (the phase space of 2D) for the studied time series in normalized coordinate axis shown in Fig. 4. Here you can see the region of attraction, which is a dense "core". At the same time for a random sequence, as already noted above, the points recovered pseudoattractor form a structureless cloud in lag space.

On the basis of the investigated time series it is possible to build the correlation entropy K, which shows the degree of divergence of close phase trajectories and allows to estimate the amount of information necessary to predict the behavior of the process in the future. Correlation entropy defines the lower bound of the entropy of Kolmogorov-Sinai K (determines the speed of loss of information about the system state and allows you to judge how dynamic system is chaotic) (Kuznetsov 2002).

In Fig. 5 shows the dependence of correlation entropy from the space dimensions attachment n of the investigated time series.

Correlation of entropy with points n=4 and n=6 (Fig. 5 a & b, respectively) is non-increasing, which indicates the presence of chaotic component. Value of K in both cases is small enough (K1=2,221; K2=1,276), what determines a good trendologist and predictability of the process for 4-6 steps forward.

One of the main and the most informative characteristics of chaotic processes is correlation dimension of the restored attractor D, which shows the degree of complexity of the system, generating the observed process. The more complex a system is, the more equations required for its description, the greater the correlation dimension, but the process is closer in its characteristics to the white noise. Thus, this value can also be seen as a measure of stochasticity of the process. In our case D1 = 7,467 and D2 = 6,353 (Fig. 6). In (Kuznetsov 2002) it is shown that the correlation dimension more than five implies a significant influence of random components. Therefore, you can leave the hypothesis that the investigated series are deterministic chaotic with stochastic components, and their phase portrait is a strange attractor.

228

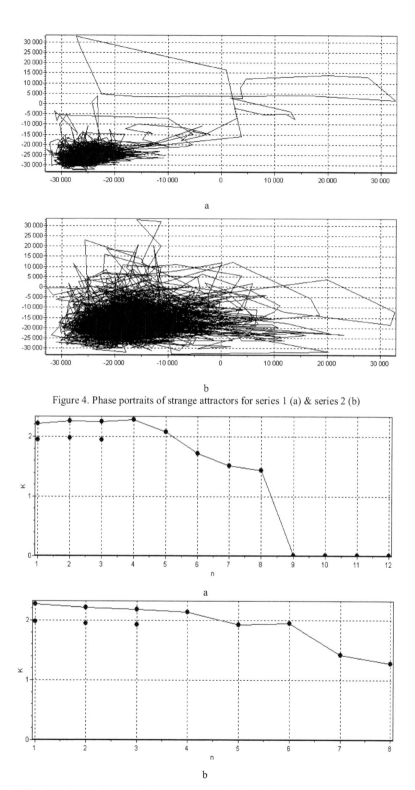

a

b

Figure 4. Phase portraits of strange attractors for series 1 (a) & series 2 (b)

a

b

Figure 5. The dependence of the correlation entropy of embedding dimension for series 1 (a) & series 2 (b)

229

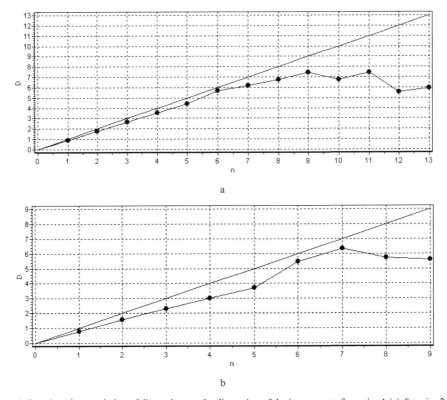

Figure 6. Based on the correlation of dimensions to the dimension of the investments for series 1 (a) & series 2 (b)

5 CONCLUSIONS

It was established experimentally that the time series representing the percentage of silicon in cast iron in the production, are fractal in nature and possess the property of self-similarity.

Using methods of the theory of nonlinear dynamical systems and theory of fractals allow us to estimate the basic fractal characteristics of time series and get an idea about the change of their properties and evolution.

It is shown that the investigated time series are persistent (trendaptive), which has a long memory, and have strange attractor with a single zone of attraction.

The presence of long memory in time series proves the possibility of developing a system of medium - and long-term forecasting. The system needs to use a fractal filter-predictors, the characteristics of which adequate the nature of the predicted time series.

REFERENCES

Kornienko V.I., Gulina I.G. & Budkova L.V. 2013. *Complex estimation, identification and prediction of difficult nonlinear processes* (in Ukrainian). Scientific bulletin of National Mining University, Issue 6.

Boffetta G., Cencini M., Falconi M. & Vulpiani A. 2002. *Predictability: a way to characterize complexity*. Phys. Rep., Issue 356.

Tyurin Yu.N. & Makarov A.A. 1998. *Statistical analysis of the data on the computer* (in Russian). Moscow: INFRA-M: 528.

Kornienko V.I., Gerasina A.V. & Gusev A.Yu. 2013. *Methods and principles of control over the complex objects of mining and metallurgical production.* Taylor & Francis Group: CRC Press: Energy Efficiency Improvement of Geotechnical Systems: 183-193.

Mandelbrot B. 2002. *The Fractal geometry of nature* (in Russian). Moscow: Institute of computer science: 656.

Feder E. 1991. *Fractals* (in Russian). Moscow: Mir: 262.

Sychev V. *Fractal analysis. The Program Fractan 4.4.* [See http:// impb.ru/~sychyov/]

Kuznetsov S.P. 2002. *Dynamical chaos* (in Russian). Moscow: Fizmatlit.

Grassberger P. & Procaccia I. 1983. *Characterization of strange attractors*. Physical Review Letters, Volume 50.

Takens F. 1981. *Detecting Strange Attractors in Turbulence*. New York: Springer: Lecture Notes in Mathematics, Vol. 898.

Sauer T., Yorke J. & Casdagli M. 1991. *Embedology*. Journal of Statistical Physics, Vol. 65, Issue 3.

Power Engineering and Information Technologies In Technical Objects Control – Pivnyak, Beshta
& Alekseyev (eds)
© 2016 Taylor & Francis Group, London, ISBN 978-1-138-71479-3

The continuous problems of the optimal multiplex partitioning an application of sets

L. Koriashkina
Dnipropetrovsk National University named after Oles Honchar, Dnipro, Ukraine

A. Cherevatenko & O. Mykhalova
State Higher Educational Institution "National Mining University", Dnipro, Ukraine

ABSTRACT: The continuous problems of the optimal multiplex-partitioning of sets (OMPS) for the fixed centers or those to be located with a variety of the partitioning quality criteria are formulated. The practical application of these problems in economics, as well as in the image analysis and segmentation, are presented. In particular, the problems of multiplex-partitioning are presented as mathematical models of problems for service centers' location within a limited area simultaneously with partitioning of this area into regions that cover customers with the same set of k nearest centers. As this takes place, the formation of partitioning-location quality criterion to specify the located centers, along with a type of provided services, is considered. By means of the presented models for the OMPS problems the competition between service centers, determined real activity sphere of each center according to its capacities, etc can be studied.

1 INTRODUCTION

The first mathematical models for the continuous problems of optimal multiplex-partitioning of sets are linear (Koriashkina 2015). These problems are an extension (generalization) of linear continuous problems of optimal set partitioning from n-dimensional Euclidean space, which are non-classical problems of the infinite dimensional mathematical programming with Boolean variables (Kiseleva 2005; Kiseleva 2013).

In addition to mathematical models for the continuous linear OMPS problems we formulate some non-linear (so-called minimax) problems of the multiplex set partitioning.

We show that the choice of the multiplex partitioning optimality criterion is defined from the type of the located centers, along with their specificity.

In linear case we deal with optimal partitioning of a given area into regions, each of which consists of points (customers) with the same k nearest neighboring centers. It is assumed that customers of each region can be served by one of these nearest neighboring centers. When partitioning quality criterion for such problems, the minimization of the total distance from the center to all clients served by it is usually selected. The examples of pairs "service center – clients" can be: enterprises and consumers, post-offices and subscribers, analysis collection stations or other medical institutions and patients, etc.

These objectives of optimal set partitioning will be useful to study the competition between service centers, to identify each center's real sphere of activity, according to its capacity, as well as additional information about the possibilities of competing service centers, the demand at each point of a given area etc.

If the centers extremely need enterprises or services (emergency, police, medical facilities, etc.) and customers who are continuously distributed in the area Ω, the optimality criterion can minimize the distance (or time) from the service point to an innermost point of the area Ω. Hence, in non-linear case, we need to optimize the worst case. The most common problem is to locate a few of such service points, and to consider that every customer can be serviced by any of k closest service points (for example, when the closest one cannot provide the service for some reason).

In either case we adduce economic interpretation of partitioning problems. Another application models and methods of optimal multiplex-partitioning of sets are indicated. Among them is the construction of higher order Voronoi diagrams and image segmentation.

2 GENERAL FORMULATION OF CONTINUOUS PROBLEM OF OPTIMAL MULTIPLEX SET PARTITIONING AND ITS VARIANTS

Let us formulate general mathematical formulation of continuous problem for the optimal multiplex set partitioning, where the partition is necessary to find the unknown coordinates of centers, regarding to their capacities. We will present different special cases further.

Let the Ω be bound Lebesgue measurable closed set in E_n; $\tau_i = \left(\tau_i^{(1)},...,\tau_i^{(n)}\right) \in \Omega$, for all $i = \overline{1,N}$, are some points called «centers» (they can be fixed or subjected to determination).

We introduce the following notations: $N = \{1, 2,..., N\}$ is a set of all indexes of centers; $M(N,k)$ is a set of all k-elements subsets of the set N, $|M(N,k)| = C_N^k = L$; $\sigma_l = \{j_1^l, j_2^l,..., j_k^l\}$, $l = \overline{1,L}$, are elements of the set $M(N,k)$. We associate each element σ_l from the set $M(N,k)$ with some subset Ω_{σ_l} of points from Ω, $l = \overline{1,L}$. In its turn, the subset of Ω_{σ_l} is associated with a set of centers $\{\tau_{j_1^l}, \tau_{j_2^l},..., \tau_{j_k^l}\}$.

Let us call the collection of Lebesgue measurable subsets $\Omega_{\sigma_1}, \Omega_{\sigma_2},...\Omega_{\sigma_L}$ from $\Omega \subset E_n$ as a partition of the k-th order of the set Ω into the disjoint subsets $\Omega_{\sigma_1}, \Omega_{\sigma_2},...\Omega_{\sigma_L}$, if

$$\bigcup_{i=1}^{L} \Omega_{\sigma_i} = \Omega, \; \text{mes}\left(\Omega_{\sigma_i} \cap \Omega_{\sigma_j}\right) = 0, \; \sigma_i, \sigma_j \in M(N,k),$$

$$i \neq j, \; i, j = \overline{1,L},$$

where $\text{mes}(\cdot)$ means the Lebesgue measurement.

Under the k-subsets of the order of Ω we mean the subsets $\Omega_{\sigma_1}, \Omega_{\sigma_2},...\Omega_{\sigma_L}$ within this set.

Let $\Sigma_{\Omega}^{N,k}$ be a class of all possible partitions of the k-th order of the set Ω into disjoint subsets $\Omega_{\sigma_1}, \Omega_{\sigma_2},...\Omega_{\sigma_L}$:

$$\Sigma_{\Omega}^{N,k} = \Big\{\bar{\omega} = \{\Omega_{\sigma_1}, \Omega_{\sigma_2},...,\Omega_{\sigma_L}\} : \bigcup_{i=1}^{L} \Omega_{\sigma_i} = \Omega;$$

$$\text{mes}(\Omega_{\sigma_i} \cap \Omega_{\sigma_j}) = 0; \sigma_i, \sigma_j \in M(N,k);$$

$$i \neq j; i, j = \overline{1,L}\Big\}.$$

Under continuous problem of optimal multiplex-partitioning of sets under constraints with centers' location we will understand the next problem.

Problem A-k. Find

$$F\left(\bar{\omega}, \tau^N\right) \to \min_{\substack{\bar{\omega} \in \Sigma_{\Omega}^{N,k} \\ \tau^N \in \Omega^N}}, \tag{1}$$

under conditions

$$\sum_{\substack{l=1 \\ l:i\in\sigma_l}}^{L} \int_{\Omega_{\sigma_l}} \gamma_i^l \rho(x) dx = b_i, \quad i = \overline{1,p}, \tag{2}$$

$$\sum_{\substack{l=1 \\ l:i\in\sigma_l}}^{L} \int_{\Omega_{\sigma_l}} \gamma_i^l \rho(x) dx \leq b_i, \quad i = \overline{p+1,N}, \tag{3}$$

where $F(\bar{\omega}, \tau^N)$ is some partitioning- location quality criterion defined below; $x = \left(x^{(1)},...,x^{(n)}\right) \in \Omega$; coordinates $\tau_i^{(1)},...,\tau_i^{(n)}$ of a center τ_i, $i = \overline{1,N}$, are unknown in advance; $\rho(x)$ is bounded measurable integral on the set Ω function; $w_i > 0, a_i \geq 0, b_i \geq 0, i = \overline{1,N}$, are given numbers. Coefficients γ_j^l are such that for all $j = \overline{1,N}$, $l = \overline{1,L}$:

$$0 \leq \gamma_j^l \leq 1, \; \gamma_{j_1^l}^l + \gamma_{j_2^l}^l + ... + \gamma_{j_k^l}^l = 1. \tag{4}$$

Here $\sigma_l = \{j_1^l, j_2^l,..., j_k^l\}$ is a set of indices of centers $\{\tau_{j_1^l}, \tau_{j_2^l},..., \tau_{j_k^l}\}$ associated with the subset Ω_{σ_l}, $l = \overline{1,L}$.

A pair $\left(\bar{\omega}^*, \tau_*^N\right) = \left(\{\Omega_{\sigma_1}^*,...,\Omega_{\sigma_L}^*\}, \{\tau_1^*,...,\tau_N^*\}\right)$, that affords minimum to the functional $F(\bar{\omega}, \tau^N)$ and satisfies conditions (2), (3), is called as optimal solution of the problem A-k.

The functional $F(\bar{\omega}, \tau^N)$ for the problem A-k was earlier proposed (Koriashkina 2015) as follows:

$$F_1(\bar{\omega}, \tau^N) = F_1\left(\{\Omega_{\sigma_1},...,\Omega_{\sigma_L}\}, \{\tau_1,...,\tau_N\}\right) =$$

$$= \sum_{l=1}^{L} \int_{\Omega_{\sigma_l}} \sum_{i\in\sigma_l} \left(c(x, \tau_i) / w_i + a_i\right) \rho(x) dx,$$

where $c(x, \tau_i)$ are real limited definitions on $\Omega \times \Omega$ functions measurable at x for any fixed $\tau_i \in \Omega$ for all $i = \overline{1,N}$; $w_i > 0, a_i \geq 0, i = \overline{1,N}$, are given numbers.

We propose three more functional types:

$$F_2(\bar{\omega},\tau^N) = F_2\left(\left\{\Omega_{\sigma_1},...,\Omega_{\sigma_L}\right\},\left\{\tau_1,...,\tau_N\right\}\right) =$$

$$= \sum_{l=1}^{L} \int_{\Omega_{\sigma_l}} \max_{i\in\sigma_l}\left(c(x,\ \tau_i)/w_i + a_i\right)\rho(x)dx,$$

$$F_3(\bar{\omega},\tau^N) = F_3\left(\left\{\Omega_{\sigma_1},...,\Omega_{\sigma_L}\right\},\left\{\tau_1,...,\tau_N\right\}\right) =$$

$$= \sum_{l=1}^{L} \sup_{x\in\sigma_l} \max_{i\in\sigma_l}\left(c(x,\ \tau_i)/w_i + a_i\right)\rho(x),$$

$$F_4(\bar{\omega},\tau^N) = F_4\left(\left\{\Omega_{\sigma_1},...,\Omega_{\sigma_L}\right\},\left\{\tau_1,...,\tau_N\right\}\right) =$$

$$= \max_{l=1,L} \sup_{x\in\Omega_{\sigma_l}} \max_{i\in\sigma_l}\left(c(x,\ \tau_i)/w_i + a_i\right)\rho(x),$$

where functions $c(x,\ \tau_i)$, $\rho(x)$ and parameters $w_i > 0, a_i \geq 0, i = \overline{1,N}$ are the same as in functional $F_1(\bar{\omega},\tau^N)$.

Usually, in practical applications as $\rho(x)$ performs so-called "density" of customers at the point x and as $c(x,\ \tau_i)$ is a function of distance between points x and τ_i.

The problem A-k with functional $F_1(\bar{\omega},\tau^N)$ can be called as a continuous linear problem of optimal partitioning of the k-th order of set $\Omega \subset E_n$ into disjoint subsets $\Omega_{\sigma_1},\Omega_{\sigma_2},...,\Omega_{\sigma_L}$ among which can be empty ones, under constraints in the form of equalities and inequalities with centers $\tau_1,...,\tau_N$ location.

If, for example, centers $\tau_i, i = \overline{1,N}$ in the problem A-k are fixed, or constraints (2), (3) are absent, then we get different particular cases of problems of optimal multiplex set partitioning.

Let us introduce three separate formulations of the problem A-k. All functions constraints in the functions, which are listed below, remain the same as in the problem A-k.

Problem A1-k. Continuous problem of optimal multiplex-partitioning of the set $\Omega \subset E_n$ with fixed coordinates of centers $\tau_1,...,\tau_N$ without constraints. Find

$$F\left(\left\{\Omega_{\sigma_1},\Omega_{\sigma_2},...,\Omega_{\sigma_L}\right\}\right) \to \min_{\left\{\Omega_{\sigma_1},\Omega_{\sigma_2},...,\Omega_{\sigma_L}\right\}\in\Sigma_{\Omega}^{N,k}},$$

where $F(\bar{\omega},\tau^N)$ is any of the forms $F_i(\bar{\omega},\tau^N), i = \overline{1,N}$, and vector τ^N is fixed.

The partition of the k-th order $\bar{\omega}^* = \left\{\Omega_{\sigma_1}^*,\Omega_{\sigma_2}^*,...,\Omega_{\sigma_L}^*\right\}$, that delivers minimum to

the functional $F(\bar{\omega},\tau^N)$, is called as optimal solution of the problem A1-k.

If in the problem A1-k coordinates of the centers $\tau_1,...,\tau_N$ are unknown in advance, and they have to be defined along with the k-th order partition $\bar{\omega}^* = \left\{\Omega_{\sigma_1}^*,\Omega_{\sigma_2}^*,...\Omega_{\sigma_L}^*\right\}$ of the set $\Omega \subset E_n$, then we have next problem.

Problem A2-k. Continuous problem of optimal multiplex-partitioning of the set $\Omega \subset E_n$ without constraints and with centers $\tau_1,...,\tau_N$ location. It is required

$$\min_{\substack{\left\{\Omega_{\sigma_1},\Omega_{\sigma_2},...,\Omega_{\sigma_L}\right\}\in\Sigma_{\Omega}^{N,k} \\ \left\{\tau_1,...,\tau_N\right\}\in\Omega^N}} F\left(\left\{\Omega_{\sigma_1},...,\Omega_{\sigma_L}\right\},\left\{\tau_1,...,\tau_N\right\}\right),$$

Problem A3-k. Continuous problem of the optimal multiplex-partitioning of the set $\Omega \subset E_n$ under constraints in form of equalities and inequalities with given coordinates of centers $\tau_1,...,\tau_N$. It is required

$$\min_{\left\{\Omega_{\sigma_1},\Omega_{\sigma_2},...,\Omega_{\sigma_L}\right\}\in\Sigma_{\Omega}^{N,k}} F\left(\left\{\Omega_{\sigma_1},\Omega_{\sigma_2},...,\Omega_{\sigma_L}\right\}\right),$$

under conditions (2), (3).

It is necessary to note that the right parts of inequalities (2), (3) must satisfy certain conditions, at which problems A-k and A3-k would have a solution. These conditions of problem A-k or A3-k solvability were obtained (Koriashkina 2015).

Lemma 1. Let $S = \int_{\Omega}\rho(x)dx$. In order to obtain the non-empty class of admissible partitions of the k-th order of Ω in the problem A-k (or A3-k) for any set of centers τ_i, $i = 1,..,N$, it is enough the following conditions to be fulfilled:

$$0 \leq b_i \leq S, \quad i = \overline{1,N}; \quad \sum_{i=1}^{p}b_i \leq S \leq \sum_{i=1}^{N}b_i.$$

Obviously, continuous linear problems of optimal set partitioning are particular cases of the problem A-k with quality criterion $F_1(\bar{\omega},\tau^N)$ or $F_2(\bar{\omega},\tau^N)$. Previously, a similar connection was considered for the problem with criterion $F_1(\bar{\omega},\tau^N)$ (Koriashkina 2015).

Next, we present an idea of solving method of the problem A2-k and its particular case problem A1-k, where the function $F(\bar{\omega},\tau^N)$ is chosen as $F_1(\bar{\omega},\tau^N)$.

Let us rewrite the initial problem A2-k as a problem of infinite-dimensional mathematical programming with Boolean variables.

Let $\bar{\omega} = \{\Omega_{\sigma_1},...,\Omega_{\sigma_l},...,\Omega_{\sigma_L}\}$ be some k-th order partitioning of the set Ω. For each point $x \in \Omega_{\sigma_l}$, $l = \overline{1,L}$, we introduce $L \cdot N$-dimensional vector $\lambda^l(x) = (\lambda_1^l(x),...,\lambda_N^l(x))$, the coordinates of which are determined as follows:

$$\lambda_i^l(x) = \begin{cases} 1, x \in \Omega_{\sigma_l} \ \& \ i \in \sigma_l, \\ 0 \ \text{in other cases} \end{cases} i = \overline{1,N},\ l = \overline{1,L},$$

where $\sigma_l \in M(N,k)$, $\sigma_l = \{j_1^l, j_2^l,..., j_k^l\}$ is the set of center indices $\{\tau_{j_1^l}, \tau_{j_2^l},...,\tau_{j_k^l}\}$, associated with a subset Ω_{σ_l}. Using these functions, we introduce characteristic functions of subsets Ω_{σ_l}, $l = \overline{1,L}$, that form the partition of the k-th order of the set Ω, namely for $l = \overline{1,L}$:

$$\chi_l(x) = \begin{cases} 1,\ x \in \Omega_{\sigma_l}, \\ 0, x \in \Omega \setminus \Omega_{\sigma_l}, \end{cases} \Leftrightarrow \chi_l(x) = \prod_{\substack{i=1 \\ i \in \sigma_l}}^{N} \lambda_i^l(x).$$

Hereinafter, the vector-function $\lambda(x)$ defined on the set Ω will be called as a characteristic vector-function of the subset Ω_{σ_l} included into the k-th order partition of the Ω (by analogy with discrete mathematics characteristic vector for finite set subset is given).

Thus we rewrite problem A2-k in terms of characteristic functions of subsets, that form the k-th order partition of the set Ω.

Problem B2-k. Find $\min_{\substack{\lambda(\cdot) \in \Gamma_0^k \\ \tau^N \in \Omega^N}} I(\lambda(\cdot), \tau^N)$,

$I(\lambda(\cdot), \tau^N) =$

$$= \int_\Omega \sum_{l=1}^{L} \left(\sum_{i=1}^{N} (c(x,\ \tau_i)/w_i + a_i)\lambda_i^l(x) \right) \rho(x)dx,$$

$\Gamma_0^k = \big\{ \lambda(x) = (\lambda^1(x),...,\lambda^l(x),...,\lambda^L(x)) :$

$\lambda^l(x) = (\lambda_1^l(x),...,\lambda_N^l(x));$

$\lambda_i^l(x) = 0 \vee 1$ a.e. for $x \in \Omega, i = \overline{1,N}, l = \overline{1,L},$

$\sum_{i=1}^{N} \lambda_i^l(x) = k, l = \overline{1,L}$, a.e. for $x \in \Omega \big\}.$

Optimal solution of problem B2-k can be obtained in following form: for $i = \overline{1,N}$, $l = \overline{1,L}$ and almost all $x \in \Omega$

$$\lambda_{*i}^l(x) = \begin{cases} 1,\ \text{if } (c(x,\tau_{*i})/w_i + a_i) \le \\ \qquad \le (c(x,\tau_{*j})/w_j + a_j), \\ \qquad i \in \sigma_l, j \in N \setminus \sigma_l, \\ 0 \ \text{in other cases,} \end{cases}$$

under the capacity of $\tau_{*1},...,\tau_{*N}$ the optimal solution of problem is

$$G(\tau) \to \min_{\tau \in \Omega^N}, \tag{5}$$

where

$$G(\tau) = \int_\Omega \min_{\sigma_l \in M(N,k)} \sum_{i \in \sigma_l} [c(x,\tau_i)/w_i + a_i]\rho(x)dx. \tag{6}$$

is chosen.

When the centers are fixed, the problem A2-k converts to a problem A1-k. For the last one the following necessary optimality conditions of multiplex-partitioning were obtained.

Theorem 1. A possible k-th order partition $\bar{\omega}^* = \{\Omega_{\sigma_1}^*,...,\Omega_{\sigma_L}^*\} \in \Sigma_\Omega^{N,k}$ of the set Ω is optimal for problem A1-k, so it is necessary i.e. for each $l = \overline{1,L}$ and $x \in \Omega_{\sigma_l}^*$

$$(c(x,\ \tau_j)/w_j + a_j) \le (c(x,\ \tau_i)/w_i + a_i), \tag{7}$$

$$\forall j \in \sigma_l \ \& \ \forall i \in N \setminus \sigma_l.$$

To solve the finite dimensional problem (5) by means of a non-differentiable function (6) we can use any subgradients' algorithm (Shor 2003).

3 HIGHER ORDERS VORONOI DIAGRAMS AS PROBLEMS A1-k AND A2-k SOLUTION

If $c(x,\tau_i)$ is Euclidean metric, $a_i = 0$, $w_i = 1$, $i = \overline{1,N}$, $\rho(x) = 1 \ \forall x \in \Omega$, then optimal solution of the problem A1-k is determined by vector-function $\lambda(\cdot) = \lambda^*(\cdot)$ as (7) it turns out from Voronoi diagram of the k-th order, known in the computational geometry (Preparata 1985), i.e. such partition of set Ω into subsets $\Omega_1,...,\Omega_M$, that:

$$\bigcup_{i=1}^{M} \Omega_i = \Omega \ ; \ mes(\Omega_i \cap \Omega_j) = 0,\ \forall i \ne j, i,j = \overline{1,M},$$

$$\Omega_m = \{x \in \Omega : \forall j \in T_m \ c(x,\tau_j) < c(x,\tau_i), i \in N \setminus T_m\},$$

where $T_m = \{i_1^m, i_2^m,...,i_k^m\}$, $m = \overline{1,M}$ are all possible of k-element subsets of indices set N.

Voronoi Diagram is a mathematical object of

computational geometry widely applicable in different fields of science and technology. For instance, some developers (Mashtalir 2013; Mihnova 2014) propose to use the higher order Voronoi diagrams for video segmentation. It allows to stabilize the representation of the frame content and to use small number of generator points (so-called subsets "centers") in order to monitor changes of the frame boundaries. Increasing of the order in Voronoi diagrams leads to the area number raise with the same number of generator points, in its turn the computational complexity of segmentation algorithm decreases about 1.5 times. Furthermore, the authors show that the Voronoi diagrams of the highest order can be applied for video segmentation

in time, i.e. the partition into frames, as well as to search for repeats and unique frames.

In Table 1 we present some solution results of the continuous problems A1-k and A2-k for square area of 10×10 points and nine subset centers. Each area therewith is partitioned into subareas using Voronoi diagrams of first, second, third, and eighth order (from left to right).

Coincidently with the solution of problem A1-k we received the solution of problem A1-k with quality criterion $F_4(\bar{\omega}, \tau^N)$, that represents radius r of spherical coverage of k-th order. In Figures radius is represented as a thin line.

Table 1. The solutions of OMPS problem using Voronoi diagrams of first, second, third, and eighth order

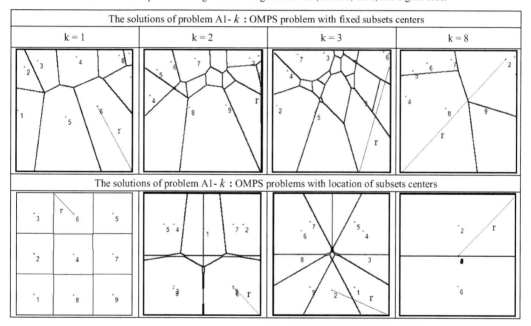

4 THE ECONOMIC INTERPRETATION OF PROBLEM A-k WITH VARIOUS CRITERIA OF PARTITIONING QUALITY.

The choice of optimal multiplex partitioning criterion is determined from the type of located centers, as well as from the specificity of services they provide. If the centers are extremely needed enterprises or services (emergency, police, medical facilities, etc.) and customers are continuously distributed on the area Ω, then the optimality criterion can be to minimize the distance (or time) from the service point to the innermost point of Ω, i.e. to optimize the worst case. In the most common

problem it is required to locate a few of such service points and to consider that every customer can be served by any of k nearest centers (for example, when the closest one for whatever reasons cannot provide the service). The problems of multiplex partitioning with fixed (or to be located) centers with the functional of the form F_i, $i = 2,3,4$ arise in such case. By analogy with optimization problems of the location of centers on graph (Christofides 1975; Farahani 2009), described problems will be called *minimax* problems of multiplex partitioning-location.

Other location problems goals are to minimize the sum of distances from customer to service center (or, as we propose, to the nearest k centers). Such

237

criterion is the most appropriate if as centers, for instance, represent shops, warehouses, and customers, that continually fill some area Ω, can be served by one of k closest to them centers. Hence, while describing mathematical models of arising problems of partitioning of the customers set into areas to be served by a set of k centers we can use the criterion F_1. The problem of centers location on graphs with similar criteria is called *minisum* location problem. We also sometimes will use this specification for problems A-k or A2-k with criterion F_1.

Using the presented models for the OMPS problems it can be studied the competition between service centers, identified relevant market for each center according to their capacity, received additional information about the possibilities of competing service centers and quantity demand at each point of a given region.

Figures 1 and 2 show how the relevant market of the fourth and second service centers respectively extends with an increase in partition order of the specified region into the service areas by only one, two or three nearest centers.

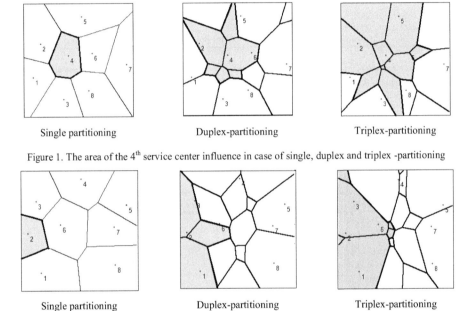

Single partitioning   Duplex-partitioning   Triplex-partitioning

Figure 1. The area of the 4th service center influence in case of single, duplex and triplex -partitioning

Single partitioning   Duplex-partitioning   Triplex-partitioning

Figure 2. The area of the 2th service center influence in case of single, duplex and triplex -partitioning

As for restrictions (2), (3), suppose we are considering a coexistence of similar service centers (providing similar in price and quality services or providing the same product), whose customers are some society stratum with roughly the same income, living conditions and advantages scale.

The demand for this service at every point x of the area Ω can be defined with a help of marketing researches. Let $\rho(x)$ be a demand function, and values $b_1, b_2, ..., b_N$ be «capacities» of service centers, that determine the maximum of services or products which appropriate center can offer.

Constants γ_j^l in the left parts of restrictions (2), (3), show centers' part of market, which the j-th

company occupies on the territory Ω_{σ_l}, among enterprises $\{\tau_{j_1^l}, \tau_{j_2^l}, ..., \tau_{j_k^l}\}$ that provide service in this area. If we presume, that service (product) market is shared among enterprises throughout the region Ω in proportion to their capacities, then for all $l = \overline{1, L}$ and for all $j = \overline{1, N}$, such that $j \in \sigma_l$, the value γ_j^l can be defined as $\gamma_j^l = b_j \Big/ \sum\limits_{q: q \in \sigma_l} b_q$ .With the above assumptions, constraints-equalities (2) indicate that all opportunities of centers τ_i, $i = \overline{1, p}$, should be fully implemented, and constraints-inequalities (3) indicate that the possibilities of the centers τ_i, $i = \overline{p+1, N}$, are restricted.

It should be noted that the problems of the multiplex set partitioning with the criterion $F_4(\overline{\omega}, \tau^N)$ are similar in their mathematical formulation to continuous problems of multiple spherical coverage (Kiseleva 2014), which have a wide range of applications.

5 CONCLUSIONS

Thus, continuous problems of the multiplex set partitioning are represented.

As well as during solving continuous problems of the optimal sets partitioning (Kiseleva 2005; Kiseleva 2015; Kiseleva 2015) the resulting partition of the given bound set is often Voronoi diagram, where points-generators are service centers, some problems of multiplex partitioning lead to higher order Voronoi diagrams. The given linear OMPS problems and their solving algorithms open the possibility for constructing the higher order Voronoi diagram and its various generalizations (weighted, with restrictions on power of Voronoi cells, with capacities of generator points, with optimal placement of points-generators in a given bounded area), which have not been mentioned in the literature before. As a mathematical object of computational geometry Voronoi Diagram is widely applicable in different fields of science and technology. The authors hope that the continuous problems of the multiplex set partitioning and methods of their solving will be useful as an algorithmic apparatus for constructing Voronoi diagrams and their generalizations.

One of the possible directions for further research of solving methods for continuous OMPS problems and similar to them problems of multiple spherical coverage is, for example, their use for optimal location of radiation sources (Blyuss 2015).

REFERENCES

Blyuss O., Koriashkina L., Kiseleva E., Molchanov R. 2015 *Optimal Placement of Irradiation Sources in the Planning of Radiotherapy: Mathematical Models and Methods of Solving.* [See http://www.hindawi.com/journals/cmmm/aa/142987/cta/]

Christofides N. 1975. *Graph Theory – An Algorithmic Approach.* New York: Academic Press: 415.

Farahani R.Z. 2009. *Facility location. Concepts, models, algorithms and case studies.* Berlin, Heidelberg: Springer-Verlag: 530.

Kiseleva E.M., Koriashkina L.S. 2013. *Models and Methods for Solving Continuous Problems of Optimal Set Partitioning: Linear, Nonlinear, and Dynamic Problems* (in Russian). Kyiv: Naukova Dumka: 606.

Kiseleva E.M., Koriashkina L.S. 2015. *Theory of Continuous Optimal Set Partitioning Problems as a Universal Mathematical Formalism for Constructing Voronoi Diagrams and Their Generalizations. I. Theoretical Foundations.* Kyiv: Cybernetics and Systems Analysis, Vol. 51, Issue 3: 325 – 335.

Kiseleva E.M., Koriashkina L.S. 2015. *Theory of Continuous Optimal Set Partitioning Problems as a Universal Mathematical Formalism for Constructing Voronoi Diagrams and Their Generalizations. II. Algorithms for constructing Voronoi diagrams based on the Theory of Optimal Set Partitioning.* Kyiv: Cybernetics and Systems Analysis, Vol. 51, Issue 4: 489 – 499.

Kiseleva E.M., Koriashkina L.S., Mykhalova A.A. 2014. *Constructive Algorithms for Solution Problems of Multiple Continuous Covering.* (in Russian) Dnipropetrovsk: NMAU: System Technologies, Issue 4(93): 3 – 13.

Kiseleva E.M., Shor N.Z. 2005. *Continuous Problems of Optimal Set Partitioning: Theory, Algorithms, and Applications* (in Russian). Kyiv: Naukova Dumka: 564.

Koriashkina L.S. 2015. *Extension of one class of infinite-dimensional optimization problems.* (in Ukrainian) Cherkasy: Cherkasy University Bulletin: Scientific journal, Issue 18 (351): 28 – 36.

Mashtalir S.V., Mikhnova O.D. 2013. *Stabilization of key frame descriptions with higher order voronoi diagram.* Bionics intelligence: scientific and technical journal, Issue 1 (80): 68 – 72.

Mikhnova O.D. 2014. *Video analysis based on Voronoi diagrams of different orders* (in Russian). Charkov: Proceedings of Kharkiv Air Force University, Issue 1(38): 142 – 145.

Preparata F.P., Shamos M.I. 1985. *Computational Geometry: An Introduction (Texts and Monographs in Computer Science).* New York: Springer-Verlag New York, Inc: 390.

Shor N.Z., Zhurbenko N.G., Likhovid A.P., Stetsyuk P.I. 2013. *Algorithms of Nondifferentiable Optimization: Development and Application.* Kyiv: Cybernetics and Systems Analysis, Vol. 39, Issue 4: 537 – 548.

Power Engineering and Information Technologies In Technical Objects Control – Pivnyak, Beshta
& Alekseyev (eds)
© 2016 Taylor & Francis Group, London, ISBN 978-1-138-71479-3

Automation of mine-type dryer at the enterprises engaged in grain storage and processing

S. Tkachenko, D. Beshta & M. Prosianik
State Higher Educational Institution "National Mining University", Dnipro, Ukraine

ABSTRACT: Working operations as well as equipment for mine-type grain dryer has been analyzed. General production algorithm of grain-dryer operation has been developed. Problems of grain-dryer equipment control have been formulated. Both area and ways to solve problems concerning Automatic Process Control System in the form of functional structure and note to it has been proposed. Results of APCS implementation for grain dryer in accordance with the proposed functional structure have been shown.

1 INTRODUCTION

Moisture and associated with it storage temperature and intensity of biologic processes are the most critical for successful conservation of technological and application properties of grain.

First of all, designated purpose of required grain moisture depends on drying process when either direct (i.e. online) or indirect random control of current moisture values is accomplished. It is desirable to dry grain in automatic mode to minimize human influence which is not always qualified; that can result in considerable losses. Grain drying process in automatic mode is possible if only adequate mathematical models are available. That is why, despite a number of models known from scientific sources, neither of them describes grain drying process in automatic mode.

Currently, a problem of grain dryer automation has been solved in some measure: either with the help of development of separate control circuits for the equipment of grain dryer with the help of a complex of separate means and control circuits which involves maintenance by producer company or specialists certified by the company. As a rule, the means are narrow-oriented at specific model range of dryers positioning as mechatronic ones that is those making operation of dryer possible and full-range.

As a result, use of ultramodern and expensive technical equipment – sensors, industrial sequence controls, powerful computers and networks – can not add new features to control grain dryer and production on the whole. Thus, it is required to substantiate ways to solve problems concerning control of mine-type grain dryer at the enterprises of grain storage and processing.

In Ukraine, tendencies in the field of drying control are first oriented at complicated multistage variators. The variators are also applied for stochastic mathematical models to divide perturbing effects according to their importance for keeping within operating procedure. In this context, control of grain moisture on exit from dryer should be performed according the results of prognostic algorithms application (Khobin, Stepanov 2003). Admitting the future of such an approach for ДСП-32 dryers modernization which were involved in corresponding research with the help of simulation modeling techniques (Stepanov, Yeremin & Veridusov 2014) it should be noted that the results of the research cannot be applied for the dryers of other types as they are currently. Such impossibility can be explained by following reasons:

- Dryers by different manufacturers have their own design features and as a result operational ones. For example, different types of burners (either long-flame or short-flame), use of heat exchangers instead of heat generators, various recuperation facilities, moist air exhaust chambers being different in design (with either horizontal or vertical position of cyclofans) and aerodynamic characteristics (such a component is not available in ДСП-32);

- Incompletely processed or unclear demands for technical operating regulations of dryers to be often for new, experimental and foreign types of dryers;

- Target controlling function is not always understood as it can be oriented both at efficiency improvement and acceleration of the process (for example, meeting the requirements of time electricity tariff span), maybe even with conscious deterioration of grain quality.

As a result, it is required to perform operations concerning simulation modeling and identification of parameters either for all or for the majority of current models of mine-type dryers for different purposes. Today it depends on long-term and high-cost operations at the enterprises for grain storage and processing. Moreover, at this stage modeling errors will result in unstable future operation of multistage controls. It can also result in grain damage and emergency operation mode of a dryer due to inadequate control.

We believe that under the conditions of lack of reliable data of technology and parameters of control objects nowadays it is more expedient to solve the problem of automation by means of situational control when control is reached at the level comparable with qualified operator. Anyway, drying foreman performs key functions; he/she should directly monitor status of equipment and technological parameters; obtain periodically laboratory data concerning moisture content and take independent decision regarding effect on equipment or control circuit of a dryer. Direct formalization and logging of drying foreman decisions within enterprise as well as following analysis of results aimed at the development of open undetermined knowledge model and final recommendations makes it possible to solve automation problem without complicated research concerning identification of parameters and simulation of a dryer. In this context, a system of fuzzy control will use the most successful decisions to develop a system of smart support for operator to make decisions (Sosnin 2015) for future implementation of fully automated mode to dry grain enabled to be implemented by current element base.

Formalization of efforts by drying foreman should involve previously developed system of automated control; the system should put a priority on the development of general functional decisions and basic algorithms. Then the problem of control quality improvement will be reduced to a process of measuring devices upgrade – grain temperatures within substantiated characteristic points, moisture content of input-output grain, drying agent temperature, position of final control elements, meters, counters – and means to produce main effects.

2 FORMULATION PROBLEM

On the basis of analysis of processing procedures performed with the help of mine-type dryers we will substantiate both direction and ways to solve a problem of APCS for mine-type dryer able to introduce new features into control of drying processes with their implementation in automated mode and quality of grain storage.

3 MATERIALS FOR RESEARCH

From the viewpoint of production control, a dryer is involved into operational procedure of grain processing being a part of technology including actions to change and determine state of a product (GOST 3.1109-82). It performs grain drying, displacement, cooling; besides it can perform storage operation.

To determine transition of the operations consider general list of technological equipment to be used in mine-type dryer. According to functional features it is possible to make following list of equipment:

- equipment to feed a dryer: bucket conveyors and feeding fasteners, over-the-dryer hopper, equipment being terminal as for dryer, transportation and technological equipment (hoppers, fasteners, conveyors etc);
- equipment for active drying: zones of grain heat within dryer columns;
- equipment for grain cooling and moister redistribution: zones of grain cooling within dryer columns;
- heat generation equipment: gas, solid-fuel, oil-fired or other heat generator or heat exchanger and column of drying agent;
- equipment for ventilation and transfer of drying agent: pressure or exhaust fans or cyclofans, chamber of moist air, means to control flow velocity;
- energy-saving equipment: moist air chamber, recuperation means;
- equipment for dryer unloading: unloading mechanisms, under-the-dryer unloading hopper, transportation and technological equipment being initial as for dryer (conveyors, bucket conveyors, hoppers etc.);
- recirculation equipment: feeding and unloading conveyors, bucket conveyors, reversing valves;
- auxiliary and anti-damage equipment: alarms, alerters, local means of safety locks, fire-fighting means, compressor for pneumatic valves, fasteners, unloading tables.

On the basis of the equipment list divide each of the operations into processing steps; as a result we obtain enlarged technological algorithm of the dryer operation shown in Fig.1.

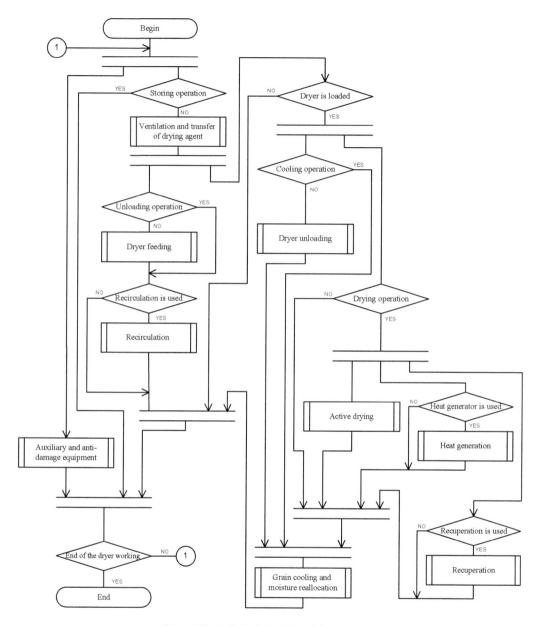

Figure 1. Technological algorithm of dryer operation

The technological algorithm shows that the majority of processing steps is parallel to the others depending on current technological operation and physical availability of the equipment used in each specific model of a dryer, for example, available recuperation equipment. The technological algorithm is not sufficient to describe drying process as it concerns only sequence of processing steps rather than functions of dryer equipment control. Taking into account technological algorithm and available technological equipment identify purpose:

- control of dryer feeding for its safe operation and required level of grain in columns necessary for standard operation mode of heat generation equipment and drying agent transfer with simultaneous prevention from abnormal operation mode of feeding equipment and grain spillage within production area;

- control of ventilation and drying agent transfer aimed at standard drying agent transfer within mines and dryer chambers with the velocity making it possible to remove grain moister without its blowing beyond mine area;

- control of heat generation required for standard and emergency control of heat generation equipment aimed at maximum heat transfer to grain mass within the zone of mine heat without any losses of its technological or consuming properties being one of the factors which help the dryer to be operate with maximum efficiency;

- control of grain unloading in terms of batch mode or streaming mode to unload grain from the dryer with simultaneous purposeful moister removal and keeping within the required temperature range;

- control of dryer recirculation which can be used for initial start of contiguous dryer or for operation of recirculation one;

- control of auxiliary and emergency equipment supporting continuous operation of the main equipment, signaling while changing technological operations and occurrence of abnormal operation modes; initial fire-fighting protection of equipment.

Fig.1 shows functional units as well as technological algorithm making it possible to make up enlarged scheme of functional structure for mine-type dryer control.

Functional structure in Fig.2 helps understand only general principle of dryer control without demonstrating a role of drying foreman in the process of control. In fact each of the abovementioned control units can be implemented in various forms – from a complex of manual control means which should be performed by a drying foreman personally up to fully automated circuits. The latter helps obtain automated system of dryer control where role of operator falls to making general decisions: when, how long, and what type of grain should be dried as well as what drying temperature and target moisture should be applied.

Fig.2 helps draw following conclusions:

- complete automatic control is expedient for auxiliary and emergency equipment. This circuit does not involve signals which processing will depend on such variable conditions as moisture, grain grade, and temperature parameters being noncritical from the viewpoint of fire-fighting safety. Moreover the circuit can be manufactured even separately with the help of local automation means;

- control circuit "feeding-heat generation-unloading" is the most complicated, multilinked, and important. To maintain it drying foreman should apply maximum of his/her skills and attention; thus the issue of automation becomes topical right in this case;

- if local mechatronic control system plays a role of control means of heat generator then they should be integrated into general control system and. In particular, into "feeding-heat generation-unloading" circuit as automation control of heat generation requires extra data concerning state of equipment for feeding, unloading, ventilation, technological characteristics of grain drying process in terms of temperature and moister;

- successful solution of grain moisture measurement has direct influence on control efficiency in terms of "feeding-heat generation-unloading" control circuit. Taking into consideration the fact that input grain moisture may be of jump-type nature varying even for one batch (which can be out of laboratory control) automatic water content measurements of grain within a stream is a required condition to develop automated system of dryer control;

- standard ways to solve control problems by means of direct use of PID-regulators are complicated. We have complex circuit of functional units of "feeding-heat generation-unloading"; each of them should solve its own control tasks: grain level within over-the-dryer hopper, output grain moister, and drying agent temperature. Moreover there are also control problems arising at the levels of functional units interaction: grain temperature within active drying zone and on exit from a dryer; input grain moisture and within intermediate drying stages when regulator of unloading unit cannot cope with the task.

We have designed and implemented APCS for "Brice Baker" dryer with the capacity of 300 t/h in its low-current part and in the part of automated process control.

Means of automated control include industrial control box mostly on Siemens element base, grain moisters within a stream of own design at the input and output of a dryer, automated working place of operator (drying foreman) on the basis of WinCC V13 (TIA Portal V13), databases of equipment checkout and process protocols. The work was performed on request of "Karlivsky mashynobudivnyi zavod" PJSC for the elevator of Alebor Group company in a village of Voronovytsia, Vinnytsia region. Fig.3 (photo) demonstrates general view of interface of automated working place of APCS dryer operator.

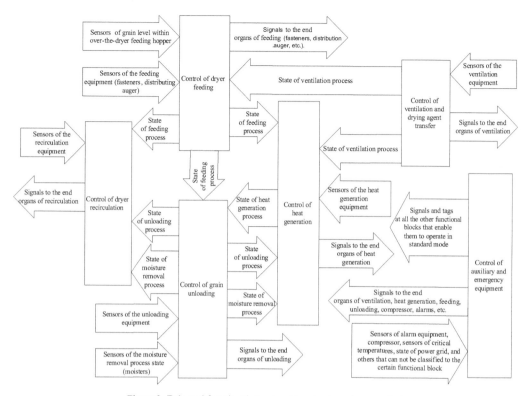

Figure 2. Enlarged functional structure for mine type dryer control

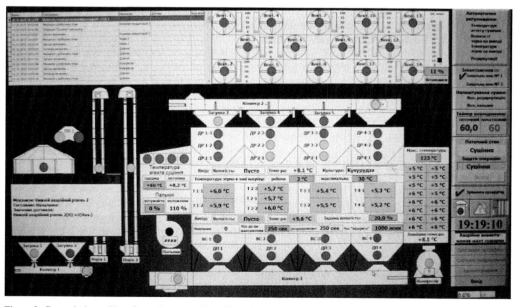

Figure 3. General view of interface of automated working place of APCS dryer operator of "Brice Baker" grain dryer with the capacity of 300t/h

The system has been implemented according to functional structure shown in Fig.2. Its functions are:

In the context of automated mode:

- obtaining of output grain moisture target value;

- control of both transportation and technological equipment of a dryer including ventilation, gas burner Tecflam, compressor, pneumatic unloading tables, feeding fasteners while performing all required technological operations – drying, transferring, cooling and grain storing – in the context of complete automation of operation steps. For example, if operator selects "drying" operation in automatic mode APCS initializes ventilation, feeds dryer up to required level with its subsequent automated control, initializes unloading equipment, and then initializes heat generator with its following control. If operator makes decision to transfer to dryer unloading the system automatically stops unloading and disconnects heat generator. Then during 40 minutes it cools grain down (operator may vary the period). Only after all obligatory operations have been performed a drying automatically shifts to unloading algorithms. Law-Denis dryers (Law-Denis Engineering Ltd., 2012) use similar principles of transitions between operations;

- control of feeding, unloading and recirculating of dryer; however, currently the operations are controlled by earlier implemented ACS;

- control of auxiliary equipment – compressor for pneumatic unloading tables, signal for transfer between technological operations and transitions, emergency unloading after local button of emergency unloading has been pressed, ventilation and burner blocking after critical temperature sensor within wet air chamber has been actuated. All the time temperature values within wet air chamber are displayed for operator (see Fig. 3);

- control of grain level within over-the-dryer feeding hopper;

- control of drying agent temperature and grain temperature within heating zone. When dryer transits to temperature mode, in terms of grain overcontrol was not higher than 1^0 C; in terms of drying agent it was not higher than 2^0 C. In this context, operator displays desired temperature individually;

- keeping a database of the process protocol according to changes in the state of a dryer and equipment. Current changes are automatically displayed at AWS interface (see Fig. 3); earlier introduced data are available on demand of operator from supplementary AWS tags. Extraordinary situations are followed by audio signals from columns. The signals should be obligatory confirmed by operator;

- grain unloading according to settings by operator with the number of unloaded portions; monitoring of emergency situations of pneumatic columns: incomplete opening, incomplete closing, incomplete discharging of under-the-drying hoppers between portions which can depend on both improper calibration of pneumatic table and waste which blocked technological equipment;

- identification of grain availability at the input/at the output of a dryer involving registration of its moisture content and providing of recommendations to operator as for adjustment of subsystem for grain discharge. Trends containing moisture indices are also displayed at AWS of operator.

In the context of automated mode:

- operator can specify air flow rate for drying agent with the help of control of fasteners of cyclofans of a dryer using means of automated working place interface.

The implemented decisions are based upon successful solution by the staff of "Eldorado" Association. The problem has concerned the control of operation-routing sequences of grain within elevator. It experienced numerous confirmations in the context of other enterprises (Prosianik, Tkachenko, Gorbunov and Prosianik 2010). Currently implementation of the task is trivial.

4 CONCLUSIONS

Basing upon analysis of technological operations performed with the help of mine-type dryers, technological algorithm explaining mine-type grain dryer has been proposed. The algorithm differs in its nonlinearity; that is it has several parallel branches; besides they may have different control efficiencies. This concept means the use of synchronized control algorithm methods that is finally reflected both in the proposed functional structure and in specific implementation of APCS.

Basing upon the proposed technological algorithm decomposition of the dryer equipment and its functions has been performed. As a result enlarged functional structure of mine-type grain dryer control system has been proposed as the trend to solve the problems of APCS design. In addition to functional control units they should be manufactured in the form of independent hardware or software modules and functional structure allows understanding the grouping of input (from sensors) and output (to end organs) signals according to their purpose as for functional units. Besides, functional structures are added with sensors of moisture removal process control or moisters as automated control of a dryer is impossible without them. However, it does not concern drying process as currently moisture content of grain in a stream is the only direct control technique. The matter is that in addition to commercial quality control it allows operative and automated response to disturbance factor being step-like moisture deviations while grain batch drying (Prosianik, Klabukov & Sosnin 2002).

The functional structure has been confirmed and proved its efficiency as a result of its implementation in the context of "Brice Baker" dryer. Despite complexity, actual nonavailability of input grain stream control as well as rather nontraditional combinations of automated control circuits concerning temperature and grain level with mathematical methods of parallel processes synchronization satisfactory results have been obtained including those concerning stability of control system.

Nowadays control of grain discharge according to the factor of its temperature and moisture content is performed in a mode of recommendations to an operator.

Automated control of grain discharge has been implemented with bench tests. Its implementation in the context of operating objects needs accumulation of knowledge base as the number of factors will be involved – from individual design features of a dryer and its heat generator to qualitative characteristics (including its moisture content) of grain itself plus weather conditions.

Moreover effect of automated control of moister removal on stability of control circuits of heat generator is also needs additional field tests as it is constant extra disturbance effect not only in the context of moister increase at the input.

Further development of the solution concerning mine-type dryers control is planned to be implemented to improve functional units of "feeding-heat generation-unloading" control circuit to transfer from the mode of recommendations to an operator as for loading to automated control of moisture removal that is drying process. Moreover it is planned to develop prescription base, i.e. ready-made settings of units being used and control to work with different cultures under different conditions. That is why operating APCS has implemented database of the process protocol in including those depending upon temperature, moisture content and discharge settings.

REFERENCES

Khobin V.A., Stepanov M.T. 2003. *System of reliable control of mine grain dryer ДСП-32от* (in Russian). Odesa: Odesa National Academy of Food Technologies: Collection of scientific papers, Issue 25: 137 – 142.

Stepanov M.T., Yeremin M.A., Veridusov P.A. 2014. *Mathematical simulation modeling of grain dryers as the basis to develop efficient control systems* (in Russian). Automation of technological and business-processes, Issue 3 (19): 16-19.

Sosnin K.V. 2015. *System of intellectual support of decision-making while controlling grain drying* (in Ukrainian). Kirovograd: Dissertation abstract.

GOST 3.1109-82. Unified system for technological documentation. Terms and definitions of main concepts.

Law-Denis Engineering Limited. Instructions on operating dryers of SCN and SRN series. 2012. Peterborough: Law-Denis: 91c.

Prosianik A.V., Tkachenko S.M., Gorbunov M.Yu., Prosianik M.A. 2010. *Functional structure to control manufacturing route of grain transportation with the use of scada-system* (in Russian). Dnipropetrovsk: NMU: Collection of scientific papers, Vol. 34, Issue 2.

Prosianik A.V., Klabukov V.F., Sosnin K.V. 2002. *Grain moister in a stream - small rain lays great dust* (in Russian). Grain storage and processing, Issue 8: 44-46.

Power Engineering and Information Technologies In Technical Objects Control – Pivnyak, Beshta & Alekseyev (eds)
© 2016 Taylor & Francis Group, London, ISBN 978-1-138-71479-3

System of automated control of traffic flow in the context of direct route and bypass one

A. Bublikov, I. Taran & L. Tokar
State Higher Educational Institution "National Mining University", Dnipro, Ukraine

ABSTRACT: The application of ideas of closed automated control systems in terms of low level to develop a system for traffic flow control within the small area of transport scheme is under consideration. In this context, the system of traffic flow control is a regulator of the average traffic flow intensity with the use of feedback principle. Conditions connected with transition of traffic flows to automated control as well as a criterion of control effectiveness are substantiated. Dependences of cycles of traffic lights at the junctions on a control signal have been determined if the crossing efficiency is high. An algorithm for automated control of traffic flows has been proposed; its efficiency has been tested with the help of a computational experiment.

1 INTRODUCTION

Today developed countries apply intelligent systems of traffic control. The systems are based upon an online analysis of traffic (Xie et al. 2014; Kale et al. 2013; Chattaraj et al. 2009). As a rule, traffic is controlled within the large area of town transport scheme with a great number of junctions. Available intelligent systems for traffic control are of centralized structure; they govern transport flows with the help of controlled signboards and information boards. Various knowledge acquisitions and decision-making algorithms are applied to develop intelligent systems for traffic control. In any single case, they are characterized by their own subjectivity and uniqueness. Thus, standardized approach to develop such systems is not available.

2 PROBLEM FORMULATION

Available intelligent systems for traffic control in megacities are of complex, multilayer, spatially extended, and centralized structure. Hence, the development of such systems as well as their control service is rather expensive. On the one hand, development of such a system for large area of town transport scheme involves purchase of numerous cameras, sensors, controlled signboards and information boards. Moreover, the greater the area within which traffic flows are controlled is, the more complicated and expensive the procedure to develop algorithms for information processing and form control for intelligent system is. On the other hand, online processing of information concerning road situation and using it to form supervisory effects must be fast and reliable which means expensive channels for information transmission between local systems at junctions and main control system.

A decentralized system for traffic control is an alternative for intelligent centralized systems for traffic control in megacities. The former consists of the aggregate of relatively independent local systems to control traffic flows within small areas of town transport scheme. Under certain road conditions, such local systems may involve simple algorithms for automated control of basic parameters of traffic flows usually applied for regulators at the lower level of automated systems to control operation schedules. In this context, both development and maintenance of such a decentralized system to control traffic requires far less expenses as it is possible to implement it gradually. Besides, its operation needs less sensors, controlled signboards and information boards. Moreover, cameras and fast online channels to transfer much information are not required at all.

3 RESEARCH TOOL

The paper studies rules to form traffic flows within small areas of town transport scheme (Fig.1).

Primary traffic travels down-top and top-down on the right road (Fig.1) being the main one and having two traffic lanes for every traffic direction. On the left at the distance of 750m, parallel two-way secondary road is available; it has only one traffic lane for each direction. A one-way traffic road (right to left) with one traffic lane is located perpendicularly to the main road. Thus, a layout of the area under consideration has two junctions. Let us take up the junction with the main road as junction 1 and the junction with secondary road as junction 2. Traffic lights control traffic flows at each junction.

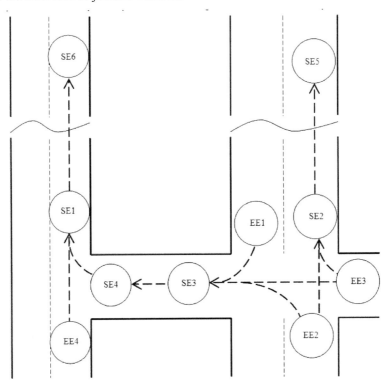

Figure 1. Structural pattern of a transport system model

The objective of the area of town transport scheme is to provide passage of vehicles through the area in any direction in minimum time. Previous observations help determine that there is a problem of periodical traffic jams on the main road (after a junction is passed) for vehicles moving down-top due to the poor highway vehicle capability in this place. This very problem stipulates the research objective – improvement of the strategy for the main traffic flow in the relevant direction (down-top) to avoid traffic jams within the main road when the first junction is behind. It stands to reason, that the improved strategy of the main traffic flow control should not result in traffic jams within other areas of the road section.

If traffic capability of the main road (after a junction 1 has been passed) is poor, redirection of certain share of traffic flow to the bypass route through junction 2 and on the secondary road is one of the problem solutions. In this context, a high automated system to control traffic competed with vehicle-mounted telematic facilities is the tool to implement

the segregation of the main traffic flow in terms of direct routes and bypass ones.

Thus, the consideration of the research objective and the problem solution help come to conclusion shown in Fig.1. Dot arrows demonstrate only those vehicle routes which govern formation of down-up traffic flows within the main road and secondary ones when junctions have been passed.

According to the objective, we formulate the criterion of transport system operation in general: the transport system is efficient if transport flow is characterized by free traffic (down-up in Fig. 1) after junction 1 or 2 has been passed. The transport system is not efficient if in terms of relevant direction (after the 1st or the 2nd junction has been passed), traffic flow is characterized by grouped traffic to be a signal of traffic jam.

To determine a type of transport flows in the context of main road and secondary one after junction 1 and 2 have been passed, it is required for the flows to be continuous. That is, transport flows should be analyzed in a spaced position from junctions after vehicle groups have been diffused; moreover, effect

250

of traffic lights within following junctions (up to the place of flow analysis) should be eliminated.

One of the key methods to improve the efficiency of complex transport objects is application of widely used research techniques with the use of modern IT. The most popular and advanced technology to analyze complex sociotechnical objects is based on system approach. According to the approach, we segregate a section of town transport scheme under consideration into areas where rules concerning formation of covered traffic flows differ greatly. "Entrances" and "exits" of junctions as well as section located far from junctions 1 and 2 are such areas. Fig. 1 shows the areas as circles; there are ten areas. Having taken the marked areas as system components, we will substantiate which of them belong to the system, and which ones belong to the environment.

In our case, we may consider boundaries of town transport scheme as boundaries of transport system section under consideration; however, it is desirable to mark the boundaries through junctions so that "entrances" transmitting environmental effect coincide with "entrances" of the junctions. If so, areas ahead of junctions locating within the boundaries can be considered as environment components (EE in Fig.1) while other areas can be considered as components of system behaviour model (SE in Fig.1). Thus, we have four components of environment and six components of transport system behaviour model with numbering (Fig.1).

Internal links of the system behaviour model components and external ones of the components with those of environment determine vehicle routes within the section of town transport scheme under consideration. Thus, routes identified with dot arrows in Fig. 1 help develop structural diagram of the system behaviour model and environmental model.

Models of environmental components are developed using the results of test measurements of the number of vehicles ahead of junctions; they have both a deterministic component and a random one. The deterministic component explains time change of the number of vehicles ahead of the junction; the component is time function in the form of degree polynom. The degree polynom is determined as a result of averaging of instaneous values of the number of vehicles which took place at a given time during several days. Fig. 2 exemplifies the definition of the deterministic component of element 1 of the environment in the context of MATLAB Program.

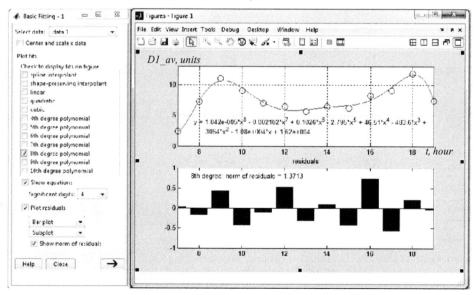

Figure 2. Results of approximation of test data by means of an 8th degree polynom

The random component of environmental element explains time variation of instantaneous value of the number of vehicles ahead of junction and its deviation from the average value; the component is a randomizer, parameters dispersion of which (dispersion of random value and its distribution law) is determined as a result of statistical processing of test measurements data. As an example, Fig. 3 represents a scheme of a model of the 1st element of the environment implemented in Simulink application of MATLAB Program. Models of other elements of the environment are determined analogously.

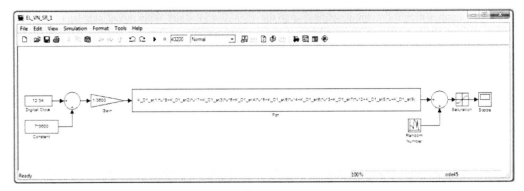

Figure 3. Structural scheme of a model of the 1st component of the environment in Simulink application

The model of system components describing the behaviour of traffic flows just after they have passed the junction is based upon a differential equation in an operator form (1) simulating dynamics of the acceleration of a cluster of vehicles ahead of a junction when the light turns green

$$\overline{V} \cdot s = \frac{K}{T} \cdot \overline{U} - \frac{1}{T} \cdot \overline{V} \tag{1}$$

where K and T are an amplification factor and time constant of the system component, respectively; V is

space velocity of traffic flow, km/g; U is action on traffic light by an automatic control system (one is green light; zero is either red light or yellow one).

K and T parameters of equation (1) are determined using the test curve of a cluster of vehicles after they have passed a junction with the help of "System Identification Tool" application of MATLAB Program. Fig. 4 demonstrates an example of K and T parameters of equation (1) for another junction (component of SE1 system) if six vehicles are available ahead of it.

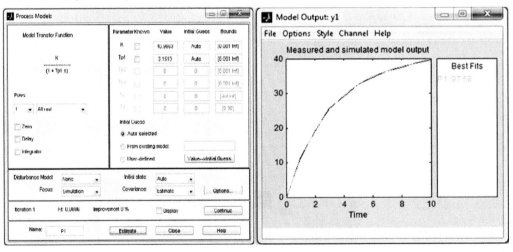

Figure 4. Results of parametrical estimation of a dynamic model explaining time variation of traffic flow spatial velocity if six vehicles are available ahead of the 2nd junction.

We should mention that the time constant T of the system component is determined for different numbers of vehicles ahead of the junction being represented as the value function.

Fig. 5 demonstrates a scheme of dynamic model explaining time variation of traffic flow spatial velocity within the 2nd junction (Simulink application of MATLAB Program). The model consists of three main parts shown as dashed lines in Fig. 5. Part one

is a unit of control action simulation; control signal to be transmitted to traffic light is generated on its exit (one is the green light; zero is either the red light or the yellow one). Part one is a unit implementing the functional dependence of time constant $T1$ of dynamic model of component 1 of transport system on the number of vehicles ahead of the 2nd junction (a case is assumed when twelve vehicles are ahead of a junction). Part three is an implemented equation (1). Moreover, dynamic model output (Fig. 5) is multiplied by control action to simulate sudden stop of traffic flow when the light turns red. In this context, dynamics of traffic flow within transitional period of signal cycle is neglected.

"Counter Limited" unit in Fig.5 generates saw-like signal; its period is equal to a signal cycle period being 100s. To do that, we introduce number 99 using menu of "Counter Limited" unit within "Up-

per limit" field; we introduce 1s sampling interval of outcoming signal within "Sample time" field.

Then saw-like signal is compared with a constant being a period during which the green signal is lightened in relevant direction. The constant is formed as a result of the period share multiplying by the whole period of a signal cycle. In Fig.5 share of the period when the green signal is lightened is 0.5; it means that green light takes 50% of the whole signal cycle (that is 50s).

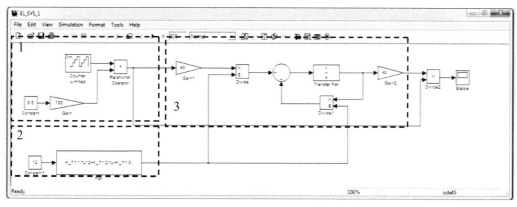

Figure 5. Structural scheme of a dynamic model explaining time variation of space velocity of traffic flow within the 2nd junction.

In exemplification of that, Fig. 6 demonstrates the structural scheme of the 1st component of the system (SE1 in Fig.1) implemented in Simulink application of MATLAB Program.

The model of the 1st component of involved transport system consists of two basic parts shown in dotted lines in Fig.6. Part one figures the intensity of transport flow for the conflict direction of the 2nd junction. Part two figures the intensity of transport flow for the main direction. As for the conflict direction, the number of vehicles ahead of the 2nd junction is taken up as equal to 12 ("Constant1" unit in Fig.6); as for the main direction, it is 40 ("Constant3" unit in Fig.6). Since the green light as well as red lights for the main direction and the conflict one are of antiphase nature, effect of control signal inversion on part one of the model is control effect on part two of the model of the 1st component. Inversion operation is implemented with the help of "Logical Operator" unit.

Within the model of the 1st component of the involved transport system (Fig.6), in units "Fcn1" and "Fcn3" transition from space velocity of transport flow to its intensity takes place by the formula

$$I = n \cdot \frac{V \cdot \frac{1000}{3600}}{L+2}, \, u/s,$$

where I is the intensity of transport flow as regard to the stop line ahead of a junction, u/s; n is the number of traffic lanes in relevant direction; and L is the average length of a vehicle in relevant direction, m. The average distance between vehicles during the whole period is taken up as a 2m constant. In this context, it is taken into consideration that the number of traffic lanes for all directions of the 2nd junction is one, and the average length of a vehicle in both directions is 4.6m.

Time integration of signals in each part of the model in Fig. 6 is performed with the help of "Integrator" unit. It differs from "Transfer Fcn" unit as follows: its input signal may be reset to zero in response to the injected signal being control action as it defines the start of each new light cycle. It should also be noted that the intensity of transport flow may become equal to zero before the cycle is over when the light is green if all the vehicles having been ahead of the second junction in the relevant direction have already crossed the stop line before the cycle is over. To foresee this situation, it is required to compare a time integration result of the traffic flow intensity with the number of vehicles which have been ahead of the 2nd junction in the relevant direction. As soon as the time integration result of the traffic flow intensity prevails the number of vehicles which have been ahead of 2nd junction, the intensity should be taken up as zero. Note that at the end of each light cycle, the time integration result of traffic flow intensity should be reset to zero.

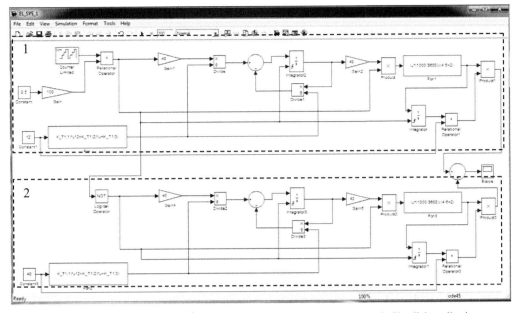

Figure 6. Structural scheme of the 1st component of the involved transport system in Simulink application

Models for other components of transport system behaviour model describing traffic flows right after a junction has been passed are developed analogously.

The 4th component of involved transport system (SE4 in Fig.1) explains the process of forming the line of vehicles taking the main direction ahead of the 2nd junction. The model of the system component is implemented basing upon functional dependence of the number of vehicles ahead of D junction on the intensity of traffic flow I

$$D = \int_{o}^{Tc} I \, dt, u,$$

where T_c is a period of the light cycle for the second junction, s.

As soon as the light turns red within the 2nd junction (the main direction), a result of traffic flow intensity integration within the direction is reset to zero. Thus, the situation at the end of the cycle, where the green signal is lightened for the main direction, is neglected to be admissible for free traffic.

The intensity of the traffic flow at the beginning of the bypass route is an input signal for the 4th component of the involved transport system to be an output signal of the system 3rd component (Fig.1). However, the traffic flow passed the 1st junction in the direction of the bypass route needs time to get to the 2nd junction. Taking into account the fact that the distance between junctions is 750 m and the average space velocity of the traffic flow within the section is 45 km/h, the time delay of traffic flow is

$$\frac{750}{45 \cdot \dfrac{1000}{3600}} = 60, \text{s}.$$

The time delay of the signal within Simulink application is implemented with the help of "Transport Delay" unit. The value of time delay is introduced into "Time delay" field of the unit menu.

Note that depending upon a short distance, the diffusion of a cluster of vehicles within the section between the 1st and the 2nd junctions is not involved.

Taking into consideration the description of 4th component model of the involved transport system, its structural scheme within Simulink application will be similar to that in Fig.7.

In the model in Fig.7 the first input signal ("In1" unit) is a traffic flow intensity at the beginning of bypass route; the second one ("In2" unit) is a control signal for the conflict direction of the 2nd junction (if integration results are reset to zero according to the transition from zero to one).

254

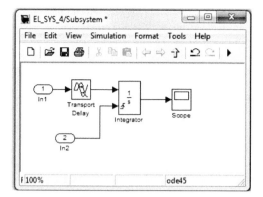

Figure 7. Structural scheme of the 4th component model of the involved transport system within Simulink application.

The 5th component of the involved transport system describes a process of traffic flow formation at the distance of 1500 m from the 1st junction along the direct route (SE5 in Fig.1). The 5th component model is based upon the functional dependence between intensities of noncontinuous traffic flow within a zone of the signal-controlled junction I and traffic flow after the diffusion of vehicle clusters at the distance of I_{diff} from the junction

$$I_{diff} = \frac{\int I(t)dt - \int I(t - T_{av})dt}{T_{av}}, u/s$$

where T_{av} is the time of sliding averaging of noncontinuous traffic flow intensity, s.

The functional dependence between intensity of noncontinuous traffic flow and traffic flow after vehicle clusters diffusion is true if the algorithm of coordinated traffic light control called "green wave" has been implemented in the main direction.

The distance between junctions is 750 m and of traffic flow within the section is 45 km/h, the time delay of traffic flow is

$$\frac{1500}{71 \cdot \frac{1000}{3600}} = 76, s.$$

Thus, structural scheme has been developed basing upon the description of the 5th component model of the involved transport system within Simulink application of mathematical package MATLAB (Fig.8)

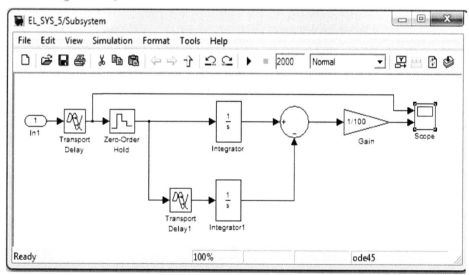

Figure 8. Structural scheme of the 5th component model of the involved transport system within Simulink application

The model in Fig. 8 implements the time delay of the input signal "In1" to take into consideration the time required for a vehicle to pass 1500 m from the 1st junction with the help of "Transport Delay" unit. The value of the time delay (76 s) is introduced into "Time delay" field of the unit menu.

Before performing sliding averaging of input signal "In1", its time sampling takes place with 1 s pitch. To do that, "Zero-Order Hold" unit is used. 1

s time sampling pitch is introduced into "Sample time" field of the unit menu.

The functional dependence between the intensity of noncontinuous traffic flow and traffic flow after the vehicle cluster diffusion is implemented on the basis of two integrators ("Integrator" and "Integrator1") within which integration results resetting is not available ("None" option is selected within "External reset" field of the menu). A sliding averaging period of the input signal is selected as that equal to the period of the light cycle (100 s) and introduced into"Time delay" field of "Transport Delay1" menu and "Gain" unit (Fig.8).

The 6th component of the involved transport system describes a process of traffic flow formation at the distance of 1500 m from the 2nd junction in terms of the bypass route (SE6 in Fig.1). The 6th component model of transport system coincides with the 5th component model of the system (Fig. 8); however, the difference of the average space velocity of traffic flow in the context of the bypass route after the 2nd junction is taken into account.

4 THE RESEARCH RESULTS

Numerical criterion of transport system efficiency is determined basing upon the research objective – the availability of free traffic flows at the distance of 1500 m from junctions in the context of the direct and bypass routes. Thus, the efficiency criterion for the involved transport system should reflect the nature of traffic flow. To determine the criterion, we have to apply one of the basic rules of traffic flow to monitor traffic flow nature. This rule describes the dependence between the average intensity and the average space velocity of traffic flow (Fig. 9).

Free traffic flow can be observed between points A and B in Fig. 9. Cluster and unstable vehicle traffic is observed between B and D points when traffic capability is close to its maximum (point C in Fig. 9); however, in this context transition to intense flow is quite possible (graph section between D and E points). Thus, the efficiency of transport system should clarify where the point determining current nature of traffic flow (within A-B section or B-D one) is (Fig.9).

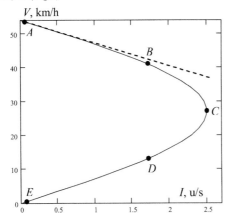

Figure 9. Dependence between the average intensity and the average space velocity of traffic flow.

The boundary between free vehicle traffic and cluster one (point B in Fig. 9) is characterized by disturbance of the linear dependence between the average intensity and the average space velocity of traffic flow. Thus, if initial part of A-B section is approximated with a straight line (dashed line in Fig. 9) and deviation of a straight line from the coordinates of the dependence graph between the intensity of traffic flow and its velocity is analyzed, it is possible to determine roughly the boundary between free traffic of vehicles and cluster one.

Let us introduce the boundary relative deviation of a straight line coordinated from coordinates of the dependence graph between the intensity of traffic flow and its velocity. If it is higher, we can conclude that the linear dependence between the intensity and velocity is disturbed and the nature of vehicle traffic experience changes from free traffic to cluster one (transition from section A-B to section B-D in Fig.9). Let us take up the deviations as those equal to 10 %.

If so, the formula of the involved transport system efficiency criterion referring to traffic flow intensity is

$$K = \begin{cases} \dfrac{|I(V)-F(V)|}{I(V)} \cdot 100 \leq 10,\% & - \\ \text{free vehicle traffic;} \\ \dfrac{|I(V)-F(V)|}{I(V)} \cdot 100 > 10,\% & - \\ \text{cluster vehicle traffic,} \end{cases}$$

where $I(V)$ is a function describing the dependence of the average intensity of traffic flow on its average space velocity, u/s; $F(V)$ is a straight line equation approximating initial part of the dependence graph between the intensity of traffic flow and its velocity, u/s.

If in terms of the involved road section, the dependence between the average intensity and average space velocity of traffic flow remains constant, then formula for the efficiency criterion can be simplified as follows

$$K = \begin{cases} I(t) \leq I_{cr} & - \quad \text{free vehicle traffic;} \\ I(t) > I_{cr} & - \quad \text{cluster vehicle traffic,} \end{cases}$$

where $I(t)$ is the current intensity of traffic flow within the involved road section, u/s; I_{cr} is the critical value of the traffic flow intensity fro road section involved, if the value is higher than the linear dependence between the intensity of traffic flow and its velocity is disturbed, u/s. Critical value of I_{cr} intensity is calculated only once according to a complete formula of efficiency criterion being *I(V)* value when the following equation is true

$$\frac{|I(V) - F(V)|}{I(V)} \cdot 100 = 10 \text{ , \%.}$$ (2)

For the involved section of town transport scheme being considered in the article, equality in formula (2) for the direct route is met if the critical value of the traffic flow intensity is 2u/s, it is 1.4 u/s for the bypass route. Let us take up the ratio between the number of vehicles in the main direction moving along the direct route and the total number of vehicles in the main direction ahead of the 1st junction as control action on the involved transport system marking the action as *R*. Moreover, the involved automated system of traffic control foresees a possibility to control periods of main light cycles within

the 1st and the 2nd junctions. In this connection, we take up control actions of periods ratio as 2nd *R1* and 3rd *R2* when the green signal is lightened in the conflict direction to a period of the light cycle within the 1st and the 2nd junctions respectively.

The model of transport system has been applied to determine experimental dependences of *R1* and *R2* control actions on control action of *R* when traffic capability of junctions along the main direction is maximal. The criterion of maximum traffic capability of junctions along both the main direction and the conflict one is flow interrupting along relevant direction before "allowing" period is over. Fig. 10 demonstrates the results of approximation of experimental dependences of *R1* and *R2* control actions on *R* control action with the help of power polynoms.

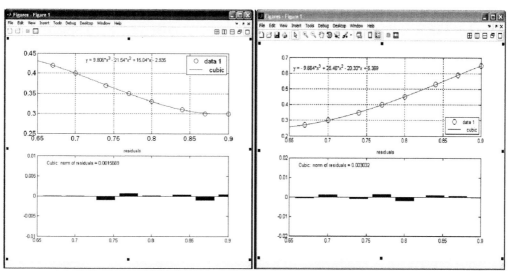

Figure 10. Results of approximation of experimental dependences of *R1* and *R2* control actions on *R* control action.

Fig.10 shows that the dependence of control action *R1* upon control action *R* (left graph) is described with the help of power polynom of the 3rd degree. In this context, the square root of a sum of square deviations of experimental dependence points upon the approximating curve is 0.0016. The dependence of *R2* control action upon *R* control action is also described with the help of power polynom of the 3rd degree (right graph in Fig. 10). If so, the square root of square deviations of experi-

mental dependence points on approximating curve is 0.003.

Taking into account a proposed criterion of the system efficiency, the algorithm scheme of control actions formation on the transport system of decision-making model has been developed (Fig. 11).

In the case of the very first use of decision-making model, an algorithm for the formation of control action effecting the transport system starts from unit 1.

In this context, the initial value of control action R is set up in unit 2 corresponding to standard traffic mode when the distribution of vehicles is not regulated along either direct or bypass routes ($R = 0.9$).

For initial value of control action R, control actions $R1$ and $R2$ are calculated in unit 3 according to the determined power polynoms.

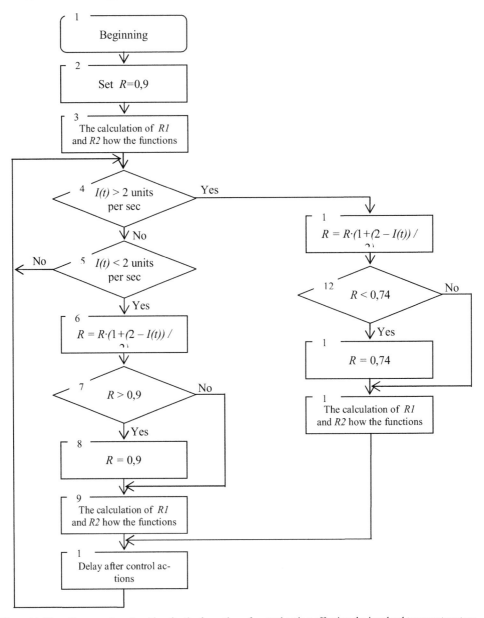

Figure 11. Flow diagram of an algorithm for the formation of control action effecting the involved transport system.

Then unit 4 is used to check the condition of accessing the average intensity of traffic flow at the distance of 1500 m from the 1st junction along the direct route $I(t)$ of its critical value 2 u/s. If the condition is not met in unit 4, then its inverse condition is being checked. If the condition in unit 5 is not met either, it means that the current average intensity of traffic flow $I(t)$ is equal to its critical value and transition to unit 4 takes place while control actions on the system remain constant.

If the condition is met in unit 5, it means that there is a reserve of traffic capability of direct route and more vehicles can move along it. To do that, a new value of control action R is calculated in unit 6. According to the calculation formula from unit 6, the value increases proportionally to the relative deviation of the current value of traffic flow $I(t)$ intensity from the critical one. Thus, changes in control action R increase by the deviation of the current value of traffic flow $I(t)$ intensity from the critical one.

Then a condition when control action R excesses its upper boundary 0.9 is being checked in unit 7. If the condition is met, the value of control action $R = 0.9$ is set in unit 8; if the condition is not met the value remains unchangeable. After control action R has been changed, control actions $R1$ and $R2$ are recalculated in unit 9.

The distribution of the number of vehicles following either the direct route or the bypass one is controlled within the 1st junction when the efficiency of transport system is estimated according to the traffic flow intensity at the distance of 1500 m from the 1st junction. Hence, after control action of transport system has been corrected, outcomes of the process can be estimated after a certain period when vehicles within the 1st junction pass 1500 m from the junction.

That is why the time delay is required after each correction procedure of control effects upon transport system. During the delay control actions remain unchangeable and the system efficiency is not analyzed. Taking into consideration the average space velocity of traffic flow (71km/h) within 0 – 1500 m from the 1st junction along the direct route, the time delay is 76 s. Let us take up the time delay after correction of control action on the transport system as that equal to 100s (it is desirable to adopt it as divisible by periods of light cycles within junctions). The time delay is implemented in unit 10 (Fig.11). Then transition to unit 4 takes place. After that the efficiency of the transport system is estimated again for new control actions.

If the condition in unit 4 is met, then according to the adopted efficiency criterion, it means that the nature of vehicle traffic varies at the distance of 1500 m from the 1st junction along the direct route (from free to cluster). Thus, to avoid intense traffic flow within the road section, it is required to direct certain share of vehicles ahead of the 1st junction to the bypass route. For this reason, the new value of control action R is being calculated in unit 11. According to the calculation formula in unit 11 the value experiences decrease proportionally to the relative deviation of the current value of traffic flow intensity $I(t)$ from the critical one.

Further condition of control action R decrease below 0.74 value, being a control bottom, is being checked in unit 12. If the condition is met, the value of control action $R = 0.74$ is set in unit 13; if not, it remains unchangeable.

After control action R has been changed, $R1$ and $R2$ control actions are being recalculated in unit 14. Then transition to unit 10 takes place to implement the time delay when control actions on transport system were corrected.

Basing upon the model of transport system, let us analyze the behaviour of the uncontrolled transport system and control system if the above algorithm of control action formation is applied (Fig.11). The efficiency of transport system is analyzed according to the criterion adapted for comparative analysis of various strategies to control a system. The criterion is relative deviation of the current value of the traffic flow intensity $I(t)$ from the critical one to increase

$$K = \begin{cases} \dfrac{I(t) - I_{cr}}{I_{cr}} \cdot 100, & if \quad I(t) > I_{cr}, \%; \\ \qquad 0, & if \quad I(t) \le I_{cr}, \%. \end{cases}$$

It follows from Fig.12 that in the context of standard mode of motion and lack of vehicle distribution in terms of the direct route and the bypass one ($R = 0.9$), the average intensity of traffic flow at the distance of 1500m from the 1st junction excesses its critical value (2 u/m) from 8.30 a.m. to 10.30 a.m. and from 4.00 p.m. to 6.30 p.m. During rush hours the controlled transport system shows the decrease in traffic flow intensity $I(t)$; at that very time it periodically excesses its critical value 2u/s with 9% relative deviation (Fig.13). While simulating, the average value of efficiency criterion K was 4.32% for the uncontrolled system and 3.06% for the controlled one. Hence, thanks to the application of algorithm to control traffic flows, average value of efficiency criterion K experienced 1.41-fold decrease which means the increase in transport system efficiency.

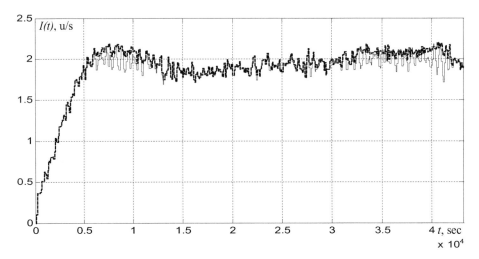

Figure 12. Time variation of the traffic flow average intensity at the distance of 1500 m from the 1st junction in terms of the direct route: a – for uncontrolled system (thick dashed line); b – for controlled system (thin full line)

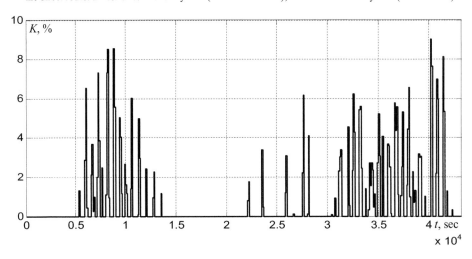

Figure 13. Time change in relative deviation of the current value of the traffic flow intensity $I(t)$ from the critical one towards the increase.

5 CONCLUSIONS

1. A condition when the current intensity of traffic flow excesses its critical value with the interruption of the linear dependence between the average intensity and the average velocity of traffic flow meaning changes in traffic (from free to cluster) is taken up as a criterion of transport system efficiency. Experimental data processing has shown that for the direct route within the section of town transport scheme involved, the critical intensity of traffic flow is 2u/s; it is 1.4u/s for the bypass route.

2. The research has shown that to provide the maximum traffic capability of junctions in terms of

the main direction, ratios of basic periods of light cycles within junctions should be changed in a function of ratio of the number of vehicles taking the main route and the bypass one. Meanwhile, the dependences are precisely described by means of power polynoms of the 3rd order.

3. The calculation experiment basing upon a model of the transport system has shown that introduction of a proposed algorithm to control traffic flows resulted in 1.14-fold decrease in the average value of transport system efficiency. It means the similar increase in transport system efficiency. However, the intensity of traffic flow in terms of the direct route demonstrates periodical excess of its

260

2u/s critical value with relative 9% deviation. That's why further improvement of the algorithm to control distribution of traffic flow in terms of the direct route and the bypass one is required on the basis of the proposed model of transport system.

REFERENCES

Xie X.-F., Smith S.F., Chen T.-W., and Barlow G. J. 2014. *Real-Time Traffic Control for Sustainable Urban Living*. IEEE 17th International Conference on Intelligent Transportation Systems (ITSC), Qingdao, China: 1863-1868.

Kale S.B., Dhok G.P. 2013. *Embedded system for intelligent ambulance and traffic control management*. International Journal of Computer and Electronics Research, Vol.2: 137-142.

Chattaraj A., Bansal S., Chandra A. 2009. *An intelligent traffic control system using RFID*. IEEE: Potentials, Vol. 28: 40-43.

Power Engineering and Information Technologies In Technical Objects Control – Pivnyak, Beshta
& Alekseyev (eds)
© 2016 Taylor & Francis Group, London, ISBN 978-1-138-71479-3

Software diagnostics for reliability
of SCADA structural elements

O. Syrotkina, M. Alekseyev
State Higher Educational Institution "National Mining University", Dnipro, Ukraine

ABSTRACT: The experimental research results for SCADA diagnosis software through a specialised DgnMethod toolkit are proposed for consideration in this paper. The task of determining the reliability for SCADA structural elements on the basis of calculation of the current and average coefficients of work instability for each unit is decided. A brief description of functional options and the operating algorithm of the DgnMethod toolkit are displayed. Figures are given to illustrate the DgnMethod interface windows with detailed analysis of displayed information.

1 INTRODUCTION

Modern SCADA systems are distributed multi-level and multi-tasking real-time hardware–software complexes. These systems are subject to strict requirements to ensure their reliability and safety. To meet these requirements, SCADA systems have powerful built-in auto diagnostic tools displaying diagnostic messages about the occurrence of faults in the system (Siemens 2009). Unfortunately, currently employed methods for SCADA diagnostics contain a great deal of low-level diagnosis information, and are, as a rule, designed for a manual system recovery by repair and maintenance personnel. The following articles (Syrotkina 2014; Syrotkina 2015) analyse the methodology of SCADA diagnostics with respect to the change in values of diagnostic features of Technological Control Object (TCO) parameters in the data streams running through SCADA structural elements.

This methodology allows us to evaluate the efficiency of the system's structural elements in real time. The experimental studies of the diagnostic methodology using a specialised DgnMethod toolkit, developed on the basis of the methodology in Borland C++ Builder, are proposed for consideration in this paper.

2 MAIN PART

The work of the DgnMethod toolkit includes the following steps:
– Configuring the structure of the SCADA system;

– Formation of the Diagnostic Matrix (DM) using a random number generator;
– Detection, localisation and distinction between independent and dependent failures for the current Diagnostic Matrix;
– Calculation of the current coefficient of instability for the work of each system's structural element.
– Determining the most vulnerable system's structural elements (in terms of their SCADA reliability).

Using the input data, DgnMethod creates a configuration diagram of the structural elements in the system (see Fig.1). The index numbers of the controlled parameters are located on the abscissa axis. The levels of SCADA hierarchy are located on the ordinate axis. The index number of a structural element that corresponds to the level of hierarchy is defined in the square located over each point on the chart. If you press with the left button of your mouse on the point of the chart CP TCO, it creates a chart Polyline that shows a path of a picked controlled parameter from a lower hierarchy level of the system to an upper one.

When forming the diagnostic matrix, it is necessary to take into account the allowable and non-allowable state changes of the controlled parameter $x_{iC}(t)$ as it passes through the levels of the hierarchy of the SCADA system, described with the help of the vector $V_{iC}(t)$ that corresponds to the column iC of the Diagnostic Matrix $D(t)$.

$$\begin{cases} V_{iC}(t) = [d_{iL,iC}(t)] \\ d_{iL,iC}(t) \in E_3 \\ iL = l(S)+1-l, \quad 1 < l \le l(S) \\ 1 \le iC \le n(X) \end{cases} \quad (1)$$

where iL is an index number of the vector element $V_{iC}(t)$ that corresponds to the row of the matrix $D(t)$; iC is an index number of the controlled parameter $x_{iC}(t)$ that corresponds to the

column of the matrix $D(t)$; $E_3 = \{0,1,2\}$ is a three-element set that corresponds to the states {Absent, False, True} of controlled parameter $x_{iC}(t)$; l is a hierarchy level in the SCADA system; $l(S)$ is a hierarchy level of the server in the SCADA system; $n(X)$ is a number of controlled parameters of Technological Control Object (TCO).

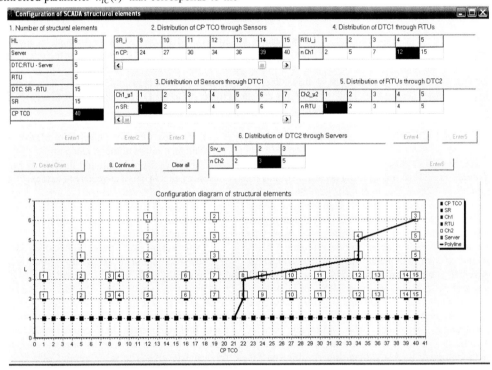

Figure 1. Window of the DgnMethod software interface for the configuration of SCADA system structure

The following symbols are presented in Fig.1: HL – Hierarchy Level; RTU – Remote Terminal Unit; DTC – Data Transfer Channel; SR – Sensor; CP TCO – Controlled Parameter of the Technological Control Object (TCO); DTC1 - Data Transfer Channel between the Sensors and the RTUs; DTC2 – Data Transfer Channel between the RTUs and the server; SR_i – the index number of the sensor; n_CP – the number of controlled parameters that are measured by the appropriate sensor; Ch1_i – the index number of DTC1; n_SR – the number of sensors that are connected to the appropriate data transfer channel; RTU_i – the index number of the RTU; n_ Ch1 – the number of data transfer channels from sensors that are connected to the appropriate RTU; Ch2_i – the index number of DTC2; n_ RTU – the number of

RTUs that are connected to the server through the appropriate data transfer channel; Srv_i – the index number of the server; n_Ch2 – the number of data transfer channels from the RTUs that are connected to the appropriate server;

Let's define n_0^{iL}, n_1^{iL}, n_2^{iL} as a number of combinations of values of vector $V_{iC}(t)$ with the corresponding value $d_{iL,iC}(t)$. Because $d_{iL,iC}$ is a member of the three-element set $(d_{iL,iC} \in E_k, k = 3)$, then we can derive the following dependencies:

$$n_0^{iL} = k^{\lfloor (iL-1)/2 \rfloor}, \quad (2)$$

$$n_1^{iL} = n_0^{iL-1} + 2^{(iL\%2)} * n_1^{iL-1}, \quad (3)$$

$$n_2^{iL} = n_0^{iL-1} + n_1^{iL-1} + n_2^{iL-1}, \qquad (4)$$

where $\lfloor \ \rfloor$ is the floor function, % is a remainder of division.

Therefore, when forming the diagnostic matrix $D(t)$ in the DgnMethod toolkit, we need to set a random number generator to an appropriate range of possible combinations.

The window of the DgnMethod software interface with the diagnostic matrix $D(t)$, formed by using a random number generator and taking into account the ranges of possible combinations for the correct values $V_{iC}(t)$, is shown in Fig. 2.

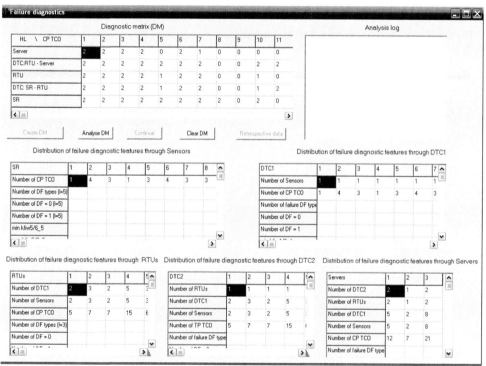

Figure 2. Window of the DgnMethod software interface with the formed Diagnostic Matrix

D. H. Lehmer's linear congruential generator (Knuth 2000), which is included in the standard library of the Borland C/C++ compiler, was applied in order to generate a sequence of random numbers. The sequence of random numbers X_n is calculated by using formula (5).

$$X_{n+1} = ((aX_n + c) \bmod m) \bmod_{mU}, \qquad (5)$$

where m is a module $(m \geq 2)$; a is a multiplier $(0 < a < m)$; c is an increment $(0 < c < m)$; X_0 is the initial value; m_U is the user module that limits the variation range of the generated random values.

The following values were employed in Borland C/C++ for a random number generator: $m = 2^{32}$; $a = 22695477$; $c = 1$;

In the developed toolkit for diagnostics, X_0 is initialised by the value of the current computer time (in sec). This value is the number of seconds from 01/01/1970 00:00:00 GMT, the starting point for UNIX/POSIX time, to the moment that the random number generator launches.

In general, the value of the user module m is calculated as follows:

$$\begin{cases} m_U(0, iL) = n_0^{iL} \\ m_U(1, iL) = n_1^{iL} \\ m_U(0 \lor 1, iL) = m_U(0, iL) + m_U(1, iL) \end{cases}, \qquad (6)$$

The values of the user module $m_U(d_{iL,iC}(t), iL)$ are shown in Table 1.

265

Table 1. The values of the user module m_U

iL	$d_{iL,iC}(t) =$		
	0	1	$0 \vee 1$
1	1	1	2
2	1	2	3
3	3	5	8
4	3	8	11
5	9	19	28

The window of the software interface DgnMethod with the analysis results of the diagnostic matrix using the method outlined in (Syrotkina 2014; Syrotkina 2015) is shown in Fig. 3. The values of the failure detection function $g2(iL, j)$ in SCADA are displayed in the analysis log field, where $[iL, j]$ are coordinates of the SCADA structural module. When analysing the current diagnostic matrix, the failures are also sorted into independent and dependent ones. The diagnostic features of dependent failures are marked with the symbol "*" in the field of diagnostic matrix.

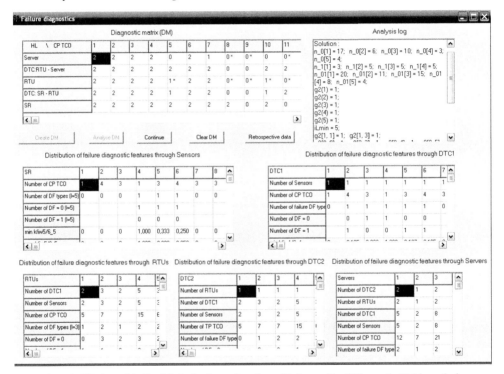

Figure 3. Window of the DgnMethod software interface with the results of Diagnostic Matrix analysis

The coefficient of work instability $Kw_{iL,j} \in [0;1]$ of the SCADA structural element, for which independent failure has been detected, is calculated in terms of the data flow that passes through it. The calculated values for the coefficients of work instability (CWI) for the SCADA structural elements are shown in Fig.4; the retrospective changes in the values of CWI over time are shown in Fig.5.

266

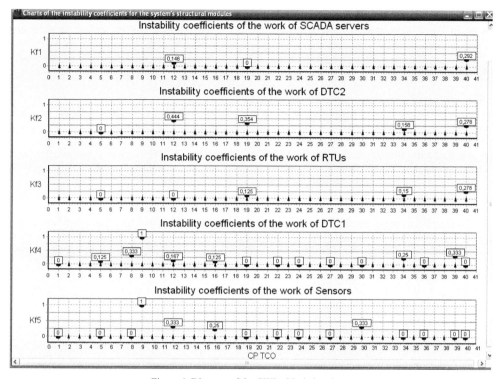

Figure 4. Diagram of the CWI with their values

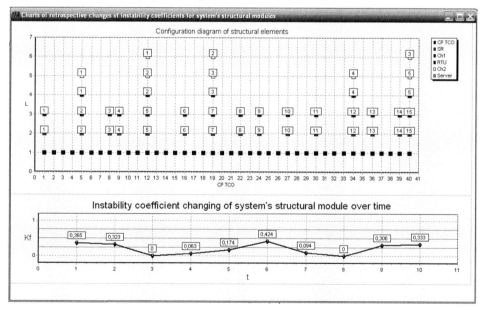

Figure 5. Chart of the CWI changes for S1 over time

The user interface allows for the prompt analysis of the operational readiness of the whole system. It also illustrates the most vulnerable SCADA structural elements.

The SCADA current structure is displayed at the top of the user interface. If you left click with the mouse on a system's structural element, the chart of the CWI appears for the selected structural module in the bottom part of the user interface. The chart shown in Fig. 5 demonstrates this process for the server S1.

3 CONCLUSIONS

The results of the experimental investigations carried out using the toolkit DgnMethod allow us to make the following conclusions:
- the predicate calculus system, which was suggested in this method (Syrotkina 2015), provides a proper description of the logic of the diagnosis algorithm for the toolkit DgnMethod;
- when forming the diagnostic matrix, the allowable and non-allowable state changes of the controlled parameter are taken into account as it passes through the levels of the system's hierarchy (according to the method of diagnostics);
- when analysing the diagnostic matrix, the independent and dependent failures are correctly determined (according to the method of diagnostics);
- SCADA diagnostics using the DgnMethod interface is intuitive and transparent, and it can be adapted to any structural configuration;

- On the basis of the aforementioned, the DgnMethod can be applied to determine the efficacy and reliability of SCADA structural modules arranged in an arbitrary configuration.

REFERENCES

IEEE Std C37.1™. 2008. *IEEE Standard for SCADA and Automation Systems*. New York: 143.

Siemens. 2009. *User guide SPPA-T3000. Diagnostic system.* Munich: Siemens AG: 251.

Knuth D. 2000. *The art of computer programming* (in Russian). Moscow: Williams: Sorting and searching, Vol 3: 832.

Syrotkina O.I. 2014. *Structural and logical model of SCADA system faults diagnostics* (in Russian). Dnipropetrovsk: NMU: Scientific bulletin of National Mining University, Issue 4: 52-57.

Syrotkina O.I. 2014. *SCADA analytical model of fault detection and localisation* (in Russian). Dnipropetrovsk: Metallurgical and Mining Industry, Issue 5: 112-115.

Syrotkina O.I. 2015. *Automatic diagnosis method for SCADA operability.* Ivano-Frankivsk: Methods and Devices of Quality Control, Issue 1: 19-26.

© 2016 Taylor & Francis Group, London, ISBN 978-1-138-71479-3

Social engineering threat for SME and European Cybersecurity Implementation program in Ukraine.

A. Sharko
State Higher Educational Institution "National Mining University", Dnipro, Ukraine

ABSTRACT: The current tendencies and scale of worldwide cyber threats for governmental and corporate sectors were analyzed. Significant impact of social engineering type of the registered attacks and incidents was highlighted. Details and new combinations of social engineering and technology-based attack methodologies were described in the context of corporate intellectual property and infrastructure threats. The current state of European Cybersecurity Implementation approach by ISACA was presented at the levels of European Union and ISACA Kiev Chapter.

1 INTRODUCTION

In view of the increasing scale of cybercriminal incidents, the growth of total budget loss in corporate and governmental sectors around the globe, the level of corporate employees and users' awareness and cybersecurity knowledge remains on the relatively low level. With the increase in the defined and applied defense practices for typical technology-only based approaches, the number of incidents and attacks with applied social engineering practices have acquired a noticeably predominant trend. Additionally, the corporate sector suffers of decrease in information security budgets, while overall cyber-security landscape highlights the lack of executive and implementation activities for building the security framework, which is stable and sufficient for the countries and corporate market needs. The discrepancy is especially striking between the preventing, monitoring and handling activities dealing with security risks based on human factor and malicious or non-malicious personnel-related attacks and the growing number and complexity of attacks of this type.

In this context, as far as the European countries and Ukraine specifically are concerned, the need in well-coordinated, systematized and feasible program of cybersecurity education, awareness and defense can hardly be underestimated.

2 FORMULATION OF THE PROBLEM

In the actual political and economic situation in Ukraine, the ability of the country and corporate sector to handle and overcome the growing risk and damage inflicted by cybercriminal attacks and cyber-war has become the minimal critical requirement for upcoming years.

In addition to the high external interest and cybercriminal targeting, the high level of software engineers education, wide share of software development markets, dramatically low level of information security awareness and training programs as well as massive usage of legally invalid copies of public and corporate software makes Ukraine a highly active scene for potential cyber-attacks and attacks initiated and held from within the country.

3 MATERIALS FOR RESEARCH

Employees are the most-cited culprits of incidents

The Global State of Information Security reports by companies like PwC, HP and security institutes like ISACA for 2014 and 2015 highlight that even with the growth of complexity and number of technology defense and attack approaches, the main security threat is still based on the human factor. Thus, according to the Global State of Information Security Survey of 2016, nearly 50% of all the security incidents confirmed by SME sector companies in Europe were fully or partially sourced by current or former companies' employees. This fact evidently corresponds to similar worldwide tendencies described in 2015 PwC Security State Survey published by ISACA showing that more than 65% of security incidents among surveyed companies around the globe were initiated by former and current employees. In a financial sector, this number reaches the mark of 72%. With this, the reason for employees' participation in security incidents is not always maliciousness, financial or social crime goals, but the lack of basic information security education and training at general social and internal corporate levels. Thus, according to ISACA State of Cybersecurity Report of 2015, only 29% of all the incidents are connected with the company insiders participation in a crime on a malicious basis, while nearly 41% of the incidents were caused by non-malicious employees.

Non-malicious employee participation in cyber-crime against the company can usually be associated with the absent or outdated basic cyber defense security tools and frameworks, low quality of cyber threats monitoring, preventing and management subsystems like intrusion prevention and detection tools, malicious code and unauthorized access monitoring systems, vulnerability scanning practices and data loss prevention tools. This correlation is grounded on the mentioned ISACA Cybersecurity Report of 2015, which stresses that only 59% (at most) of the surveyed companies have the activities of these types set up and handled on a decent level. Some of the most critical of them, like data loss prevention tools and patch management activities are exercised by just 52% of the companies. Besides, only 51% of the companies have any kind of security awareness training program.

In this perspective, another aspect of the incidents analysis appears to be highly important - the significant amount of social engineering attack approaches. Thus, according to the report, at least 65% of all the incidents had social engineering practices, "watering hole" and "man-in-the-middle" approaches involved. In 69% of these cases, the so-called "phishing" practices were used as the attack initiation steps.

What was the estimated source of security incidents?

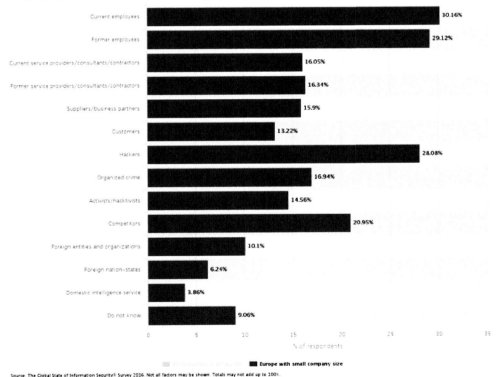

Figure 1. Sources of security incidents within European SME by Global State of Information Security report

Which of the following attack types have exploited your enterprise in 2014?

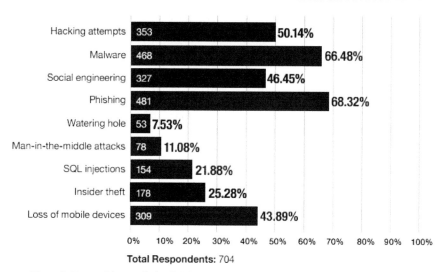

Total Respondents: 704

Figure 2. Types of the attacks by ISACA and RSA State of Cybersecurity: Implications for 2015 report

Social engineering evolution and technical aspects

With the evolution of attack and technological aspects, the growth of the overall connectivity, massive social networks and physical life penetration with interconnected devices within trending concepts of the Internet of Things, the social engineering practices evolve and adapt for the new reality. The classical set of the social engineer attacks defined more than 30 years ago and used since the growth of the corporate sector.

In a quick view, this set would contain the following online and offline approaches that were defined and deeply analyzed first by Kevin Mitnik (social engineer and hacker) in his books back in 2002.

- *Baiting* - manipulating the interest of the authorized personnel to use or install the infected hardware
- *Phishing* - usage of false emails, IM, or websources that impersonate the trustworthy originals provoking target employee to enter the personal sensitive data
- *Pretexting* – approach which is usually based on capturing information by phone calls based on face-to-face impersonification of the authorized or trusted co-worker, company partner or client company employee
- *Quid pro quo* - capturing sensitive information in a form of informational or

social exchange that "benefits" the target employee in a certain way

- *Tailgating* – getting the physical access to a secured office or hardware areas and resources by physical following ("tailing") the authorized employee entering these areas or usage of his mobile devices that have or may potentially have the needed level of systems' access.
- *Reverse social engineering* - provoking target employee to request attacking engineer for help or support to gain the needed level of trust and access

Most of these approaches are still more than actual and usable in a nowadays corporate sector. Especially in the field of SME businesses that do not have the ability or strategic desire to spend sufficient resources on the information security programs.

However, in each of these and other new methods of socially based attacks there are certain changes and innovations that correspond to the modern social and commercial paradigms.

Here are the most prominent of these changes:

The area of Phishing attacks has dramatically widen with time and is seen via the security reports data highlighted above, as one of the most effective and thus common practices.

Falsing the messages and emails that every corporate employee processes on a daily basis, makes initial level of trust high enough to make this kind of attack an effective entry point to most of the

271

companies secured framework. Current massive interaction of an average user with banking, commerce, legal, and support online and offline services gives a diverse range of possible messaging subjects that may attract the victim to interact with the infecting resources from entering the sensitive data up to installing and launching the malicious software.

Due to this level of effectiveness and relatively low technical complexity of implementation, the number of this kind of attacks is increasing dramatically. Thus, according to APWG Phishing attack trends report by April 2015, the total number of unique phishing campaigns registered within 2014 comes up to 704,178 which is more than 4 times bigger than 173,063 registered by this organization in 2005.

This also led to appearing of such sub-types of this kind of attacks as the so-called Spear Phishing (phishing attacks that are dedicated to the specific person or group of people) and Whaling (name comes from "whale" for highly dedicated and prepared attack on the company's executives and C-Level managers).

Both of these techniques require from the attacking party gathering of additional personal, corporate and subject-related information to increase the level of the attack personification.

Current availability of personal information through social networks and sharing of the information via multiple public sources like Bing and Google profiles, LinkedIn, Facebook and other social networks, makes this process much easier and more efficient. Professional or just highly skilled cyber-criminals execute this preparation phase by automating or semi-automating the aggregation of DNS, mail and PGP servers, personal, accounting and corporate information, using such publicly available tools as "theHarvester", "recon-ng" and analogues or proprietarily created scripts for gathering of target specific information.

For example, with minimal time and configuration involved, "theHarvester" tool has gathered information about 825 hosts, 450 virtual hosts, 96 TLDs (top-level domains) and 91 Internet connected devices (using Shodun search engine) associated with cisco.com

After analyzing the gathered information, the attack is usually based on delivering to the victim the content that visually and contextually imitates the one that may be interesting or provoking to the targeted employee and then, using misspelled or spoofed URL, cross-site scripting, hidden redirects and web browser address bar manipulation, it directs the victim to the website that supports the deception by imitating typical UI, content and

behavior of the original web resource down to dynamic imitation of two-factor authentication, live support of web chats and other implementation of MITM (man-in-the-middle) attack idea.

No doubt, such a sophisticated approach and variety of available information make the socially based attack highly powerful, dangerous and successful, especially for target employees at governmental or executive level. Bright examples of impactful Spear Phishing and Whaling are the cases described in 2013 by Bruce Schneier (cryptography and security technologist) who wrote about the group of US government officials that were compromised with the help of professionally targeted Spear Phishing campaign during their trip to Copenhagen to debate climate change related issues. A sample of Whaling cited by Schneier is the case of Coca-Cola's acquisition deal of $2.4bn that was canceled due to the information revealed after the successful email Phishing attack on the deputy president of Coca-Cola's Pacific Group, Paul Etchells, who opened the email that had seemed to be sent by the company's chief executive.

Other social engineering approaches have become more efficient through using the publicly available information about the victim. Thus, Pretexting and Reverse social engineering approaches are mostly based on a hacker social skills and information about the target. In fact, these techniques exploits the approach known as OSINT (Open source intelligence) – the practice that is commonly used by government intelligence services since World War II and based on the idea that most and dominating part of the information is available within public media domains and platforms. A classic example of WWII OSINT analysis was the case of US government Foreign Broadcast Information Service (FBIS) monitoring of oranges market prices in Paris as an indicator of whether railroad bridges had been bombed successfully. During the year, the technique was constantly evolving, utilizing more and more advanced tools for information gathering and processing.

An additional factor that exposes the company's property even more is diversity and extension of the mobile devices from laptops, smartphones to wearable devices and Internet connected infrastructure elements. The above mentioned Tailgating approach in one of its forms is based on the idea of compromising the mobile device owned by an authorized employee. For example, it is deploying the malicious software on a personal cell phone of the victim asking to make a call from this device or use the corporate laptop in a public network prepared in advance for sniffing the data. To the same group we can relate the attacks based on a simple theft of such authorized devices that were not followed up by appropriate security procedures of deactivation or de-authorizing actions. Thus according to the PwC Global State of Information Security Report, the number of attacks targeting mobile devices increased by 36% in 2015 compared with 2014. Another 44% of the incidents relate to attacks associated with the lost mobile devices.

Social engineering threat awareness and prevention

As a global answer to these increasing security risks, corporate sector reacts with increasing budgeting for security coverage of mobile devices management and assets utilization and for the IoT (Internet of things) concept, covering the wearable and connected devices. But so far only 36% of companies surveyed within 2015 PwC report to have IoT security strategy implemented.

However, within last decade, the global information security community and particular organizations have been putting visible efforts into increasing awareness of social engineering as a threat for the corporate intellectual property and resources. On a regular basis, the organizations like Social-Engineer, Inc. and its founder Christopher Hadnagy conduct the activities aimed to educate public communities interested in social engineering subjects and companies that want to increase their level of internal cyber-security and educate their personnel. Christopher Hadnagy is the professional penetration tester, coach and the author of the "Neuro Linguistic Hacking (NLH) research that combines technological skills, theory of nonverbal communications and Paul Ekman's ideas of Microexpressions used to identify a person's untrustworthiness. Providing the services of "ethic hacking" and security penetration tests for the companies around the globe and running series of conferences and events, some of them are held in the format of the so-called "Social-Engineer Capture the Flag Competition" as a sub-track of DEF CON conferences. Within this competition, organized with the support of EFF (Electronic Frontier Foundation), the attendees are competing in

"ethic" social engineering attacks on pre-selected major US private-sector companies like Macy's, Wallmart and others. The format of attacks and competition correspond to Pretexting approach. Professional and non-professional engineers start with conducting the OSINT research based on the information which can be obtained from Google, LinkedIn, Flickr, Facebook, Twitter, WhoIs, etc. social networks. The second phase of the attack was performed in a format of phone calls executed inside the transparent soundproof booth in front of conference and track attendees. The aim is to gather certain pieces of information through both phases that in a situation of "black hat" attack would be meaningful, but not business critical information (see Table 1. below). No physical (i.e. facility) or technical (i.e. network) penetration into target companies was allowed. Contestants were also forbidden to collect any confidential data such as credit card information, social security numbers, and passwords or visit any location of their target or interact with any person from the target before the call at DEF CON. The only source of information should have been the OSINT one. The main goal of the competition was to perform within legal and educational boundaries the penetration testing and highlight that the companies with significant budgets and serious attitude to information security are still vulnerable to the social engineering attacks. The secondary goal was to indicate any correlations in sex, experience and social background of contestants with the success gained during OSINT and calling phase of the competition. While the audience, organizers and contestants were highly motivated for positive reactions on the situations when the attacks were failing due to the strict procedures correctly performed by the target company employees, only in 2 cases of 121 calls the contestants were shut down from the call without reaching the needed contest goal. Figure 3 below demonstrates the overall score based on the number of pieces of the confidential information captured from the companies through all the calls performed by the contestant teams.

Table 1. Sample of the sensitive information pieces captured from private-sector companies during SECFC contest on DEF CON 22 in Las Vegas

DEFCON 22 SECTF Flag List		
Logistics	**OSINT**	**CALLS**
Is IT Support handled in house or outsourced?	3	6
Who do they use for delivering packages?	3	6
Do you have a cafeteria?	4	8
Who does the food service?	4	8
Other Tech		
Is there a company VPN?	4	8
Do you block websites?	2	4
If websites are blocked = yes, which ones? (Facebook, Ebay, etc)	3	6
Is wireless in use on site? (yes/no)	2	4
If yes, ESSID Name?	4	8
What make and model of the computer do they use?	3	6
What anti-virus system is used?	5	10
Can Be Used for Onsite Pretext		
What is the name of the cleaning/janitorial service?	4	8
Who does your bug/pest extermination?	4	8
What is the name of the company responsible for the vending machines onsite?	4	8
Who handles their trash/dumpster disposal?	4	8
Name of their 3rd party or in house security guard company?	5	10
What types of badges do you use for company access? (RFID, HID, None)	8	16
Company Wide Tech		
What operating system is in use?	5	10
What service pack/version?	8	16
What program do they use to open PDF documents and what version?	5	10
What browser do they use?	5	10

Current information on security situation in Ukraine and European Cybersecurity Implementation program

As related directly to the current political and economic situation in Ukraine, the number of cyber-attacks and threats has risen significantly during the last 2 years. On the one hand, there is the cyber-warfare as unavoidable part of the global geopolitical situation that Ukraine is overcoming at the moment. There are several factors that aggravate the situation: extremely low level of security threats awareness in major, SME corporate sector and governmental structures, lack of information security specialists that have sufficient level of expertise to handle the threats of APTs (Advanced Persistent Threats) and country-wide level attacks. And on the other hand, speedy integration of the country to US, European and global IT markets, big number of proficient software engineers, developing of the national software and hardware products and active participation of Ukrainian players in the markets of cryptocurrency. These factors together become a ground for the dramatic increase in cyber-attacks and risks in the Ukrainian digital landscape. According to the reports of the Kaspersky Lab research center, Ukraine has appeared in TOP-15 of the countries that were encountering the most massive DDos attacks during the first quarter of 2015. Another prominent example occurred on Feb 16, 2015 when the forensic investigation of one of Ukrainian banks revealed the worldwide operating cyber-criminal group that managed to successfully execute a series of highly sophisticated attacks on banks of Ukraine, Russia, USA, China, Germany and other countries for the total amount of $1bn loss.

In such situation, the continuous and effective improvement can only be achieved by applying structured, organized and internationally supported programme of cyber-security. One of important steps made on the way in this direction is related to the set up in 2007 of CERT-UA (Computer Emergency Response Team of Ukraine). Later in 2009, it was formally accepted on FIRST (Forum for Incident Response and Security Teams) which enabled the Ukrainian team to effectively communicate with other 326 CERT teams around the globe. Another event to be mentioned is the opening in Kiev of the ISACA Kiev Chapter. ISACA's impact as a global educational, regulatory and security standards development organization that also drives and supports the European Cybersecurity Implementation program, cannot be overestimated.

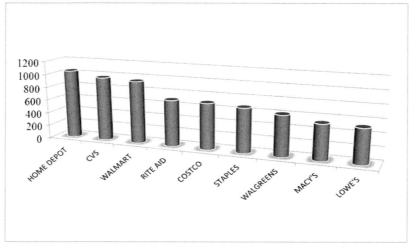

Figure 3. Companies penetration rating (the higher score – the more vulnerable company is) by the results of SECFC contest on DEF CON 22 in Las Vegas

Just recently, on December 9, 2015, the American Chamber of Commerce in Ukraine hosted the event where Alexey Yankovski, (President of ISACA), Irina Ivchenko, (Vice President on Information Security, ISACA) and other speakers presented the Ukrainian translation of the ISACA European Cybersecurity Implementation Series.

ISACA European Cybersecurity Implementation program is a Series of papers that addresses cybersecurity implementation from a European perspective, including the European Union (EU) and its associated countries that will overview and define the layout of implementing cybersecurity good practice in line with existing laws, standards and other guidance focusing on risk guidance, resilience and assurance in cybersecurity targeted for C-level and senior management of the business sector and Cybersecurity practitioners and auditors.

According to European security landscape, the main tendency addressed in the program is the fact that Security breaches have evolved from opportunistic attacks by individual perpetrators to targeted attacks that are often attributed to organized crime or hostile acts between nation states. Hence, EU and its member states have launched a number of programs and initiatives, like forming the European Network and Information Security Agency (ENISA), publishing formal Cybersecurity Strategy issued by Euro Comission, Digital Agenda for Europe and launching the Horizon 2020 Research and Development program that is dedicated to enforcement of cybersecurity union wide. To analyze, co-ordinate and apply all of these sources of valuable information, enterprises need practical implementation guidance for cybersecurity in the European context that Series may support using recognized frameworks and standards to provide targeted insight about implementing cybersecurity.

In this context, the programme will be addressing different levels of attacks from unsophisticated up to the APT sponsored by the countries regarding political, economic, social, technical and environmental aspects of cybersecurity taking into account the level of efforts to be involved in each of them (see Figures 4 and 5). The message actively sent to the corporate sector with this document is that Business can no longer consider itself as "not interested" to participate in the global cybersecurity. Development of more and more advanced and automated attack methods and overall vulnerability of network participants create and will create bigger "botnet" and "dragnet" networks involving every entity connected to cybercriminal activity.

Series will define the strategy, definition, and breakdown of the goals and actions from the governmental level to Enterprise, IT-related goals and Change Enablers activities to perform the necessary transformations. The goal of the program and its support on the national level is to make sure the performance of these activities for an enterprise will embed cybersecurity, as an integral part, into its overall governance, risk management and compliance (GRC) frameworks. The framework will include governance that is in line with existing principles of corporate governance, comprehensive management of cybercrime and cyberwar fare risk and threats that are aligned with existing enterprise risk management (ERM) systems, comply with existing or planned EU-level and national laws and regulations, are resilient to organizational infrastructures and personnel and assurance for information, processes and related controls.

Another important part of the program will be completion of actions and fulfilment of recommendation actions defined by the Digital Agenda for Europe document.

Once all these activities are declared, fully defined and explained, the Ukrainian business sector and IT development activists will have to take their turn implementing it for the overall strengthening the cybersecurity level of Ukraine and their own interest in particular.

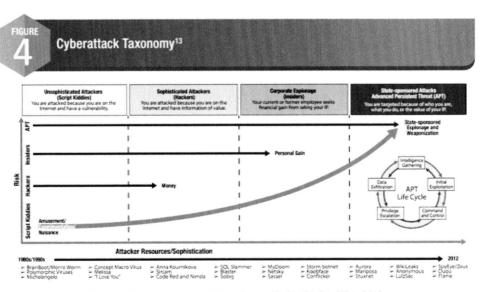

Figure 4. Responding to Targeted Cyberattacks by ISACA, USA, 2013

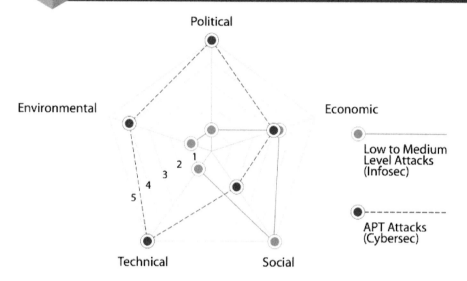

FIGURE 5 Information Security and Cybersecurity Focus (PESTLE)

Political

Environmental

Economic

Low to Medium
Level Attacks
(Infosec)

APT Attacks
(Cybersec)

Technical   Social

Figure 5. Information Security and Cybersecurity Focus by ISACA, USA, 2013

4 CONCLUSIONS

The world technological trends and monitoring show in a definite way that overall number of cyber-threats and attacks is growing at much higher speed than informational security budgets and awareness. This discrepancy unveils critical vulnerabilities in security structure of governmental, financial and enterprise sectors around the globe. Human factor, non-malicious participation of insiders and spreading of smart and connected mobile devices with low level of corporate information security education make social engineering approaches most effective and damaging. Modern automation and technology means make them even more powerful and dangerous.

However, on a global scale, European and Ukrainian security landscapes can be clearly seen at the wide range of activities, programs and initiatives intended to bridge this gap and provide the defense from cyber-threats in a coordinated and united way. Ukraine is now to make a breakthrough increasing the level of national cyber-security which is critically important for the country and will be more and more significant in upcoming years.

REFERENCES

Gray A.M. 1990. *Global Intelligence Challenges in the 1990s.* American Intelligence Journal.

Kingsbury A. 2008. *Spy Agencies Turn to Newspapers, NPR, and Wikipedia for Information. The intelligence community is learning to value "open-source" information.* U.S.News

Hadnagy C. 2011 *Social Engineering: The Art of Human Hacking.* Indianapolis: Wiley Publishing: 380.

Hadnagy C., Fincher M. 2015. *Phishing Dark Waters: The Offensive and Defensive Sides of Malicious Emails.* Indianapolis: John Wiley & Sons: 192.

Ekman P. 2012. *Emotions Revealed: Understanding Faces and Feelings.* UK: Hachette: 304

Ekman P., Friesen W.V. 2015. *Unmasking the Face: A Guide to Recognizing Emotions From Facial Expressions.* US: Paperback: 308

Mitnick K.D., L.W. Simon, Wozniak S. 2002. *The Art of Deception.* New York: John Wiley & Sons: 304.

The Global State of Information Security® Survey 2016. Turnaround and transformation in cybersecurity. 2015-2016. Pricewaterhouse Coopers LLC (PwC).

2015 Global Cybersecurity Status Report. Cybersecurity Nexus (CSX): ISACA, 2015.

Phishing attack trends report by April 2015. Anti-Phishing Working Group (APWG), 2015.

Schneier B. 2013. *Phishing Has Gotten Very Good.* New York: John Wiley & Sons: 384.

AO Kaspersky Lab. 2015. Kaspersky Security Bulletin 2015.

European Cybersecurity Implementation: Overview. ISACA, 2014.

Adaptable model of OLTC regulating transformer in MATLab software

L. Zhorniak, O. Volkova & K. Shapka
Zaporizhzhia National Technical University, Zaporizhzhia, Ukraine

ABSTRACT: The issues of the voltage quality improvement for electric power users in voltage supply systems were investigated. MATLab software (Simulink) was used in order to investigate the power transformer voltage regulation system of OLTC type power regulation on the RNTA basis – 110/630 OLTC.

1 INTRODUCTION

Modern advanced industrial processes are often sensitive to outages and voltage distortions. On the other hand, the growing use of non-linear loads generates interference and contributes to the increased levels of voltage pollution in power networks. This situation can affect normal operation of other customers connected to those networks. Thus, in recent years the sensitivity towards power quality issues has been increasing mainly as a result of the demands posed by many consumers.

Currently, the power systems use a large number of devices, which ensure the maintenance of the required voltage level. First of all, transformers with an adjustable transformation ratio under load, capacitors, secondary batteries, reactors, synchronous compensators, generators, power plants, and so on are used. These devices are equipped with regulators that maintain the voltage at a predetermined level. The main means of regulating the voltage in electrical networks are power transformers with on-load tap changers (OLTC).

2 ADAPTABLE MODEL

To increase the regulation stability for (1), a model that allows consideration of the sign of the derivative enveloping, the regulated voltage is proposed. It allows not to make additional transformer tap switch if the regulating parameter moves into this zone under the influence of external factors, but is out of the deadband.

Assessment of the stability of the investigated automatic control system (ACP) +

(1) is combined with the assessment of the quality of its transient (the latter damp over time in case of stable system). In turn, the transition process in the

system is measured by its response to the greatest possible jumps setting or exciting effects of voltage. Obviously, the synthesized system refers to the essentially nonlinear, and the investigation of its stability should be carried out by computer simulation. The mathematical model of the ACP-voltage power transformer with on-load tap changer of RNTA - 60/630 (eight switching steps) can be implemented by the dependencies (2):

$$u(t) = K_1 \left[\left(U(t) - U_y \right) - K_2 \left(I(t) - I_{min} \right) \right]; \quad (1)$$

$$K_m = \begin{cases} \dfrac{U_{i+1}}{U_{rat}}, if \begin{cases} u(t) \le u_{1.t}; \\ u(t - \tau_1) \le u_{1.t}; \\ \dfrac{dU_b}{dt} \le 0; \end{cases} \\[4mm] \dfrac{U_i}{U_{rat}}, if\ u_{1.t} \le u(t) \le u_{u.t}; \\[4mm] \dfrac{U_{i-1}}{U_{rat}}, if \begin{cases} u(t) \ge u_{u.t}; \\ u(t - \tau_1) \le u_{u.t}; \\ \dfrac{dU_b}{dt} \ge 0; \end{cases} \end{cases} \quad (2)$$

Simulation model of three-phase three-winding power transformer with on-load tap changer, as shown in Fig. 1, is created in the software package MATLab (Simulink 4) to solve the provided dependencies. Since the voltage control channels in the direction of its decrease or increase work uniformly, only one channel was implemented in the model, which simplifies the calculations significantly. In addition, we can take into account that at significant deviations of voltage, regulation system provides OLTC switching without delay (at least for two positions). Therefore, to simplify the model, it was assumed that the device has only two tap-changer section. A transformer with OLTC was simulated in MATLab package using controlled transmission system (FACTS). This was possible because the OLTC switch is used to switch the load and carry current in the windings of the transformer without breaking the circuit. Since the real time of the windings switching (30 - 60 ms) is much smaller than the actual time of selection taps (3 - 10 s), this block can be represented by vector model to investigate the stability of the power system in range from several seconds to several minutes.

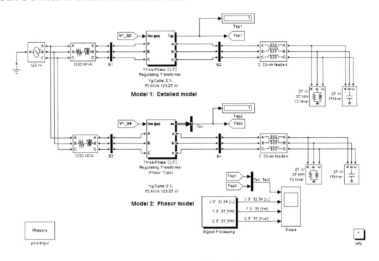

Figure 1. Model used

The parameters of external control of the tap changer are defined in the block menu. Regulation of the voltage-controlled single-phase transformer takes place at changing of transformation coefficient V2/V1 by OLTC device (Fig. 2). OLTC device can be connected both to the winding 1 and to the winding 2, and also any number of taps can be chosen (starting from zero - for correction without turns and ending with the greatest possible number Ntap - with a maximum coil correction). OLTC has the reverse that allows connecting the regulating winding oppositely or cumulatively. The multiplication factor Vnom2/Vnom1 (Fig. 2) is called the correction of the voltage and is defined as $1/(1 + N \cdot \Delta U)$ - for OLTC on the winding 1 or $(1 + N \cdot \Delta U)$ - for on-load tap changer on winding 2. Here, N is number of tap; ΔU is voltage increment on the branch line with respect to voltage winding 1 or 2. A negative value N corresponds to reversing switching (dashed line).

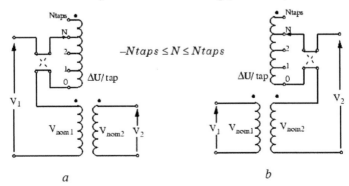

Figure 2. The circuit of voltage regulation in one phase of a three-phase regulating transformer

a) on the first winding:

$$\frac{V_2}{V_1} = \frac{1}{\left(1 + N \cdot \Delta U\right)} \cdot \frac{V_{rat2}}{V_{rat1}}$$

b) on the second winding:

$$\frac{V_2}{V_1} = \left(1 + N \cdot \Delta U\right) \cdot \frac{V_{rat2}}{V_{rat1}}$$

A 25 kV distribution network consisting of three 30km distribution feeders connected in parallel supplies power to a 36 MW/10Mvar load (0.964 PF lagging) from a 120 kV, 1000 MVA system and a 120kV/25 kV OLTC regulating transformer. Reactive power compensation is provided at load bus by a 15Mvar capacitor bank. The same circuit is duplicated in order to compare the performance of two different models of OLTC transformers:

- model 1 is a detailed model where all OLTC switches and transformer characteristics are represented. This model can be used with either continuous or discrete solvers to get detailed wave shapes or with the phasor simulation method to observe variations of phasor voltages and currents.

- model 2 is a simplified phasor model where the transformer and OLTC are simulated by current sources. This model can be used only with the phasor solution method. It is much faster to execute and it should be the preferred model for transient stability studies, when several such devices are used in the same system.

Both OLTC transformer models implement a three-phase regulating transformer rated 47MVA, 120kV/25 kV, Wye/ Delta, with the OLTC connected on the high voltage side (120 kV). The OLTC transformers are used to regulate system voltage at 25 kV buses B2 and B4.

Voltage regulation is performed by varying the transformer turn ratio. This is obtained by connecting on each phase, a tapped winding (regulation winding) in series with each 120/sqrt(3) kV winding. Nine (9) OLTC switches allow selection of 8 different taps (tap positions 1 to 8, plus tap 0 which provides rated 120kV/25 kV ratio). A reversing switch included in the OLTC allows reversing connections of the regulation winding so that it is connected either additive (positive tap positions) or subtractive (negative tap positions). For a fixed 25 kV secondary voltage, each tap provides voltage correction of +/-0.01875pu or +/-0.875% of rated 120 kV voltage. Therefore, a total of 17 tap positions, including tap 0, allow voltage

variation from 0.85pu (102 kV) to 1.15pu (138 kV) by steps of 0.01875pu (2.25 kV).

The positive-sequence voltages measured at buses B2 and B4 are provided as inputs to the voltage regulators (input 'Vmeas' of the transformer blocks). The voltage regulators were the following:
    - 'Voltage regulator' parameter = 'on'.
    - The relative voltage is set to 1.04%.
    - In order to start simulation with 25-KV voltages close to 1.04% at buses B2 and B4, the initial tap positions are set at -4, so that the transformers are increasing the voltage by a factor 1/(1-4*0.01875)=1.081.

The detailed model is built with a fixed number of taps (8). Note that the phasor model provides more flexibility as it allows selection of primary and secondary winding connections (Wye or Delta) as well as changing the number of taps and using the OLTC either on primary or secondary side.

The tap transition is performed by temporarily short-circuiting two adjacent transformer taps through resistors (5 Ohm resistances and 60ms transition time as specified in the block menu). The phasor model is built with current sources emulating the transformer impedance, which depends on winding resistances, leakage reactances and tap position. Both models use a voltage regulator that generates pulses at the 'Up' or 'Down' outputs and orders a tap change either in the positive or negative direction. The voltage regulation depends on the specified dead band (DB = two times the voltage step or 0.0375pu). This means that the maximum voltage error at buses B2 and B4 should be 0.0165 pu. As long as the maximum tap number is not reached (-8 or +8), voltage should stay in the range: (Vref-DB/2< V<1.04+DB/2) = (1.021< V< 1.059).

As tap selection is a relatively slow mechanical process (4 sec per tap as specified in the 'Tap selection time' parameter of the block menus), the simulation Stop time is set to 2 minutes (120s). The Three-Phase Programmable Voltage Source is used to vary the 120 kV system voltage in order to observe the OLTC performance. Initially, the source is generating its rated voltage. Then, voltage is successively decreased (0.95pu at t = 10s) and increased (1.10pu at t = 50s).

After the simulation started, OLTC operation is observed on the Scope (Fig. 3).
− Graph 1 shows the tap position.
− Graph 2 shows superposition of positive-sequence voltages at 120 kV bus B1, at 25 kV bus B2 and bus B4.
− Graphs 3 and 4 show the active and reactive powers measured on 120 kV side (buses B1 and B3).

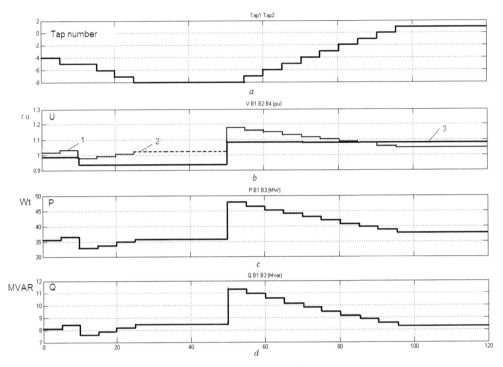

Figure 3. Waveforms of tap changing process

These graphs allow estimating how quickly transient processes when exposed to step excitation are damped in the synthesized system of automatic voltage regulation of power transformer with on-load tap changer RNTA - 60/630. Furthermore, whether additional tap switching of OLTC is required until the transition process ends.

3 CONCLUSION

Simulation of control voltage in programming software MATLab (Simulink 4) can significantly improve not only the quality of delivered energy in energy-producing industries, but also the efficiency of production. The results obtained in the simulation and related calculations suggest how rapidly synthesized in the automatic voltage regulation in the power transformer on-load tap changer (in this example RNTA - 60/630) is damped transients-step winding, and whether such an additional tap-tap-changer until the transition process is ended. Prediction of the number and speed of the switching power transformer windings will optimize not only the regulation of the power consumption, but also ultimately provide the requested process.

REFERENCES

Govorkov F.P. 1993. *By the Question about Regulation of Voltage in Electrical Urban Networks* (in Russian). Energetic and electrification, Issue 4: 42-44.

Rozanov Y.K., Ryabchinskij M.V. 1998. *Contemporary Methods of Quality Improvement Electrical Energy (Analytic Analisys)* (in Russian). Energetic and electrification, Issue 3: 10-17.

Mokin B.I., Vigovskij Y.F. 1985. *Automatic Regulator in Electrical Network* (in Russian). Kyiv: Tehnika: 104.

Grabko V.V. 2005. *Models and Methods of Voltage Regulation with the Help of Transformers with OLTC. Monograph* (in Russian). Vinnitsa: UNIVERSUM-Vinnitsa: 109.

Zhizhelenko I.V. 1981. *Quality Factors of Eelectrical Energy and Their Control in Production Plants* (in Russian). Kyiv: Tehnika: 160.

Pospelov G.E., Such N.M. 1981. *Power and Electrical Energy LossesiIn Networks* (in Russian). Moscow: Energoizdat: 216.

Orlov V.S. 1985. *Additional Power and Electrical Energy Losses during Voltage and Frequency*

Deviate (in Russian). Izvestia vuzov: energetics, Issue 6: 19-23.

Venikov V.A., Idelchik V.I., Liseev M.S. 1985. *Voltage Regulation in Electrical Systems* (in Russian). Moscow: Erergoatomizdat: 216.

Dyakonov V.P. 2002. *Simulink 4. Special Reference Book* (in Russian). Saint-Petersburg: Piter: 5

Printed and bound by CPI Group (UK) Ltd, Croydon, CR0 4YY

24/10/2024

01778286-0001